INFORMATION-
THEORETIC
ASPECTS *of*
NEURAL
NETWORKS

INFORMATION-THEORETIC ASPECTS *of* NEURAL NETWORKS

Edited by

Perambur S. Neelakanta, Ph.D., C. Eng.

CRC Press
Boca Raton   London   New York   Washington, D.C.

Library of Congress Cataloging-in-Publication Data

Catalog record is available from the Library of Congress.

No claim to original U.S. Government works
International Standard Book Number 0-8493-3198-6
Printed in the United States of America 1 2 3 4 5 6 7 8 9 0
Printed on acid-free paper

INFORMATION-THEORETIC ASPECTS OF NEURAL NETWORKS

PREFACE

This book is a compilation of concepts behind the information-theoretic considerations as applied to the neural complex. It offers a focused insight into newer perspectives of information-processing associated with real and artificial neural networks.

Information-handling is ubiquitous with every conceivable system and specifically, it offers a vivid portrayal of the functioning of so-called complex systems—a subset of which is the neural (real and artificial) network.

Within the science of neural complex exists a generous blend of synergetic subjects, namely:

- Information thermodynamics
- Stochastical processes
- Extremum information entropy

Hence, neural informatic considerations are often built upon a macroscopic framework structured to form a pyramid with layers of entropy concepts, thermodynamic principles, and probabilistic axioms.

However, only a few of the relevant considerations (such as the relative entropy concepts) have been widely advocated in the existing literature towards modeling the neural complex in the information-theoretic point of view. A comprehensive outlook of deploying information-theoretics to study the neural complex is still at large.

Information-theoretics vis-à-vis neural networks, in general, embodies parametric entities and conceptual bases pertinent to:

- ❖ Memory considerations and information storage
- ❖ Information-theoretic based cost-functions
- ❖ Neurocybernetics and (self)-organization

Existing studies largely cover the memory characteristics in the neural complex, but sparsely address the entropy and/or cybernetic aspects of neural information while algorithmizing the neural training and deducing optimal neural (prediction) performance. Hence, in order to comprehend and explore this virginal discipline, the present book is developed with the following segmented topics laid out cohesively in eight chapters:

- Information-theoretics, Shannon information and information thermodynamics: An overview.

- Neural complexity as an information processing system: Descriptive, declarative and pragmatic aspects of neural information.

- Memory and information storage in the interconnected neural web.

- Extremum (maximum and minimum) information entropy: A plausible cost-function for artificial neural network (ANN) learning.

- Neural network training: Minimization of cross-entropy functions (Kullback, Jensen and Csiszàr family of functions).

- ANN training via Maximization of negentropyy error-metrics.

- Statistical distance-measures such as Bhattacharyya and Mahalanobis metrics for neural network optimizations

- Symmetric and asymmetric characteristics of information-theoretic error-metrics: Discrimination, distance or divergence measures?

- Algorithmic complexity based representation of neural information-theoretic parameters.

- Genetic algorithm (GA) versus neural information.

- GA-based optimization strategies using ANNs.

4   Dynamics of neurocybernetics in terms of nonlinear, information-theoretic transfer function of the neural cellular units.

- Statistical mechanics, neural networks and information theory: A synergetic view.

- Semiotic framework of neural information processing and neural information flow.

- Fuzzy information and neural networks: Fuzzy nonlinearity of neural transfer function specified in the information theoretic plane.

- Neural dynamics conceived via fuzzy information parameters.

- Neural information flow dynamics

The chapters have been written with appropriate titles and contents to include one or more of the aforesaid topics sequenced in a streamline of relevant details. The projected audience who will find this book as an exhilarating visit to Shannon's world in the universe of neurons are engineers and physicists who model the neural complex (real and artificial) as well as biologists who would like to see the intricacies of communication theory and protocols as managed by the morsels of the brain.

The concepts of information and entropy as neural entities as portrayed in this book open new avenues hitherto unexplored and relieve the bottleneck in realizing a common platform staged to analyze the neural complex and develop advanced artificial neural networks thereof.

The authors are confident that this book will be most useful for all those interested in associating new techniques of formalism with the neural complex systems via classical principles of information theory.

Perambur S. Neelakanta   Boca Raton
Editor

CONTRIBUTORS

S. Abusalah, Ph.D.
Department of Electrical Engineering
University of West Florida
Pensacola, Florida

T.V. Arredondo, M.S.E.E.
Siemens Information and Communications Networks, Inc.
Boca Raton, Florida

D. De Groff, Ph.D.
Department of Electrical Engineering
Florida Atlantic University
Boca Raton, Florida

P.S. Neelakanta, Ph.D., C.Eng.
Department of Electrical Engineering
Florida Atlantic University
Boca Raton, Florida

J.C. Park, Ph.D., P.E.
2d3D Inc.
Delray Beach, Florida

TABLE OF CONTENTS

Chapter 1

Introduction

P.S. Neelakanta

"...where knowledge is abundant, uncertainty is sparse."

Unknown

This book is a compendium of studies on the information-theoretic aspects of the neural complex. It has evolved to include the new and unfamiliar considerations in the existing field of studies on neural activity from the information theory point of view. Conjectured lucidly as a part of the text are relevant extensions on sparsely known analyses pertinent to the performance of artificial neural networks (ANNs) as viewed in the framework of information-theoretics. The content of this book has a multi-layered structure. The core is spanned by the following objectives, specifically relevant to the artificial neural networks and broadly linked to the real neural architecture:

The need and the scope of the subject matter as conceived and deliberated in this book stem from certain obvious lacuna and unexplored "black-holes" prevailing in the existing literature [1.1-1.10]. Emerging theories pertinent to comprehending the information-theoretic perspectives of the methodologies associated with the development of artificial neural networks (as computational and pattern-recognition tools) "in the image of" real neural complex are also considered.

The collective responses and statistically unified activities of (real) neurons have given ample directions to ponder over newer considerations to refine the existing heuristics in developing artificial neural paradigms. As rightly stated, "research under a paradigm must be a particularly effective way of inducing a paradigm change." The wealth of knowledge that persists and evolves in studies associated with neural networks (real and artificial) offers intuitively newer directions. These directions achieve a better understanding, refined modeling and faster computational capabilities of artificial neural networks vis-à-vis that have similar aspects of the real neural complex.

This book is concerned essential with the modeling of artificial neural networks in the information-theoretic plane. It is consistent with the conceivable architectures of such networks and spatiotemporal activities of the interconnected cellular units deliberated towards maximizing the networks' ability to learn from the examples presented to them. The performance considerations addressed in this book refer to both crispy as well as vague (or fuzzy) attributes of the ANNs supplemented by appropriate biological considerations where necessary.

Further, the primary goal of the endeavors presented here is to nurture a new avenue of research in the area of neural complex modeling. The studies addressed thereof are tailored to characterize the neural-network based optimization algorithms in the information-theoretic plane and apply such algorithms in the artificial neurocomputational efforts. The relevant scope of the studies is to achieve better performance characteristics and broaden the neural computational and information-processing abilities so as to characterize and recognize the complex input data and/or patterns robustly.

1.1 Neuroinformatics

The cross-discipline that emanates from the synergetic considerations related to the functioning of the neural complex and the concepts behind acquisition, processing, storage, transmission, retrieval and dissemination of neural information leads to coining a concise word "neuroinformatics." It includes in a broad sense, the information-theoretic methods to describe and analyze the real as well as artificial neural networks and embraces the associated processes and methods pertinent to information transfer, storage, retrieval and dissemination.

What are the specific aspects of the neural complex which may come within the purview of neuroinformatics? A response to this query dwells in the twin-kernel elements of informatics: The structural or classificatory characteristics of the neural information and the semantic part of (or, the meaning carried by) the complex neural data. At the cellular level, the neural information is dictated by the purposive selection of a dichotomous (synaptic) state with the intention of influencing another synaptic state facilitated *via* dendritic interconnections. On an extensive scale, it depicts a spatiotemporal entity proliferating across the neural complex. In its structural as well as semantic attributes, neural data are stochastical due to the entropic contributions introduced in its characteristics as a result of inevitably coexisting neural disturbances.

A neural network is a conglomeration of several subsets which are in part its constituents and in part interact with it through massive interconnections. Further, the subspaces of a neural domain have feedbacks which portray explicit influence of one subsystem over its neighbors. A neural network also represents a self-controlling set of automata, a memory machine, and a homeostat.[1] It is a domain of information-conservation—a process in which a chain of elements (cells) takes part in the collection, conversion, storage, and retrieval of information pertinent to the neural complex. Neurons

---

[1] *Homeostatis:* Refers to internal processes and behavior of living systems which cause physiological entities to be maintained within certain limits conditioned by the adaptability of the system regardless (within certain range) of changes in the environment.

essentially perform binary decisions. Their instantaneous behavior equations and memory processes take place according to mnemonic or evolution equations depicting the dynamics of changes occurring in the coupling coefficients of neurocellular interconnections with respect to time.

Upon excitation, the neural network exhibits transient and reverberatory modes. The decision equation leads to obtaining practical means for the revision and computation of diffuse reverberations of desired general characteristics with the exclusion of all other unwanted components (noise). They indicate a direction towards minimizing entropy.

In regards to mnemonic aspects of neural behavior, an adiabatic learning hypothesis can be assumed which enables a coupling of instantaneous weight changes in intra-neural coupling coefficients from the sustained, slower learning processes or the so-called long-term memory considerations.

The neural system is amenable to isolation into individual subsets such as cells, interconnections, synapses, etc. Each subsystem may in turn be broken down into microscopic structures (parts) participating in the neural activity. Thus, the neural system represents a complex automation—a *cybernetic system,* and if one is "to judge from Wiener's book [1.11], (a cybernetic system) includes information theory." The association of its subsystems via feed-forward and/or feedback information constitutes an optimum control strategy prescribed for *self-organization.*[2] In order to achieve efficient, cybernetically controlled self-adapting activities, the neural system strives to minimize the uncertainties arising from the inherent noise or (spatiotemporal) random characteristics of its activities. Such activities, in general, refer to the state-space of the units (cells) comprising the neural complex.

Compilation and storage of data on the state of the system, transmission of the data (among the interacting units), and retrieval of data where and when needed constitute the information-processing tasks in the neural system. Pertinent to neurocybernetics, the concerned data and the associated information can be quantified in a wider sense through a *measure of the disorganization.* This refers to a metric of uncertainty removed from the cellular complex by receiving knowledge from the neural environment which is *pragmatically* utilized as actions of self-organizational procedures. The neural information not only portrays a representative value available to depict the neural state of disorganization, but also it lends itself for necessary actions by the existing means toward self-organization.

---

[2]*Self-organization*: Progressive formation within the system of sequential, ordered (coherent) relationships between the interacting dynamic variables. A self-organizing system under the influence of appropriate stimuli (or, information about the state of the environment) is capable of changing its organization so as to mold into a structure which is able to perform a specified function (according to certain criterion) in an optimum manner. Self-organizing structures are invariably hierarchical.

The associated disturbances in the neural complex decide the extent of uncertainties in the neuronal states, thereby augmenting the system entropy which influences the amount of information needed to be processed. To realize the optimal self-organizing performance, the neural system warrants a minimum amount of information. The minimal information required is assessed implicitly via entropy considerations by depicting the neural complex as an information-processor which could, in general, be nonideal and lose information (*null-information*) as well as gain *false*   and/or *redundant* information. Further, the minimum information at pragmatic level has a value to the extent that it is useful as a resource for purposeful activity in overcoming the disorganizing influences. It can also correspond to a set of data of value in decision-making towards disorganizational uncertainties.

Open neural systems have entropy characteristics generally similar to that of closed systems. However, the entropy value is dependent on the temporal pattern of the input energy. This dependence relies on the following hypothesis: The difference between the state of system and the state of maximum entropy grows with the input information per unit of input energy. The above hypothesis is consistent with the concept of "neural energy" frequently used to describe the changes in the neural activity that produces the behavior controlled by the brain. Bergström [1.12] enumerates the following as the characteristic of neural energy:

- The state of the neural system can be described by the total neural energy, and by its distributor, the entropy.

- The total neural energy fulfills the common notion of physical energy. That is, the total energy is the sum of its parts which remains constant in a closed system.

- The distribution of neural energy can be calculated for a given system which maximizes the entropy.

- In a closed neural system, the following approximation for the entropy is derived:

$$H(E, n, N) \cong \left(\frac{4E}{nN}\right)\left[1 - \left(\frac{E}{nN}\right)\right](n-1)\ell og(N+1) \qquad (1.1)$$

where E is the total energy, n is the number of neuronal elements and N is the efferent branches in cells.

- The above relation of Eqn.(1.1) indicates the dependence of neural performance on the (stochastic attributes of) excitatory activity of the neurons. This performance capacity, is therefore, dependent on the entropy of the system. As a result, the capacity of the neural complex

4

handling information attains a maximum at a certain optimum level of neural energy.

The discussions above indicate a direct link between the neural energy sustained *via* neural activity protocols and the treatment of information capacity (and information transfer properties) of simple neural systems.

To a query on what sort of information one can access from neural entropy formula namely, Eqn.(1.1), Bergström [1.12] has replied that the neural entropy function which reveals the entropy of a neural system depends on the energy content of the system, so that it shows a maximum with a medium degree of energization. Since information capacity behaves in the same manner as entropy (in Shannon's sense), the information handling capacity of a neural system exhibits a maximum only for an optimal degree of energization effected, for instance, through sensory channels. The information handling capacity decreases if the energy content of the system decreases or increases from this optimum level.

In real (biological) neural systems, *mossy fibers* which originate from several brain sites carry information of a diverse nature concerning the *declarative* aspects of internal states (of the interconnected neurons) and the *descriptive* message about the physioanatomical environment. The items of information so transmitted or delivered as an input to the neural communication network are largely sensory data. In addition, there are also information-bearing commands of the central motor system which coexist in the interneural communication links.

The information discharge rates of mossy fibers are modulated over a wide range which permits them to transmit detailed parametric information. Therefore, information-theoretic analyses of the neural complex fall within a domain which represents an information-parametric or entropy-dictated spread-space.

The self-sustaining activity of the neural network arises from the dynamics of neural mechanisms and is due to feedbacks of recurrent circuits in the interconnected complex. Such self-regulating or self-sustaining activities refer to the generation of a mode (activity) pattern performed by cerebellar modules which are matched with a set of patterned outputs. Realization of a matching refers to self-organization and corresponds to achieving the *objective goal*. The self-regulation process is, therefore, an adjunct accomplishment of the associative memory problem, namely, the network's response in delivering one of the stored patterns most closely resembling that which is presented to it. The closeness of resemblance approaching infinity represents a close realization of the objective function.

1.1.1 *Neural memory: Neural information storage*

Application of informational analysis to the neural complex in the literature [1.13] has so far been limited largely to the information capacity heuristics of structures like Hopfield's net model of associative memory for the purpose of quantifying network's performance in storing information. A brief summary of relevant considerations is as follows:

In simple models, the information capacity of a standard memory is specified explicitly by the number of memory bits, but where complex dynamics of activation patterns (such as in the neural complex) are encountered, probabilistic estimates of the information capacity are indicated on the basis of certain simplifying assumptions. The estimate of information storage capacity of Hopfield's network, for example, has been formulated *via* statistical methods and extended with more rigorous stochastical techniques.

Considerations pertinent to *memory* and *thought-process* are decided essentially by the associative properties of the collective neuronal aggregates and the underlying entropy considerations. In this perspective, Caianello [1.14] outlined a theory of thought-process and *thinking machines* in terms of so-called *neuronic equations* depicting the instantaneous behavior of the system and *mnemonics equations (*or *evolution equations)* representing the permanent or quasi-permanent changes in the neuronal activities. Such permanent changes, with the passage of time in neural functioning (caused as a result of learning experience), depict explicitly the changes in neural coupling coefficients with respect to time. These changes can also be a degree of plasticity which decides the *fixation of memory* or *memory storage* involving time-dependent neural activity. The instantaneous behavior can be decoupled from the sustained, slower learning or mnemonic processes. This decoupling refers to an adiabatic learning hypothesis.

Shannon's concept of *information* or *entropy* [1.15] in stochastic systems can be extended to neuronal associative memory as a strategy towards information-theoretic modeling of neural activity. It refers to the computation of an entropy function vis-à-vis a neuronal potential train of spikes. Accordingly, the temporal and spatial statistics of this neuronal aggregate of signal elements have featured dominantly in the information-theoretic approaches formalized in the 1970s [1.10]. For example, in 1972 Pfaffelhuber [1.16] used the concept of entropy to describe the learning process as one in which the entropy is depicted as a decaying function of time concurrent to the progress of gaining information.

Why should an entropy model be advocated to depict neural information processing? Is it because of the similarity between Shannon's concepts of information content, or because of the general property of the neural system (like any other physical systems) of maximizing entropy?

Bergström [1.12] offers an elegant reply to the above question: The primary need to develop an entropy model depicting the neural system is to evolve an operational definition of the *macrostate* of the neural system. (The term macrostate here refers to the same adopted in physical thermodynamics.) The macrostate corresponds to the "over-all-state" of a neural system describing certain (neuro) physiological states. Further, Bergström [1.12] indicates that the sort of information that can be ascertained from an "Entropia" formalism (such as Eqn. (1.1)) corresponds to soliciting explanations on certain behavioral characteristics of species with different degree of complexity of their neural system.

The concept of entropy constitutes the basis in elucidating the *information capacity* of neural networks. Further, considering the associative memory as a plausible model for biological memory having a collective dynamic activity, the information capacity of Hopfield's network can be determined quantitatively as outlined below.

Characterization of information in a neural network via classical stochastic system considerations (as conceived by Shannon and Weaver) simply defines a specification protocol of message transmission pertinent to the occurrence of neuronal events (firings) observed (processed) from a milieu of possible such events across the interconnected network. Thus, in the early 1970s, as stated earlier, the information-processing paradigms as referred to the neural system were derived from the stochastic considerations of (or random occurrence of) neural spike trains. Relevant studies were also devoted to consider the probabilistic attributes of the neural complex; and, considering the randomly located cellular elements, the storage capacity of information in those elements (or the associated memory) was estimated. That is, the information-theoretic approach was advocated to ascertain the multiple input-single output transfer-function of information in the neuronal nets. The information capacity of a neural network was then deduced on the basis of the dynamics of activation patterns. Implicitly, such a capacity is the characteristic and ability of the neural complex viewed as a collective system. This capacity was represented as a memory (that stores information) and was quantified via Hartley-Shannon's law as the logarithm of the number of strings of address lines (consisting of memory locations or units) distinguished. Memory locations in this context refer to the number of distinguishable threshold functions (state-transitional process) simulated by the neurons.

Concurrent to the elucidation of optimal (memory-based) storage capacity of neural networks, the following are the basic queries posed:

- What is the maximum number of pattern examples that the neural complex can store?

- For a given set of patterns (less than the maximum value), what are the different functions that could relate the network inputs to the output?

- How do the statistical properties of the patterns affect the estimation of the network information capacity?

Further, the notions of information capacity of the neural complex (associated memory) are based on:

- Binary vector (dichotomous) definition of a neuronal state

- Matrix representation of the synaptic interconnections

- Identifying the stable state of a neuron

- Defining the information capacity of the network as a quantity for which the probability of the stored state-vector patterns (as per Hebbian learning) being stable, is maximum.

Elucidating the information capacity in the neural complex requires a model that depicts the information flow across the interconnected neurons. Such an information proliferation, in general, would face an axonal bottleneck. That is, the input arriving to a neuronal cell is not as information-starved as the output. On the input-end neurons are more ajar than on the output-end. Often they receive information at a rate three orders of magnitude higher than they give it off. Also, the neurons always tie together to form a group and, hence, permit an anisotropic proliferation of information flow across the neural complex. These groups were termed as *compacta* by Lengendy [1.17]. It is the property of compacta (due to massive interconnections) that, whenever a compactum fires, the *knowledge* imparted by the firing stimulus is acquired by every neuron of the compactum. Thus, neuronal knowledge proliferates from compactum-to-compactum with a self-augmentation of information associated in the process. Each successive structure is able to refine the information available to it to the maximum extent that the information can possibly be refined. Due to this self-augmentation of information, McCulloch and Pitts called the neural nets *networks with circles*. Further, routing of information across the interconnected compacta is a goal-pursuit problem. That is, the self-organizational structure of the neural complex dictates a goal-directed neural activity and the goal-seeking is again monitored adaptively by the associated feedback (self-)control protocols.

Pertinent to conventional memory, information storage is an explicit quantity. For example, in a Random Access Memory (RAM) with M address lines and 1 data line (2^M memory locations, each storing 1 bit of information), the storage capacity is 2^M bits. That is, the RAM as an entity distinguishes 2^{2^M} cases (in respect of setting 2^M bits independently as 0 or 1) and thereby stores a string of 2^M bits. This definition enables the entire contents of the RAM as one entity to encode a message—the logarithm of such different messages measures the information content of the message.

Along similar lines, how can the capacity of a neural network be defined? The information content in a neural net refers to the state-transitions, the associated weights, and thresholds. How many different sets of weights and thresholds can then be distinguished by observing the state-transitions of the network? Abu Mostafa and St.Jacques [1.13] enumerated such threshold functions involved and have shown that there are $2^{\alpha N^3}$ distinguishable networks of N neurons, where α is asymptotically a constant. In logarithmic measure, the capacity of a feedback network is, therefore, proportional to N^3 bits. However, the above definition of capacity of a neural net is rather heuristic and not apparent. In order to elucidate this capacity explicitly, one has to encode information in the state-transitions across the interconnected network. Decoding of the message warrants observations concerning "which states go to which state". Such a format of depicting information storage is not, however, practical or trivial.

Alternatively, an energy function can be defined to indicate the state-transitions in the neural cells leading to a set of certain identifiable states. These states are vectors of bits (which correspond to words in a regular memory). Convergence to such identifiable states is the inherency of a feedback network functioning as an associative memory. The information storage in feedback networks can be specified by these identifiable states designated as *stable-state*.[3] How many such stable-states can be stored in a feedback network? The number of these stable-states can be shown as bN, consisting of N bits (depicting the individual states of N neurons), and b, an asymptotic constant. Hence, stable-state capacity of a feedback network of N neurons is proportional to N^2 bits. The reduction from N^3 to N^2 arises from the loss of information due to the restricted (selective) observations depicting only those transitions leading to stable-states. The selection of stable states used in the computation of memory storage is a rule-based algorithmic strategy that selects a set of state vectors pertinent to the network which are stable by remaining unchanged when the update threshold rule is imposed. For example, the Hebbian rule chooses selectively the matrix of interconnection weights to be the sum of outer products of the vectors to be stored. The stable-state storage capacity of the neural network is thus dependent on the type of algorithmic rule chosen. Relevant to the Hebbian rule, for example, only $gN/\ell og_e(N)$ randomly chosen stable states can be stored (where g is an asymptotic constant); and the corresponding capacity is proportional to $N^2/\ell og_e(N)$ bits.

For a feed-forward network, however, there is no definition of memory capacity that corresponds to stable-state capacity inasmuch as such networks do not have stable-states, but rather input-output relations.

The information aspects of a neural complex viewed in terms of memory capacity also penetrate the coordinated organization of firing patterns in

---

[3] *Stable-state:* Refers to a state where the neuronal state remains unchanged when the update threshold rule is applied.

recurrently connected cellular units. That is, the *memory traces* refer to systematic sequential firing of particular sets of neurons and the *variety* in memory space depicts the vagaries in traces pertinent to different sequences of different sets of neurons. In the universe of the neural complex, any cellular unit may, in general, participate in many such traces defined by the temporal relations among the participating neurons as observed by Little [1.18]. The traces are embedded in a given cellular set *via* selective modulations of the synaptic weighting among the participant neurons in accordance with the fashion of arrangement (of the neurons) in sets in the firing patterns. Temporal traces and the spatial proliferation of state-transitions are the embodiments of memory considerations in the neural system. As observed by MacGregor [1.19],"from the very onset, anatomical connections and associated temporal firing patterns are two distinct but intrinsically coupled manifestations of a unitary phenomenon."

Do the memory traces overlap? MacGregor [1.19] answers this question by considering the *cross-talk* mediated by inappropriately activated synapsis through synaptic adjustments. That is, two distinct traces may call for the same single synapse between two cells that participate in both traces, but each asks for a distinct value to be assigned to the synapse. It is likely in multiple embedded nets that a given neural cell may project to some subset of cells because of its position in one trace and to another subset of cells as a result of its position in a second trace. Implicitly, this is same as Gardner and Derrida's approach [1.20] of learning an explicit target function or extraction of a rule formalized in terms of replica symmetry consideration.

The multiple traces embedded in a random fashion can be regarded as being completely independent of each other and the cross-talk can be considered as a disruptive aspect of memory borne by a recurrently connected network. The stochastical theory of cross-talk has been comprehensively addressed by MacGregor who associates the concepts of *beds* and *realizations* in describing the memory traces. That is, the firing of specific sequences of sets of neurons which represent the items of information consists of physiological variations (or realizations) of primarily anatomical sites which are dubbed as beds. Thus, the population of excitatory and inhibitory neurons taken as an ordered sequence of sets constitutes the bed of a trace. A realization is an observable physiological manifestation of an underlying bed. It consists of an ordered sequence of subsets of active neurons, some of which are members of the corresponding sets of the bed and which fire over a given time interval in the temporal correspondence that exists among the cells of the bed.

Using the above notion in describing the neural complex, MacGregor and Gerstein [1.21] elucidated mathematical expressions to show how the memory capacity of recurrently connected nets depends upon their characteristic anatomical and physiological parameters. Considerations of overlaps of traces, disruptive cross-talks, and random background activities were judiciously algorithmized in determining the storage capacities of information (*via* traces) pertinent to large interconnected networks.

Palm [1.22,1.23] applied information theory concepts to deduce the storage capacity of neural associated memory with randomly distributed elements. He estimated that at least 0.05 bits per storage element can be stored in such systems.

1.1.2 *Information-traffic in the neurocybernetic system*

The science of information is concerned with a set of general concepts and analytical expressions regarding the flow information in any situation for which information serves as a resource. At the pragmatic level, information has a value to the extent that is useful as a resource for purposeful activity towards certain decision-making. Information, in fact, "is data of value in decision-making". Thus, the neural activity supports a data of value in making a decision to achieve self-organization.

The existing body of information-theoretics as applied to the neural complex, does not *per se* specify explicitly the *pragmatic*[4] and *semantic* utility of neural information with regard to self-organizing (cybernetic) control-goals or decision-making strategies. This is due to the fact that the techniques pursued by the statistical theory of information do not match the analysis of control problems. Classical statistical theory of information describes only the processes involved in the transmission and storage of information. It is not comprehensible to treat information *vis-à-vis* control strategies or the extent of contraction of the parameter subspace as a result of previous training.

The neural complex handles information at three stages (Fig.1.1): The *input stage,* the *processor stage* , and the *controlling stage*. The input information includes all the *objective knowledge* on the structure, arrangement, and properties (such as the synaptic anatomy, physiology, and biochemical characteristics) of the participating neurons in the control endeavor. It also covers the details on the neural environment of the spatiotemporal domain accommodating these neurons [1.24].

The processing information characterizes the relevant properties and structure of the neural activity or state-transitional events across the interconnected neurons. These events include the excitatory and inhibitory aspects, delays, threshold levels, inherent disturbances, and a variety of processing neuronal infrastructures. In essence, processing refers to all

---

[4] *Pragmatic information*: At the pragmatic level, information has value to the extent that it is useful as a resource for purposeful activity. The primary purposeful activity refers to decision-making. Hence, "information is data of value in decision-making".The pragmatic information content of a set of data or a message is equal to the difference of the value of the decision-state of the decision-maker after and before the receipt of the message, where the values of the decision-state is a function of the determinism of the decision-maker.

relevancies of the control center which strive to attain the objective function being fullfilled. Processing is a rule-based set of *algorithms*[5] which extracts useful information from the inputs and utilizes it to achieve the goal. Once the information is processed, it represents controlling information which is looped (forward or backward) into the system to refine the achievement of the goal or reduce the *organizational deficiency* that may prevail and offset the efforts in realizing the objective function. Consequentely, the controlling information is the knowledge exerted to self-control (or regulate) the automata. The controlling information process is therefore a stand-alone strategy, distinctly operating on its own as an adjunct to the main information processor.

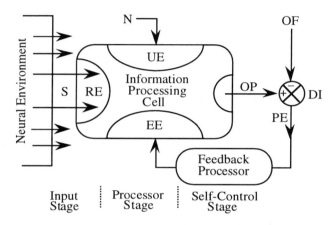

Fig.1.1: Information traffic in the neurocybernetic system
S: Sensory input information; RE: Synaptic (recognition) information of the inputs; UE: Uncertainties/entropy; N: Intra- or extracellular disturbances (noise); OP: Axonal output information; DI: Decision-making information; OF: Learning-based objective function for self-control endeavors; PE: Predicted error information; EE: Error-correcting information for self-organization

The controlling information includes the finer details of knowledge associated with sensing and recognition of the processed information from the second stage, algorithmic manipulations on the sensed data for decision-making, and strategies of predicting the errors. It supplies information to neural actuator(s) to execute the feedback or feed-forward controls through the organizational loop. The controlling information-processing works towards the realization of objectives or goals with the minimization of errors.

Thus, the concept of information-processing in the neural complex viewed from a cybernetic angle assumes an informational structure and is

---

[5] An *algorithm,* in general, refers to a sequence of steps, codes, or rules leading to specific results.

pertinent not only to perceiving knowledge from the source and analyzing it for its usefulness, but also to processing it further for applications towards achieving a self-organizing automaton. When applying information theory to neurocybernetics, the following general considerations are implicit:

- In view of the automatic and adaptive control strategies involved in neurocybernetics, an informatic-based *transfer function* can be defined which refers to information-processing algorithms concerning the attainment of a target or objective function. In terms of informational efficiency functions, they should also assess how a given processing algorithm is realized by the control strategies involved (subject to constraints on speed of response, storage capacity, range of inputs, number of interconnections, number of learning iterations, etc.).

- There are methods of elucidation by which the informational, characteristics of the neural network can be derived by strategies such as frequency domain analysis employed normally in the theory of adaptive controls.

- There exist techniques to evaluate the effectual or ineffectual aspects of classical stochastical theory of information to describe either quantitatively or qualitatively, the control activities of a neural network. In general, statistical information describes the quantitative aspect of the message and does not portray the qualitative consideration of the utility of the message in contributing the attainment of the objective function. Further, the operation of a neural system being dynamic, the statistical measure of information pertinent to the temporal neural activity is often inadequate to include the nonstatistical part of the dynamic system, such as its topological, combinatorial algorithmic features.

- The cybernetic perspective of a neural network represents the *degree of organization* of neural activity. Therefore, the corresponding informatic description of the neural system should not only address the memory considerations, but also enclave the control aspects of modeling and programming of the collective response of neurons.

- From a neurocybernetic viewpoint, the information theory should specify the *semiotic aspects* of information, covering the *syntatics* which relate the formal properties of neurons and their combinations (variety) to the amount of information they carry, and *semantics* and *pragmatics*, which define the information content and the information utility of the neural signal elements, constituted by the binary state vectors of the cellular potential transitions, respectively.

- By enunciating a relation between the degree of self-organization *versus* the informatics of *orderliness*, a new approach as applicable to neural cybernetics can be conceived. Wiener himself indicated such a futuristic trend to incorporate and extend information theory to neural *via* semiotic considerations.

- The concept of neurocybernetic informational theory rests upon the *threshold of distinguishability* of neural state variables and the amount of their *variety* pertinent to the self-control process.

- Analysis and synthesis of man-machine systems as done in the development of computer architecture and artificial neural networks mimicking the biological neurons refer to modeling and programming only at the information structural level. Such modeling and programming *via* information theory in the neurocybernetic domain could, however, broadly address:

◊ An informational description of the global neural complex

◊ Similarity or dissimilarity criteria with regard to the objective structures, information structures, and information flows specified in terms of entropy parameters of the neural complex

◊ Establishing a similarity between informational functions of processing by self-organizing (control) centers of the interconnected cells.

The bases for inculcating an informatic approach to the traffic of neural data are, in essence, the *complexity* of the system in spatiotemporal domains, *orderliness* (alternatively randomness) associated with the system due to the presence of inevitable intra- and extracellular disturbances, *degree of organization* of self-control effected by feedback control strategies, and the *entropy* of the system. In the real neural complex, the brain, for example is structurally organized in a primitive core with a random geometry surrounded by more organized shells with increasing structural order in respect to the connections between neuronal units. In proportion to the entropy content of their input and output entities, the core and the shells may exhibit in their signaling functions different degrees of randomness with negative entropy of the neural process increasing from the core to the peripheral shells. This implies that the brain represents a nonstable domain within which maximum *negentropy* (negetive entropy) extends functionally from the outer shells in decreasing value and reaching a maximum *posentropy* (positive entropy) at the core. This nonstable attribute in information-theoretic perspective is measurable as the difference in pragmatic information contents pertinent to maximum non-equilibrium and equilibrium states of the neural complex.

The *complexity* of the neural system provides a platform for the passage of neural information and is decided basically by the number of units (cells), the variety of the functional events (state variables such as the energy levels), and the complexity of occurrence of such events in the time domain. An algorithmic description of the neural complexity in the informational plane should, therefore, include all the aforesaid considerations summarized by MacGregor [1.19] as beds and realizations.

Neural complexity is a generalized estimate functionally related to the *number variety* (composition), structure, and properties of the cellular units considered in space or time or both. The overall complexity could be nonadditive of the influences arising from the number of cellular units participating in the neural activity and their *variety*. For example, the excitatory neurons, the inhibiting neurons, the neurons with different threshold levels, the neurons with varying extents of synaptic inputs, etc. constitute the "variety" features indicated above. In other words, variety is an implicit attribute of the diverse nature of beds and realizations constituting the items of the neural information traffic.

1.2 Information-Theoretic Framework of Neurocybernetics

The essential parameters deciding the information-theoretic aspects of self-control functions in a neural network viewed in cybernetic perspective are:

- Complexity of neural architecture

- Spatiotemporal randomness (disorder of the system) of neural state transitions

- Self-organization efficiency of the interconnected cells

- System entropy associated with the spatiotemporal neural events.

Corresponding informational analysis pertinent to the self-controlling or organizing characteristics of the neural system can be specified in terms of the following entities:

- Information-processing algorithms at three stages of neural information traffic depicted in Fig.1.1

- Information utility or the value of neural information

- Information efficiency resulting from loss of information due to intra- and/or extracellular disturbances

- Steady and transient (dynamic) states of neural information flow.

A major function of the neurocybernetic system is a goal-related (dictated by an objective function), self-organizing or self-regulating effort viewed within a set of bounds. The associated randomness or disorderliness, however, causes the system parameters (specified by a vector set) to veer from the system objective in realizing the target function. The corresponding deviatory response in a neural network can be quantified by an ensemble of *diversion factors* pertinent to the neural environment which can be subdivided as follows:

- External (extraneural) diversion subset due to disturbances or randomness introduced by external influences such as extracellular disturbances

- Internal (intraneural) diversion subset due to causes arising from internal randomness or cellular noise.

The gamut of information-theoretic aspects of the neural complex can be viewed from the following perspectives:

- Learning considerations and memory storage of the neural complex in holding a set of information [1.25]

- Measures of neural activity both in the real and artificial neurons

- Neural system and entropy considerations

- Information-theoretic error-measures useful for neural network training strategies

- Neurocybernetic (control) information.

The genesis of information-theoretic considerations of complex systems (such as the neural network) stems from the basic concepts of Shannon and Weaver [1.15] who in 1949 indicated the classical method of assaying the information content pertinent to stochastical signals. The basis of the Shannon-Weaver concepts is that the observed occurrence of a specific event from a milieu of possibilities in an indeterminate system constitutes a specification of information. Shannon and Weaver showed how one could quantify the amount of information in such situations. Qualitatively, the less likely the event, the more information is associated with its occurrence. That is, uncertainty associated with a stochastical system is an implicit measure of information content. Normally (from a thermodynamics point of view) *entropy* is a measure of uncertainty associated with the stochasticity of the system. Therefore, a negative entropy (or *negentropy*) portrays the explicit metric of information content. Shannon and Weaver further characterized the binary decision as the most elementary informational element and labeled the amount of information associated with a binary decision as "bit" (an acronym of "binary digit").

A number of investigators in the early 1970s discussed the application of information-theoretic ideas to neuronal spike trains. Nakahama et al. [1.26] defined the concept of "dependency" for discrete variables in a population in terms of Shannon's entropy and the mth order conditional entropy. This measure provides sufficient description of the degree of the first-order Markovian characteristics of neuronal spike trains. It is claimed by the authors to be the most concise measure for expressing the higher-order properties of time-series and to be superior to correlation or spectral measures [1.26].

In its role as an information-processor, the neural complex encounters uncertainties due to possible randomness, as well as loss of information, or acquires false/redundant data during its attempts towards self-organization. Therefore as indicated before, to realize optimum self-organizing characteristics, the neural system warrants a minimum amount of information. The question of minimum information arises because an information-processor, in general, could be nonideal and lose information (*null-information*) as well as gain *false* and/or *redundant informatiom.*

Other deliberations on the information-theoretics of the neural complex existing in the literature refer to the probabilistic issues of neural stochasticity in different pespectives. For example, the neuronal activities when viewed as pertinent to a probabilistic system have been portrayed by Palm [1.23] in the information-theoretic plane in terms of quantitative metrics measuring the "evidence" and "surprise" associated with the data.

Further, on neural information-processing, Windhorst and Schultens [1.27] have presented a method to calculate the information transfer or "transformation" in multiple input/single output neural systems. This is an extension of an approach introduced by Eckhorn and Poepel [1.28] in 1974.

Tsukada et al. [1.29] discuss Shannon's entropy concept with regard to some properties of temporal pattern discrimination in neurons. Sejnowski [1.30] discusses the significance of correlation's pertinent to neural information-processing as regards to neural information.

In [1.10] it is indicated that Borisyuk and others have proposed a new, method for analyzing several simultaneously recorded spike trains to identify and evaluate interconnections. Also indicated in [1.10] is the concept of "neurochrome" to display the neural activity patterns.

The proliferation of neural information across the interconnected set of neurons has been modeled in the information-theoretic plane by Ventriglia [1.31] who fused the relevant considerations on "information wave", neural dynamical activities and the associated memory effects.

Another information-theoretic aspect of the neural complex refers to specifying the associated information content in terms of "certainty" attributes, in contrast to the traditional evaluation of the information *via* "uncertainty" considerations. Relevant details are as follows:

Considering the spatiotemporal activity of neurons, the corresponding continuum model due to Wilson and Cowan [1.32] suggests that the neural activity can be described either in terms of the proportion of excitatory cells becoming active (per unit time) or, alternatively in terms of the proportion of inhibitory cells (per unit time) participating in the activity concerned.

Therefore, if the activity profile of excitatory cells is described in the information-theoretic domain *via* negentropic (uncertainty) parameters, it is logical that the same activity profile in terms of inhibitory cells would correspond to (pos)entropic (certainty) parameters. To accommodate these two possibilities in the information-theortic plane, a definition of information should be evolved on certainty as well as uncertainty considerations. That is, instead of the usual definition of information based on uncertainty considerations, a new information measure in terms of certainty aspects is necessary if one has to study the activity profile of inhibitory cells indicated above. This can be done by a generalized class of certainty information measures (due to Van der Lubbe et al. [1.33]) which are also based on the concept of probability.

It appears that the relation between certainty and probability (in contrast to Shannon's notion of uncertainty versus probability) was known even in 1687 to Leibnitz; Bernoulli consequently formalized an explicit relation between certainty and probability indicating that "probability is the degree of certainty and differs from the whole."[6]

In [1.33], a (ρ, σ) certainty measure (G_n or $G_n{}^*$) is defined for $P \in A_n$. That is, for P, an element of the set A_n of all complete discrete probability distribution, namely, $A_n = \{P = (p_1, p_2,..., p_n)|p_i \geq 0, i = 1,2, ..., n,$ $\sum_{i=1}^{n} p_i = 1\}$ and $(\rho, \sigma) \in D$ where $D = \{\rho, \sigma |0 < \rho < 1, \sigma > 0 \; \forall \; \rho > 1, \sigma = 0\}$, G_n and $G_n{}^*$ are given by:

$$G_n(p; \rho, \sigma) = \left[\sum_{i=1}^{n} p_i^{\rho}\right]^{\sigma} \tag{1.2a}$$

$$G_n^*(p; \rho, \sigma) = \prod_{i=1}^{n} p_i^{p_i(\rho-1)\sigma} \tag{1.2b}$$

The corresponding general classes of certainty-based information measures are given in [1.33]. Specifically, the logarithm information format of certainty measure is given by:

$$^1H_n(P; \rho, \sigma, \delta) = -\delta \log[G_n(P; \rho, \sigma)]$$

$$= -\delta \log[\sum_{i=1}^{n} p_i^{\rho}]^{\sigma} \tag{1.3}$$

---

[6] *Probabilitas enim est gradus certaintities, et ab hac differt ut pars a toto.*

where δ denotes a positive parameter which can be chosen on the basis of the base of the logarithm adopted.

Thus, probability *versus* certainty can be related to yield an information measure as a strategy parallel to the conventional probability versus uncertainty yielding the Shannon's information measure.Therefore, if activity of excitatory neurons is cast *via* Shannon's measure in the information-theoretic plane, a parallel technique to study the activity of inhibitory neurons would be to use the certainty-based information measure discussed above (or vice-versa).

As discussed earlier, the neural complex bears three domains of informatics: The first one embodies the input, the second one is a processor stage, and the third part refers to the controlling stage.

The associated knowledge or information in the first section is largely *descriptive* of the environment of the domain-space accommodating the neurons. It represents a *declarative* knowledge which reflects the structure and composition of the neural environment. On the other hand, the second section has activities which are processed in translating the neural state-transitions across the interconnected cells as per certain instructions, rules, and set of learning protocols. In other words, a set of rule-based algorithms depicting the knowledge or information useful to achieve the goal constitutes the processing sections of the neural complex. The associative memory of a neural network stores this declarative information to a significant extent.

Likewise, the controlling section of the network information is looped into the system to refine the achievement of the goal or minimize the organizational deficiency (or the divergence measure) which depicts the offsets in realizing the objective function. In this self-control endeavor, the neural automaton has the knowledge or the information which is again, largely *procedural*. It represents a collection of information which on the basis of phenomenological aspects of neurons reflects the rational relationship between them in evaluating the organizational deficiency.

Therefore, the neural complex and its feedback system combine harmoniously the declarative and procedural information. Should the control activity rely only on procedural informatics, it represents the conventional or classical model. However, due to the blend of declarative informatics, the neural controlling processor as suggested by the author [1.24], is essentially a *semiotic model*— a model that depicts a sum of the declarative knowledge pertinent to the control actions. The control actions themselves are, however, rule-based or procedural in the informatic domain.

The cohesive blend of declarative and procedural informatics in the neural automaton permits its representation by a semiotic model. That is, the neural complex is a system that could be studied *via* semiotics or the *science of sign systems*.

The semiotic modeling of the neural network relies essentially on the information pertinent to the controlled object, the knowledge on its control being present in the associated memory so that the system can be taught (or made to learn) and generate procedural knowledge by processing the stored control information.

1.3 Entropy, Thermodynamics and Information Theory

The concept of entropy was first introduced in thermodynamics where it was used to provide a statement of the second law of thermodynamics. Later, statistical mechanics provided a connection between the macroscopic property of entropy and the microscopic state of the system. This work was the crowning achievement of Boltzmann who had the celebrated equation $S = \ell og(W)$ inscribed as the epitaph on his gravestone. (*Note*: S denotes entropy and W represents energy.)

Since the time of Boltzmann, in the physical systems, quantities of the type $H = -\sum_i p_i \ell og(p_i) = -S$ have been encountered either as an expression of thermodynamic entropy or as a quantity closely related to it. Boltzmann observed that the thermodynamic entropy has something to do with *orderliness* and *disorderliness* in the system. In 1929 Sziland pointed out that a decrease of thermodynamic entropy is accompanied by the acquisition of information. In 1932 Von Neumann, using S as thermodynamic entropy, demonstrated the *irreversibility of observation* in which information about the state of the system becomes statistically inaccurate. In other words, S measures uncertainty and in a negative sense, $H = -S$ refers to an equivalence of information (that removes the uncertainty). It also measures the strength of correlation, beyond the average interaction, among the stochastical constituents of a system. A close connection between information-theory entropy and thermodynamics has also been elucidated by Brillouin.

The mathematical information theory deals with the syntactic aspects of the transmission information across a channel. Applied to a function of a single neuron, information theory, in a primitive sense is useful for the classification of coding problems of neural signals.

More considerations on the thermodynamics entropy *versus* neural information content are discussed in Chapter 8, with respect to thermo-biodynamic perspectives.

Entropy, when referred to a random variable, is a measure of the uncertainty of the variable; it is a measure of the amount of information required on the average to describe the random variable. The link between entropy, thermodynamics and the concepts of information theory was elucidated as early as 1930 by Hartley [1.34]. He introduced a logarithmic measure of information for communication. His measure was essentially the logarithm of the alphabet size. Subsequent interpretation of entropy by Shannon is that the distributions of higher (neg)entropy "assume less." Shannon was also the first to define *relative entropy* or *cross-entropy* and *mutual information*.

Relative entropy is a measure of the distance between two distributions. In statistics, it arises as an expected logarithm of the likelihood ratio. That is, the relative entropy when depicted by measure of the inefficiency of assuming that the distribution is q when the true distribution is p. For example, if one knows distribution of the random variable, then a code with average description length H(p) can be constructed. If instead, the code for a distribution p would need $H(p) + D(p\|q)$ bits on the average to describe the random variable.

Relative entropy as a statistical distance measure was first defined by Kullback and Leibler [1.35]. It is also known under a variety of names, including the *Kullback-Leibler distance*, cross-entropy, *information-divergence* and *information for discrimination*, and has been studied in detail by Csiszàr [1.36], Amari [1.37] and others [1.38]. Specifically, Kullback-Leibler entropy can be specified in the context of differential geometry as the Riemann metric in the distributions.

In addition to relative entropy, the concept of *maximum entropy* was evolved as the most unbiased representation of one's ignorance of the state of the system. It is specified by the probability distribution which maximizes the entropy, $S = -\sum_i p_i \log(p_i)$ under the condition $\sum_i p_i = 1$ where p_i (i = 1, 2, 3, ..., n) denotes the probability of the ith event; and the system under consideration permits the average value of a quantity, f_i (associated with the ith event) to be a constant equal to $\sum_i f_i p_i$. In the case of an interaction which exists between the events, f_i, in general, is a function of p_j's.

As Jaynes [1.39-1.41] defines information theory, it provides a constructive criterion for setting up probability distributions on the basis of partial knowledge, and leads to a type of statistical inference which is called the maximum-entropy estimate. It is the least biased estimate possible based on the given information. That is, it is the maximally noncommittal estimate with regard to missing information. Jaynes ingenuously has extended the maximum entropy principle to reconstruct the concepts of classical statistical mechanics. The fundamental aspects of evolving relevant equations related to information theory are presented in Appendix 1.2.

A set of associated entities pertinent to entropy versus information such as *Fano's inequality* and the notion of *sufficient statistic* are presented in Appendix 1.1 along with concepts like the *minimal sufficient statistic*. The relationship described between mutual information and sufficiency is due to Kullback [1.38].

The relationship between information theory and thermodynamics has been discussed extensively by Brillouin [1.42] and Jaynes [1.39-1.41]. Although the basic theorems of information theory were originally derived for a communication system, attempts have also been made to identify these theorems with the fundamental laws of physics. There have also been attempts to determine whether there are any fundamental physical limits to computation of such theorems as indicated in the works of Alekseev [1.43] and Bennett and Landauer [1.44, 1.45]. A summary of basic aspects of information theory is presented in Appendix 1.1.

1.4 Information-Theoretics and Neural Network Training

Pertinent to the application of information theory and the allied concepts of complexity to neural networks, Abu-Mostafa [1.46] states, "The performance of a conventional computer is usually measured by its speed and memory. For neural networks, measuring the computing performance requires new tools from information theory and computational complexity".

Considering the neural complex as an architecture with layers of cellular units which are fully interconnected, each of the cellular units can be regarded as an information-processing element. In reference to this layered architecture, Liou [1.47] suggested the use of cross-entropy (CE) principles and Ackley et al. [1.48] deployed the concepts of reverse cross-entropy (RCE) to define a *distance function* in order to depict the deviation of neural statistics in the presence of environmental inputs from that in the absence of such inputs. Minimization of this distance parameter refers to the process of attaining the objective function. Thus, both CE and RCE concepts describe implicitly an entropy-based, information-theoretic cost-function to train a layered framework of neural units.

In addition to CE and RCE techniques, a host of other information-theoretic error-measures, such as Csiszár's set of divergence measures, can be applied to the neural networks. Such preferences remove the possible constraints of working with restricted error-measures in parametric space and offer a wider scope and ample choices in selecting the most appropriate error-metric(s) for the required optimizations.

The cost-functions in the information-theoretic plane pertinent to neural networks are the error-measures representing a set of metrics based on two possible considerations:

- The minimization of relative entropy or distance function(s) in respect to two statistical sets of varieties, namely, the network output and the target values.

- As a parallel strategy, elucidation of an error-measure *via* maximum entropy considerations is also possible and will be discussed later.

A significant aspect of developing multiple error-measure strategies in the information-theoretic plane is to train the neural network in terms of global information associated with the input-output entities rather than in terms of the parametric characteristics of those entities. The parametric features are rather discrete and could be decoupled explicitly from each other. They may not portray the cohesive interrelations between themselves in dictating the neural functions trained towards yielding optimized results.

Also, elucidation of a variety of information-theoretic measures and their applications in network training, parallel to adopting conventional error-measure strategies such as quadratic cost-function, offer a broader flexibility and choice towards realizing better algorithms in characterizing the neural input-output relation more precisely and possibly with better accuracy and more robust results. Such studies would also lead to implementing the information-theoretic based error-measures in fuzzy neural networks.

In the application of information-theoretic considerations to neural network training strategies, there are two feasible methods of incorporating entropy (or information theory) based entities in the protocols of network training as illustrated in Figs.1.2 and 1.3. These methods are:

- Maximum (direct) entropy considerations (*Jaynes' formalism*)

- Minimum (cross) entropy considerations (*Kullack-Leibler formalism*)

In Fig.1.2 a multilayered neural network is trained via supervised learning strategies. The network's output O_i in response to a set of inputs $\{x_i\}$ is compared with a teacher value T_i and the difference $d_i = (O_i - T_i)$ is determined. Considering, the stochasticity associated with O_i and/or T_i, the difference entity d_i represents a stochastical variable whose direct entropy functional is given by [1.48]:

$$I_D(d_i) = -\sum_i p_{d_i}(d_i)\left[\ell og\left\{p_{d_i}(d_i)\right\}\right] \qquad (1.4)$$

The above expression of Eqn.(1.4) also represents the information content of d_i by virtue of its probabilistic attributes specified by its probability density function (pdf) denoted in Eqn.(1.4) as $p_{d_i}(d_i)$.

Considering the direct entropy I_D as a cost-function, its maximization via feedback would lead the network's performance to convergence. An example would be backpropagation using gradient-descent algorithm. That is, O_i will approach close to the target value T_i.

The feasibility of using the maximum entropy as a cost-function in elucidating the performance of a neural network stems from the simple fact that the neural network represents an interacting system. The functioning of the neural network depends on the collective behavior of the interconnecting neurons. Application of the information-theoretic considerations in terms of the maximum entropy principle to study a system of interaction elements, in general, has been elaborated by Takatsuji [1.49].

23

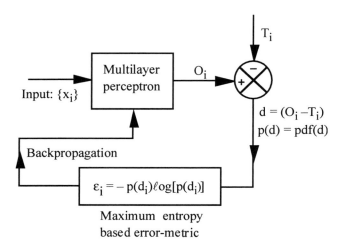

Maximum entropy
based error-metric

Fig.1.2: Entropy maximization based neural network training. ($\varepsilon_i \Rightarrow$ Direct entropy)

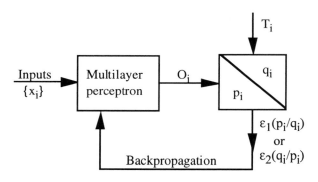

Minimum entropy (cross-entropy)
based implementation

Fig.1.3: Entropy minimization based neural network training. ($\varepsilon_{1,2} \Rightarrow$ Cross-entropy)

In contrast to the above technique, the network training can also be implemented as shown in Fig.1.3. Here, by denoting the pdf of O_i and p, and pdf of T_i as q, a relative entropy function (which measures the "distance" between the two statistics as indicated before) is written as follows:

$$D(p_i|q_i) = \sum p_i \ell og(p_i/q_i) \qquad (1.5a)$$

or,

$$D(q_i|p_i) = \sum q_i \ell og(q_i/p_i) \qquad (1.5b)$$

24

The above expressions depict the mutuality between the statistical distributions of O_i and T_i or *vice versa*. Hence, they represent *conditional* or *transinformation functions*. Either the expressions of Eqns.(1.3a) and (1.3b) or their linear combination can be adopted as a cost-function which can be minimized (*via* backpropagation based gradient-descent procedure), in order to enable the network to converge. Minimization of D(.) physically means, reducing the *distance* or *divergence* between the statistics of O_i and T_i.

The concept of "distance" measuring the statistical separation between two groups dates back to the earlier part of the century. In 1905 Mahalanobis [1.50,1.51] offered a D^2 statistical definition of "distance" in analyzing two groups in a mixture and formalized a "coefficient of likeness" parameter. It is essentially a quantification of distinguishability of sets of statistical distributions. In 1936, Fisher [1.52] prescribed a *linear discriminant function* towards establishing the distance between two statistics. In [1.53], Bhattacharyya discusses the measure of divergence as a metric of resemblance between multinomial populations. Further, he has proposed the visualization of such populations with different weights in terms of "population lines" having varying extents of "angles of divergence". Hence, he proposed a distance measure known as the *Bhattacharyya distance*, given by :

$$B(p_1; p_2) = -\ell og_e(\rho) \qquad 0 \le \beta \le \infty \qquad (1.6a)$$

where

$$\rho = \int [p_1(x)p_2(x)]^{1/2} dx, \qquad 0 < \rho < 1 \qquad (1.6b)$$

which is known as the *Bhattacharyya coefficient*.

Jeffreys [1.54] has indicated the divergence as the difference in mean values of the loglikelihood ratio under two hypotheses and is given by:

$$J(p_1; p_2) = E_1[\ell og_e L(x)] - E_2[\ell og_e L(x)] \qquad (1.7)$$

where $E_i[\ell og_e L(x)] = \int [\ell og_e L(x)]p_1(x)dx$, $i = 1, 2$; and, $L(x)$ is the *likelihood ratio*, namely, $p_i(x)/p_2(x)$ where $\int p_i(x)dx$ ($i = 1, 2$) specifies the probability of observation x when hypothesis, h_i is true.

An alternative fomalism of statistical distance is due to Kolmolgorov who defined a "variational distance" given by:

$$K(p) = \frac{1}{2} \int [\pi_i \pi_i(x) - \pi_2 \pi_2(x)]dx \qquad (1.8)$$

with $(\pi_1 + \pi_2) = 1$.

Pertinent to evolving decision rules based on the "distance" related to the problems of (goodness of) fit, Matusita [1.56] has developed a *coefficient of the affinity* similar to the Bhattacharyya coefficient, to specify a distance ($\|..\|$) between two distribution functions.

1.4.1 Cross-entropy based error-measures

The real neural complex has an entropy behavior generally similar to that of closed systems — systems in which the difference between the state of the system and the state at maximum entropy grows with the input information per unit input energy. The entropy of the neural system depends on the energy content of the system, so that it shows a maximum with a medium degree of energization (and the function is inverted U-shaped). Since the information capacity behaves in the same manner as entropy, the information handling capacity of a neural system exhibits a maximum only for an optimal degree of energization. This is effected, for instance, through sensory channels and will decrease if the energy content of the system decreases or increases from this optimum level.

The primary reason to develop an entropy model of the neural complex is due to the need for an operational definition of the *macrostate* of a neural system. As stated earlier, the term *macrostate* is used here in the same sense as in physical thermodynamic theory. It corresponds to the "overall state" of a neural system as the term is often used by neurophysiologists, in order to describe certain physiological states. The macrostate describes the collective behavior of the interconnected neurons; whereas, the microstate of the neural complex can be regarded as the localized activity state of the neurons. Any change that can be observed at the microstate level may not influence the global microscopic attributes of the neural system. The macrostate of the nervous systems refers to:

- Sensory and voluntary mapping functions

- Selection of neural input and output information

- The consciousness function

There is, so far, no comprehensive theory which explains quantitatively the connection between the *macrostate* and the *microstate* of a neural system. In [1.57], Bergström has proposed the concept of neural entropy. Considering, the microstate as equiprobable, suppose P_M depicts different possible microstates pertaining to the same macrostate. Bergström then defines the neural entropy (S_N) as:

$$S_N = K\ell og(P_M) \tag{1.9}$$

where K is a constant of proportionality.

In this entropy model, the entropy and the neural energy are parameters describing the macrostate, and the number of neuronal elements and the number of their branches are parameters describing the microstate.

Such a model shows that it may be possible to define magnitudes for the neural signaling functions, which are conserved in the system despite opposing views which may exist. In fact, it is difficult to explain certain overall performances of the brain occurring at the psychological (behavioral) level without a prescribed conservation principle. The entropy model includes the "maximum of entropy". This would mean that each disturbance of the neural system by environmental forces which decrease its entropy is followed by a maximization of the entropy. This maximization effort operates as a "statistical force", being inherent in the system, and is able to affect the behavior at the output channels of the system. Without such neural forces, it is difficult to explain the behavioral performances of neural systems. In the physiological concept of regulation of behavior there is a lack of definition of such forces. The essence of maximum-entropy principle is that it offers an information-theoretic approach to systems, where interaction between elements is encountered. Information theory works as a more powerful tool for the description of the input-output relations than many other statistical techniques. This is especially true in dealing with neurophysiological data where relevant quantities are defined and assessed in the time-course of transmitted neural information.

As stated before, by considering the entropy associated with the neural system as an entity describing the behavior of the macrostate, it represents implicitly the extent of probability of this macrostate. By using this concept, the *Weber-Fechner law* of perception and stimuli has been explained by Bergström[1.57] in the entropy framework.

From the perspective of entropy considerations, (as stated before) there are two methods of inference when given new information in terms of expected values. They are Jaynes' principle of maximum entropy and Kullback's principle of minimum cross-entropy (minimum directed divergence). As shown by Shore and Johnson [1.58], these are the only correct methods for such inductive inferences. However, more general considerations of the maximum entropy principle and the minimum cross-entropy principle are indicated as follows.

Suppose it is known that a system has a set of possible states S_i with unknown probabilities $p(x_i)$. Then the constraints on the distribution $p(x_i)$ specified either by the value of certain expectations $\sum_i p(S_i) f_k(S_i)$ or by the bounds on these values are learned. Suppose one needs to choose a distribution p that is in some sense, the best estimate of p given what is known *a priori*. Usually there remains an infinite set of distributions that are not ruled out by the constraints. Therefore, which one should be chosen?

The principle of maximum entropy states that of all the distributions p that satisfy the constraints, the one with the largest entropy $\sum_i p(S_i) f_k(S_i)$ should be chosen. That is, the most unbiased representation of our ignorance of the state of the system is provided by the probability distribution which maximizes the entropy. Entropy maximization was first proposed as a general inference procedure by Jaynes, although it has historical roots in physics. It has been applied successfully in a remarkable variety of fields, including statistical mechanics, thermodynamics, traffic networks, queuing theory and computer system modeling, system simulation, production line decision-making, computer memory reference patterns, system modularity, group behavior, stock-market analysis and general probabilistic problem solving. There is also much current interest in maximum entropy spectral analysis. More details on maximum entropy considerations are presented in a later section (Section 1.4.3).

In simple terms, the principle of minimum cross-entropy is a generalization that applies in cases when *a priori* distribution q that estimates p is known in addition to the constraints. The principle states that, of the distributions q that satisfy the constraints, the one with the least cross-entropy $\sum_i p(x_i) \ell og[p_i(x_i)/q(x_i)]$ should be chosen. Minimizing cross-entropy is equivalent to maximizing entropy when the former is a uniform distribution. Unlike entropy maximization, cross-entropy minimization generalizes correctly for continuous probability densities. Then the following functional is minimized:

$$H(p,q) = \int p(x) \ell og[p(x)/q(x)] dx \qquad (1.10)$$

The name cross-entropy is originally due to Good [1.59]. Other names, as stated before, include expected weight of evidence, directed divergence and relative entropy. First proposed by Kullback, the principle of minimum cross-entropy has been advocated in various forms by others [1.60-1.62], including Jaynes [1.63-1.64], who obtained Eqn. (1.3) with an "invariant measure" playing the role of the prior density. Cross-entropy minimization has been applied primarily to statistics, but also to statistical mechanics, chemistry, pattern recognition, computer storage of probability distributions, and spectral analysis. For a general discussion and examples of minimizing cross-entropy subject to constraints, see [1.65].

In the communications area [1.66,1.67] it has been shown that apart from maximization of an entity such as the signal-to-noise ratio, minimization criteria (for example, in reference to minimum error probability) is suitable on a context basis. In the search of such minimization criteria, the notion of a "distance" between two probability density functions is quite useful. As Kailath [1.66] points out, the two distributions can be the distributions of the observations in a binary hypothesis testing situation. The further apart one can make these distributions, the more likely the probability of mistaking one for the other will be smaller. Therefore, various distance measures have been studied as simpler substitutes for the error probability.

What one would like for such distance measures is a property of the following type: If the distance between the two distributions is greater for a signal set α than for a signal set β, then the error probability for set α is always less than the error probability for set β. This is too much to hope for and one should only be able to find *weaker* relations between distance measures and the probability of error.

In statistics, the use of such distance measures has a long history, beginning with the work of Pearson (as quoted in Tildesley [1.68]). Two widely used distance measures in statistics are the D^2-statistic of Mahalanobis [1.50-1.51] and the linear discriminate function introduced by Fisher [1.69]. These measures have been discussed in detail in many statistical texts, for example, those authored by Anderson [1.70] or Rao [1.71]. With the advent of Shannon's information theory in 1948, the divergence, a measure closely related to Shannon's logarithmic measure of information, became popular, though it had first been proposed before Shannon's work by Jeffreys [1.54].

In the engineering literature, the divergence has recently been applied to several problems with varying degrees of success. Another measure proposed in the context of selecting the criterion for signal selection is based on a statistic first introduced in a statistical context by Bhattacharyya [1.53]. It should, however, be pointed out that the Bhattacharyya distance is actually a special case of a more general distance measure introduced by Chernoff [1.72].

Pertinent to the context of this book, it is appropriate to review the extent to which cross-entropy principles have been applied to predicting the performance of neural networks. As stated in an earlier section, the concept of reverse cross-entropy was applied to neural networks by Ackley et al. [1.48]. Ackey specified a distance function to depict the deviation of neural statistics in the presence of environmental inputs from that in the absence of such inputs in the entropy plane. Minimization of this distance parameter has been referred to as the process of attaining the objective function. An alternative method based on cross-entropy was proposed by Liou and Lin [1.47] to study the learning aspects of layered networks.

Further, the geometry of the neural learning manifold has been represented *via* "information geometry" by Amari et al. [1.73] in reference to Boltzmann machines. Information geometry, which originates from the intrinsic properties of a smooth family of probability distributions. It has been considered in [1.73] to establish a natural, invariant Riemannian metric and a dual pair of affine connections on the Boltzmann neural network manifold which represents a network of stochastic neurons. The set of all the Boltzmann machines with a fixed topology forms a geometrical manifold of high dimensions, where modifiable synaptic weights of connections play the role of a coordinate system to specify networks. A learning trajectory, for example, is a curve in this manifold. It is important to study the geometry of the neural manifold, rather than the behavior of a single network, in order to know the capabilities and limitations of neural networks of a fixed topology. In [1.73], the meaning of geometric structures is elucidated from the stochastic and statistical points of view. This leads to a natural modification of the Boltzmann machine learning rule.

Windhorst and Shulters [1.27] prescribed the measures of trans-information to multiple-input/single-output neural systems. A general statistical description of learning processes is given by Pfaffelhuber [1.16] using an extension of classical information theory and the associated uncertainty (subjective entropy) considerations.

Real neural information processing has been addressed by Güttinger [1.74] and the concept of mutual information as applied to synaptic signal has been studied by Uttley [1.75] in depth. The processing of vast information received from the environment by the nervous system towards its learning and self-organization has been discussed in [1.76].

A neurocomputational model to extract information from neural responses (r) about the stimuli (s) that evoke such responses using the conditional probabilities $P(s \mid r)$ of the stimulus, given the response on *a posterior* basis, and the associated cross-entropy error-measure, has been studied by Kjaer et al. [1.77]. Usually, unsupervised learning procedures based on Hebbian principles are successful at modeling low-level feature extraction, but are insufficient for learning to recognize higher-order features and complex objects. Further, a class of unsupervised learning algorithms that are derived from information-theoretic principles also exists. These algorithms have been built on the concept of maximizing the mutual information between the outputs of different network modules, and are capable of extracting higher-order features from data. They are therefore well suited to modeling intermediate-to-high level perceptual modeling stages. This claim has been substantiated in [1.77] with some novel results for two signal classification problems, as well as by reviewing some existing results and known related approaches. Also indicated in [1.77], is the evaluation of the relevant algorithms vis-á-vis computational costs and biological feasibility.

Measuring the information carried by neuronal activity becomes difficult, particularly when recording from mammalian cells, due to the limited amount of data usually available which results in a systematic error. While empirical *ad hoc* procedures have been used to correct for such error, a direct procedure consisting of the analytical calculation of the average error, its estimation up to subleading terms from the data, and its subtraction from raw information measures to yield unbiased measures can be envisaged. The leading correction terms for both average transmitted information and conditional information can be evaluated. Further, inasmuch as one has to regularize the data first, the expressions appropriate to different regularizations need to be specified. Relevant computer simulations would indicate a broad range of validity of the analytical results.

It has been proposed in [1.78] that the backpropagation algorithm for supervised learning can be generalized. The algorithm can be put on a satisfactory conceptual footing, and very likely made more efficient by defining the values of the output and input neurons as probabilities and varying the synaptic weights in the gradient direction of the log-likelihood, rather than the "error". Correspondingly, a binary set-based relative entropy measure which can be used in the gradient-descent algorithm (applied to the neural *via* backpropagation) is given by:

$$\varepsilon_{RE} = (1/2)(1 + \tilde{T}_i)\ell og\left[(1 + \tilde{T}_i)/(1 + \tilde{O}_i)\right]$$
$$+ (1/2)(1 + \tilde{T}_i)\ell og\left[(1 - \tilde{T}_i)/(1 - \tilde{O}_i)\right]$$

(1.11)

where the *tilde* denotes the normalization of the parameters with respect to their (respective) maximum values. Baum and Wilczek [1.78] have shown that adopting the above measure (Eqn.1.4) leads to minimizing the energy or maximizing the negative entropy or the log-likelihood function.

Further, the aforesaid expression (Eqn.1.4), which is based on computing the relative entropy via normalized $\tilde{T}_i$ and $\tilde{O}_i$ parameters, has been used by Watrous [1.79] and Solla et al. [1.80] in training a layered network towards optimized predictions. In the experimental simulations performed by Solla et al. [1.80], learning in layered neural networks is posed as the minimization of error function defined over the training set. Since a probabilistic interpretation of the target activities suggests the use of relative entropy as an error measure, the merits of using this error function over the traditional quadratic function for gradient-descent learning has been investigated [1.81]. Comparative numerical simulations for the *contiguity* problem show marked reductions in learning times. This improvement is explained in terms of the characteristic steepness of the landscape defined by the error function in configuration space.

In a continued effort, Watrous [1.79] has made a study to compare the square-error and the relative entropy metric (Eqn.1.4) using several optimization algorithms. In [1.79], the convergence rates and generalization performance are compared for the square-error metric and a relative entropy metric on a contiguity problem using those optimization algorithms. The relative entropy measure converged to a good solution slightly more often than the square-error metric given the same distribution of the initial weights. However, where the results differed, the square-error metric converged on average more rapidly to solutions that generalized better to the test data. These results are not, however, in complete agreement with some results previously published.

At this juncture, it is appropriate to indicate the salient aspects of error-measure or cost-functions in general which have been prescribed conveniently towards learning strategies vis-á-vis neural networks. Cost-function, in reference to neural networks, is an *ad hoc* introduced measure, minimization of which corresponds to attaining global minimum through the adjustment of synaptic coupling. The dynamic search of such a global minimum refers to the *learning rules*. That is, the learning procedures or learning rules are, in general, synaptic modification algorithms that allow an arbitrarily connected network to develop an internal network architecture appropriate for a particular task. This goal can be achieved on the basis of direct comparison of the output of the network with known correct answers (examples). The synaptic

31

couplings are then modified in order to reproduce the examples as closely as possible. This is sometimes called *supervised learning*.

The constraints to be imposed over the cost-function are, in general, very weak, allowing enormous freedom of choice. One of the most widely used constraints is that the cost-function induce a local learning rule. This means that the variation of synapse between two neurons at a given time should depend only on the instantaneous post-synaptic potentials (PSP) received by them, and not on the PSP received by the rest of neurons. Such a requirement has a heuristic character and, although quite plausible from a biological point of view, it is not supported by concrete empirical evidence. Therefore, it is of interest to study the effect of introducing non-local learning rules in a neural network. The dynamical effect of some kind of cost-function that induces non-local rules in a simple perceptron corresponds to the Tsallis-statistics-based generalization of the gradient-descent dynamics. The effect of different choices of cost-function which generate both local and non-local rules are specific to the problem of "memorization", namely, the perceptron learning a single random pattern uncorrelated with previously learned ones.

The most popular error-measure is the so-called *square or quadratic error* (ε_{SE}) which is obtained from the mean square of the difference between the network output (O_i) and the teacher value (T_i). That is,

$$\varepsilon_{SE} = 1/2\sum_i (O_i - T_i)^2 \tag{1.12}$$

This error-measure of Eqn.(1.12) offers a convenient mode of implementation for minimization, such as application of a negative gradient-descent algorithm facilitating a step-wise procedure. It is normally formulated as an iterative "learning rule" which, for sufficiently frequent sequential "presentation" of the training pairs to be stored, gradually produces the optimal synaptic weights being adjusted in the network by means of a feedback such as backpropagation. Square-error specifies an Euclidean norm.

Another noninformation-theoretic metric used in the backpropagation-based neural network training refers to Minkowski-power metrics. Hausen and Burr [1.82] have studied the learning in connectionist models using these type of noneuclidean error-measures. The study presented in [1.82] generalizes, in particular, the backpropagation by using the so-called Minkowski-r power metrics. For small r's, a "city-block" error-metric is approximated, and for large r's the "maximum" or "supermum" metric is approached. For r = 2, the standard backpropagation model results. An implementation of Minkowski-r backpropagation and several experiments performed show that different values of r may be prescribed for various purposes. These different r-values may be appropriate for the reduction of the effects of outliners (noise), modeling the input-space with more compact clusters, or modeling the statistics of a particular domain more naturally.

They may also be used in a way to be more perceptually or psychologically meaningful— for example, speech or vision.

The aforesaid error-measures, namely, relative entropy-based metrics, mean-square error-measure and Minkowski power metrics, are all specified in terms of the parameter values of O_i and T_i and do not address the stochasticity associated with O_i and T_i.

As discussed earlier, pertinent to the stochastical environment of the neural complex, a more appropriate description of comparing O_i and T_i (that is, an error-measure based on O_i and T_i) should duly take into consideration the associated probabilistic attributes of O_i and T_i. In this perspective, it is therefore more appropriate to consider the maximization of the information-theoretic error-measure, namely, $- p_{d_i}(d_i)[\ell og\{p_{d_i}(d_i)\}]$. Here, the entity $d_i = |O_i - T_i|$ represents the magnitude of the Euclidean distance and p_{d_i} refers to the pdf of d_i. Alternatively, one could pursue the minimization of the cross-entropy based information, $p_i \ell og(p_i/q_i)$ and/or $q_i \ell og(q_i/p_i)$.

1.4.2 Symmetry aspects of divergence function

The pursuit commonly followed in using an information-theoretic based error-measure in lieu of the conventional error-metric) as a cost-function in optimization problems, refers to deducing a cross-entropy function, the simplest version of which is the Kullback-Leibler (KL) information measure. Symmetrized versions of KL-measure known as the Jensen and/or Jensen-Shannon measures, have also been considered in the literature [1.67] to specify explicitly the bidirectional influences of p and q and q and p. *Mutual cross-information measure, transinformation measure, divergence measure, dissimilarity measure, discrimination information* are the other nomenclatures used in the literature to specify the cross-entropy measure on a contextual basis. This measure described in a bilaterally symmetric form is also known as the "distance" measure. Instead of the logarithmic function, a more generalized function (Φ-function) can also be chosen to evaluate the cross-entropy values. These Φ-functions should, however, satisfy certain conditions. They may also include certain weighting constants and/or order parameters of constrained values. A class of functions known as Csiszár's error-measures [1.36] based on various Φ-functions are enumerated in Chapter 4 and are studied with respect to their effectiveness in training neural networks.

Information about an object is conveyed by the knowledge that some property is absent, as well as by its presence; for example, in recognizing a hand-printed numeral, by knowledge that there is no horizontal stroke. A similar principle can be applied to the information. If an input $F(X_i)$ occurs, a quantity $\gamma_i F(X_i)$ is added to $F(Y)$, the internal state of the neuron. Such a system can be called an unsymmetrical information system. Information with a quantity of opposites sign is added to $F(Y)$. There is a simple conclusion, based on the theorem that the long-term average information conveyed by the presence and absence of a signal tends to be zero. That is, in symmetrical information handling, the average contribution of a synapse converges to zero. This will be achieved if a synapse behaves like an electrical capacitor which, on average, passes no current; that is, if a synapse communicates only changes in its postsynaptic state $\gamma_i F(X_i)$ to the internal state of the neuron.

In terms of the above considerations, Uttley [1.82] defines the information associated with a real neural synapse as an "informon." That is, by designating the synapses obeying certain information rules as informons, he has defined the symmetrical characteristics of the informon in terms of average mutual information relevant to a set of binary signals as follows:

For two binary signal X_i and Y, the probability of Y given that X_i is absent is written as $P(Y|\overline{X}_i)$, and the mutual information function between $\overline{X}_i$ and Y is given by:

$$I(X_i : Y) = \ell og\left[\frac{P(\overline{X}_i \cap Y)}{P(\overline{X}_i)P(Y)}\right] \qquad (1.13)$$

The average mutual information given by the presence and absence of X_i over a long series of events is:

$$E[I(X_i : Y)] = P(X_i)I(X_i : Y) + P(\overline{X}_i)I(\overline{X}_i : Y) \qquad (1.14)$$

Thus, unless otherwise chosen, the mutual-information based cost-function is taken in a symmetrized functional form, as well as that corresponding entropy relations do not assume a nonadditive structure, the metrics so considered are futile in their performance requirements. Considered in Chapter 4 are ways of symmetrizing and/or balancing the error-entropy measures so as to make them suitable as usable error metrics in the test domain. The symmetric characteristics of the cross-entropy divergence measure are shown to have a significant role to play in driving the network to the global minimum. Relevant considerations addressed in Chapter 4 include typical symmetrized formulations and confirming results obtained *via* on-line simulation studies on a test multilayered perceptron architecture.

1.4.3.Csiszár's generalized error-measures
As discussed earlier, a cost-function based optimization can be specified in the information-theoretic plane in terms of generalized maximum and/or minimum entropy considerations associated with the network (see also Appendix 1.2). A set of minimum cross-entropy or mutual information error-measures, known as Csiszár's measures, can be deduced in terms of probabilistic attributes of the "guess" (output) and "true" (teacher) value parameters pertinent to neural network topologies. Their relative effectiveness in training a neural network optimally towards convergence by realizing a

realizing a predicted output close to the teacher function is discussed in Chapter 4 with simulated results obtained from a test multilayer perception. The Csiszár family of error-measures indicated thereof offers an alternative set of error-functions. These are defined over a training set, which can be adopted towards gradient-descent learning in neural networks using the backpropagation algorithm in lieu of the conventional square error (SE) and/or relative entropy (RE) error-measures. Relevant *pros* and *cons* of using Csiszár's error-measures are deliberated in Chapter 4.

1.4.4 Jaynes' rationale of maximum entropy
The concept of maximum entropy due to Jaynes has been applied successfully to interacting systems such as considering in statistical mechanics [1.65]. Maximum entropy principle is so different from that of "orthodox" statistics that it seems new and mysterious to many although, historically it dates back to Boltzmann, 1877.

Shannon interpreted entropy by stating that distributions of higher (neg)entropy "assume less". In general, the possible distributions are concentrated strongly near the one of maximum entropy. That is, distributions with appreciable lower entropy than the maximum permitted by a set of data are atypical of those allowed by the data. This can be stipulated by the *entropy concentration theorem* which runs as follows:

Suppose a random experiment has n possible results at each trial. Thus, in N trials there are n^N conceivable outcomes. Assuming each outcome yields a set of sample numbers $\{N_i\}$ and relative frequencies of occurrence { $f_i = N_i/N$; $1 \le i \le n$ }, then the associated entropy is given by:

$$H(f_1, f_2, ..., f_n) = \sum_{i=1}^{n} f_i \log(f_i) \tag{1.15}$$

Considering the subclass C of all possible outcomes that could be observed in N trials, compatible with m (m < n) linearly independent constraints of the form:

$$\sum_{i=1}^{n} A_{ji} f_i = d_j, \qquad 1 \le j \le m \tag{1.16}$$

where the matrix A_{ji} defines the "nature" or "attributes" of M's (which are interpreted conceptually as m different "physical" quantities). Further, $\{d_1, d_2, ..., d_m\}$ refers to the data set {D} considered.

A certain fraction F_C of the outcomes in the subclass C yield an entropy bound by the following relation:

$$H_{max} - \Delta H \le H(f_1, f_2, ..., H_n) \le H_{max} \tag{1.17}$$

where H_{max} is determined by an algorithm[7]. The concentration of F_c near the upper bound H_{max} can be specified as a functional relationship between F_c and ΔH, given by:

$$2N\Delta H = \chi_k^2(1 - F_c) \tag{1.18}$$

where $2N\Delta H$ is asymptotically distributed over the subclass **C** as chi-squared with $k = (n - m - 1)$ degrees of freedom regardless of the nature of the constraints. $\chi_k^2(.)$ represents the 100% significance level of the critical chi-squared distribution for k degrees of freedom.

1.4.5 Mutual information maximization

In contrast to maximizing the direct entropy (Jaynes formalism) or minimizing the cross entropy (Kullback-Leibler formalism), there is an interesting method of maximizing mutual information (cross entropy) associated with the outputs of different network modules possible. Such a method facilitates extracting higher order features from data. Normally such higher order features vis-á-vis complex objects may not be recognized by unsupervised learning procedures based on Hebbian principles. This is due to insufficiencies in learning involved, especially with respect to, a real neural complex.

Learning rules in artificial neural networks derived by differentiating global objective functions appear to lack biological plausibility, since such

---

[7] *Maximum entropy algorithm*: Suppose the *partition function* is defined as :

$$Z(\lambda_1, \lambda_2, ..., \lambda_m) = \sum_{i=1}^{n} \exp[-\sum_{j=1}^{m} \lambda_i A_{ji}]$$

Then,

$$H_{max} = \ell og(Z) + \sum_{j=1}^{m} \lambda_j d_j$$

in which the *Lagrangian multipliers* $\{\lambda_i\}$ can be deduced from:

$$\frac{\partial}{\partial \lambda_j}[\ell og(Z) + d_j] = 0, \qquad 1 \le j \le m$$

which corresponds to a set of m simultaneous equations for m unknowns. The relative frequency of occurrence that has the maximum entropy is then

$$f_i = Z^{-1} \exp(-\sum_{j} \lambda_j A_{ji}), \qquad 1 \le j \le n$$

Other distributions $\{f_i\}$ allowed by the constraints of Eqn.(1.16) will have various entropies less than H_{max}.

rules correspond to computation of spatio-temporal stochasticity arranged over an entire training pattern ensemble.

Learning equations and the governing cost-function which represent biologically more plausible forms can be specified by viewing a neuron as a communication channel. That is, a cost-function for learning can be set-up in terms of the rate of information transmission through the "channel". In other words, the mutual information between the input and output of a neuron I(x;y) is optimized.

Suppose for a particular input case α, the output of the ith unit in module A represents a variable y_{Ai}. The on-state probability of y_{Ai} is p_{Ai} and the off-state probability is $q_{Ai} = (1 - p_{Ai})$.

Suppose there are two neighboring units of different modules A and B. For N number of input samples, the probability that the ith unit in module A is in on-state is given by

$$p_{Ai} = \sum_{\alpha=1}^{N} P^{\alpha} p_{Ai} \qquad (1.19)$$

where p^{α} is the *a priori* probability of an input case α. Eqn.(1.19) denotes averaging the activity of the entire input sample distribution pertinent to $\alpha = 1, 2, ..., N$.
Similarly,

$$p_{Bi} = \sum_{\alpha=1}^{N} P^{\alpha} p_{Bi} \qquad (1.20)$$

and, the joint probability that two neighboring units are in on-state simultaneously is given by:

$$p_{AiBi} = \sum_{\alpha=1}^{N} P^{\alpha} p_{AiBi} \qquad (1.21)$$

which represents the expected value of the product of the states of the units over the sample-space.

For the set of output y_{Ai} and y_{Bi}, in terms of the probability indicated above, the mutual information, I (in Shannon's sense) is given by (See Eqns. (A 1.1.5 and A 1.1.7):

$$I(y_{Ai}; y_{Bi}) = H(y_{Ai}) + H(y_{Bi}) - H(y_{Ai}, y_{Bi}) \qquad (1.22)$$

The differential of I with respect to each incoming k^{th} weight w_{ik}, namely, $\partial I(y_{Ai}; y_{Bi})/\partial w_{ik}$ defines a weight up-date rule in biological neural machinery. The binary formalism as above can be extended to discrete multi-valued signals as well. Further, these formulations are also valid for continuous variables.

The use of mutual information cost-function, namely I, in unsupervised learning and the relevant biological plausibility considerations are discussed in [1.77] comprehensively. The relevant computational complexity is also addressed in respect to the associated gradient computation. Application of this algorithmic approach appears to be a powerful learning procedure involving extraction of the features which are coherent across different neural input channels, depicting biologically plausible models of self-organization.

1.5 Dynamics of Neural Learning in the Information-Theoretic Plane

While *the learning dynamics,* namely, the convergence characteristics of the network *versus* the temporal (epochal) iterations of error-feedback, are invariably considered in parametric space, no specific details on the convergence aspects of the neural information towards an equilibrium are portrayed in the existing literature. That is, the information-theoretic aspects of neural learning dynamics have not been addressed comprehensively in the existing literature.

The dynamics of learning, in general, indicate the temporal discourse of a network parameter (for example, the error-metric used) and the crossings of the learning curve at a specified equilibrium value which depicts the "attractor" regions. The general considerations on the size of the basin of the attraction for randomly sparse neural networks with optimal interconnection have been elucidated in [1.83]. Using the relevant results, the static properties of storage/memory have been generalized to the optimal dynamics of the neural complex. Further, the results elucidated by Gardner [1.84] indicate that for the symmetric, fully connected model of the network, the size of the basin of attraction is determined by the first time-step. This time-step is larger than the size of the basin attractions determined upon the completion of iterations when stability is obtained. In general, as Zak[1.83] points out, the failure of the so called *Lipshitz condition*[8] in unstable equilibrium points of dynamical systems leads to multiple choice response to an initial deterministic input.

By and large, in the efforts of Zak [1.83] and those of Gardner [1.84], only the parametric space of the network has been considered in analyzing the dynamics of the network's performance. The study as described in Chapter 5, however, projects the problem in the information-theoretic plane. The convergence aspect of the network is studied *via* augmentative and annhilative information processed as a *random-walk phenomenon* in the network during the epochal iterations of error-measure feedback to the network. It is considered that the methodology presented in Chapter 5 is new and has not been explored earlier. Another interesting aspect of learning dynamics refers to some non-locality (non-extensiveness) in the learning procedure, a strategy which is more akin to biological plausibility. That is, unlike the conventional method wherein the quadratic cost-function induces a local learning rule, it would be of interest to study the information-theoretic cost-

---

[8] *Failure of the Lipshitz condition*: Suppose a dynamical system is described by: $\dot{x} + F_i(x_1, x_2, ..., x_3) = 0$, where $i = 1, 2, 3, ..., n$ and F_i's denote continuous functions of all their arguments; Then, $\partial F_i/\partial x_i \to \infty$ at $x_k = x^o_k$ depicts the failure of Lipshitz condition.

functions compatible with non-local learning rules with relevant dynamics based on Tsallis-statistics as indicated earlier.

Learning in reference to real neural complex depicts progressive modifications occurring at the synaptic levels of the interconnected neurons. The presence of intraneural disturbances, inherently present, or any extraneural noise in the input data and/or in the teacher values may, however, affect such synaptic modifications specified by the set of weighting vectors of the interconnections. As discussed earlier, the noise considerations when translated to artificial neurons refer to inducing an offset in the convergence performance of the network in striving to reach the goal or the objective value *via* the supervised learning procedure implemented. The dynamic response of a learning network, when the target itself changes with time, can be studied in the information-theoretic plane. The relevant nonlinear (stochastical) dynamics of the learning process can be described by the *Fokker-Planck equation*. The dynamics are specified in terms of a conditional-entropy (or mutual information) based error-measure elucidated from the probabilities associated with the input and teacher (target) values. The logistic growth or evolutionary aspects and certain attractor features of the learning process can be described in reference to neural manifolds using the mathematical foundations of statistical dynamics as discussed in Chapter 5. On-line simulation studies on a test multilayer perceptron can be performed to portray the asymptotic behavior of accuracy, speed of learning *vis-á-vis* the convergence aspects of the test-error-measure(s) used in training the network.

Further, the learning process by which a neural network adopts itself to its environment can be analyzed by prescribing an optimal non-zero learning parameter that governs the adaptability of the network towards "accurate" learning.

The dynamics of learning specified in the information-theoretic plane can also be extended to fuzzy neural systems via the *fuzzy Fokker-Planck equation*. In extending such dynamic considerations to fuzzy networks, the probabilistic (stochastical) meaning of fuzziness should be prescribed (in the perspectives of Zadeh [1.85]). A corresponding stochastical *fuzzy differential equation* can, thereof, be evolved to describe the pdf of the associated error-measure $\varepsilon(t)$ elucidated in the information-theoretic plane, as a function of time. This fuzzy differential equation of the pdf of $\varepsilon(t)$ can be evolved based on the approach due to Kandel and Byatt [1.86] and Kandel [1.87]. The differential equation in the fuzzy error-domain (E^f) describes the stochasticity of the fuzzy network's dynamics in terms of the fuzzy Fokker-Planck equation mentioned before. Derivation and interpretations of the fuzzy Fokker-Planck as addressed in Chapter 5 are new and have not been discussed in the earlier studies known.

1.6 Neural Nonlinear Activity in the Information-Theoretic Plane

The nonlinear activity in the neural complex representing a biological or an artificial network of neurons can be described by *the Riccati differential equation* relating the input and output parameters of the network. The general solution of this equation represents a single parameter family of curves which are S-shaped (sigmoids). The underlying basis in the development of the nonlinear differential equation vis-á-vis the neural input-output with respect to a set of inputs is described in Chapter 2. Among the feasible such differential equations, it is indicated in Chapter 2 that the *Brillouin-Bernoulli function* is a generalized representation of the stochastically justifiable sigmoid as governed by the spatial long-range order of neuronal state proliferation. Further, it is shown that the sigmoidal function derived as a solution of the Riccati differential equation can be used in the algorithms specified towards the minimization of the quadratic cost-function, as well as the information-theoretic error metrics. The pros and cons of choosing a sigmoidal function arbitrarily for use in ANN paradigms are also discussed in Chapter 2.

The nonlinearity aspect of the neural input-output relation can be tied to a concept of *information interaction*. Suppose a system (Ω) is composed of nonlinear, mutually interacting dynamical entities such as neural units. There exists an interaction between these entities and the environment (Ω') which can be studied in the information-theoretic perspective [1.24]. Considering the open-system Ω energetically interacting with the environment Ω' as illustrated in Fig.1.4, the associated differential entropy $dS_i(t)$ can be specified as ($dS_i + dS_e$) where dS_i (≥ 0 always) represents the differential of the internally produced entropy; and, dS_e is the differential of the entropy pertinent to the inflowing energy from Ω'.

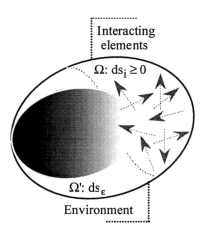

Fig.1.4: Information interaction between a set of interacting elements $\{\Omega\}$ and their environment $\{\Omega'\}$

The sign of dS_e decides the negentropic or posentropic characteristics of the interaction involved. When $dS_e \geq 0$, it refers to "disorder", or *ataxia,* entering the system from the environment. In the case of $dS_e < 0$, it facilitates an "order". In this case, the negentropic input from the environment prevails. When $|dS_e| \geq dS_i$, this condition allows the entropy of the system to decrease at the expense of the environment. That is, $dS(t) \leq 0$.

The interactions, as above deliberated in information-theoretic plane, refer to a nonlinear model representation via, the maximum entropy principle [1.88]. It is shown in [1.88] that the behavior of such an interacting system, as in ferromagnetism, follows a sigmoidal input-output relation.

The fuzzy aspects of nonlinear processes associated with the neural complex can be modeled in terms of a logistic input-output relationship at the cellular level as governed by a fuzzy Riccati differential equation. The solution of this equation represents the function approximation of overlapping output clusters resulting from the segments of input-space grouped into membership classes, each depicting a certain range of input values. Relevant analytical representation of the input-output relation in a neuron is a fuzzified sigmoid which yields a membership function, namely $^{s^0} \mu(X_i) \in [1,0]$ where $\{X_i\}$ represents the summed (and weighted) inputs. The degrees of membership $\mu(.)$ are assigned by an order parameter S^0. The logistic growth of output *versus* input in the fuzzy neural complex under consideration follows a generalized representation of a stochastically justifiable sigmoidal function as decided by the spatial long-range order of neuronal state proliferation across the network. It also captures the approximate nature of reasoning and perception associated with the "granular information" vis-á-vis the fuzzy set(s) of the variables involved. An architecture of a fuzzy neural network which deploys the fuzzy sigmoid is presented and discussed in Chapter 3.

1.7 Degree of Neural Complexity and Maximum Entropy

A neural complex, real or artificial, which is an embodiment of a massively connected set of neurons represents a cellular automaton "trained to learn" and predict *via* endeavors managed by a set of protocols. These protocols involve collection, conversion, transmission, storage and retrieval of information. The training or learning effort is to recognize and counter-balance the effects of the cellular disturbances (noise) present in the neural system which may tend to disorganize the system's convergence towards an objective function mediated through learning protocols. The extent of disorganization caused by such disturbances can be specified by a disorderliness parameter set by the maximum entropy considerations. Such an entropy functional depicts implicitly the degree of complexity of the system in spatiotemporal domains as well. Therefore, the disorderliness in the neural complex can be specified by a *complexity metric*. Using this metric-parameter as an error-measure, or cost-function, a control strategy such as the backpropagation based gradient-descent method can be developed to train a multilayered perceptron. A study presented in Chapter 6 offers relevant algorithmic considerations and simulation results.

The complexity metric depicts implicitly an information-theoretic error-measure based on maximum entropy considerations. It essentially corresponds to elucidating an error-measure towards neural training as indicated in Fig.1.3. However, the neural complexity parameter broadly includes in its definition the structural and algorithmic considerations of the neural spatiotemporal constituents, namely, the beds and realizations.

Kolmogorov [1.89] and Chaitin [1.90] have independently prescribed a theory that associates the information content to finite strings, based on the fundamental characteristics of randomness. In Kolmogorov's perspective [1.89], the quantitative definition of information falls under two categories, *combinatorial* and *probabilistic*. In the combinatorial framework, by assuming a varible x capable of taking different values in a finite set $\{\mathbf{X}\}$ containing N elements, the entropy of the variable x is given by: $H(x) = \ell og(N)$. By giving a definite value, say, x = a, one removes this entropy and communicates information, $I = \ell og_2(N)$. This approach has a logical independence of probabilistic axioms. The probabilistic definition follows the approach wherein, by assuming x as a random variable, $H(x)$ is defined as $-\sum_x p(x)\ell og[p(x)]$ where p(x) is the probability attribute of the random variable x. This probabilistic approach advocated by Kolmogorov, can be noted as the same as that of Shannon's definition of information.

Kolmogorov's complexity concept forms a basis for information theory without recourse to probability concepts. It can be interpreted as the amount of information contained in an object about itself. Since the neural complex is constituted by strings of neurons, the concepts of Kolmogorov complexity could form a basis for quantifying objectively the associated information spread across the beds and realizations. That is, the spatiotemporal neural activities are a surrogate for the total randomness of the complete complex system. One can infer that the information content in the activities cited is related to the amount of information in the complex system.

Inasmuch as the randomness or disorderliness of the neural system refers implicitly to the system complexity C_S as observed from the system exterior, and depicts the "variety" of the system subsets and/or microsubsets constituted by the neural assembly defined earlier, C_S can be represented mathematically, by:

$$C_S = \phi(N, v) \qquad C_S \in \Omega_{C_S} \Leftrightarrow \Omega_S \qquad (1.23)$$

where N is the number of cellular assemblies and v is their (associated) *variety*, namely the synaptic modifications *vis-á-vis* excitatory/inhibitory states. The function ϕ measures implicitly the uncertainty or entropy due to the complexity and therefore could be logarithmic in the perspectives of

Shannon's law. Further, Ω_{C_S} is the complexity domain equivalent of an entropy domain Ω_S. The uncertainty indicator of the complexity should be linked to the disorderliness function by a criterion that, in the event of the system tending to be well-ordered or least complex (through minimization of disorderliness) with the condition $Y_j \rightarrow 0$, it should be accomodated by an appropriate functional relation between Y_j and C_S. Further, in terms of the geometrical space Ω_{C_S} representing the disorderliness, the complexity defines not only a measure of uncertainty but also refers explicitly to the *a priori* probability of a sample within the region. In reference to the foregoing discussions, the complexity of a neural disorganization is composed of:

- Spatial factors conceived in terms of the random locations of the neuronal assemblies

- Temporal characteristics as decided by the random occurrence of state transitions across the interconnected cells in the assemblies considered

- Stochastical attributes of the neural disturbances

- Combinatorial aspects due to the number and variety of the subsets of assemblies in the neural activity.

The complexity considerations pertinent to the neural complex offer an alternative strategy to use the maximum entropy parameter as an error-metric *in lieu* of the conventional cost-functions, such as the quadratic error or cross-entropy error-metrics. A theoretical outline of deducing an inter-relation between the complexity of the network and the associated entropy is presented in Chapter 6. Hence, a complexity based cost-function is derived and the ability of such a cost-function in enabling a convergent neural network training is illustrated via simulations pertinent to a multilayered perception.

1.8 Concluding Remarks

In the world of real neurons, an inquisitive query is posed to specify a cost-function uniquely which is optimized by the cellular web of the brain complex. While the sensory data are largely redundant, it is logical to presume, consistent with the use of an optimal cost-function, that an encoding effort is rather implicitly and inevitably present in the neural complex which renders the sensory data to be encoded selectively. Such a *factorial coding* conditioning of the neural input enables the corresponding outputs to be statistically independent in conjunction with the role played by the optimal cost-function chosen.

What is the nature of a cost-function that enables the neural network to learn from a factorial code? An answer to this stems from the inherent characteristics of the neural complex — which is essentially an information processing unit — wherein a major task involved is the information-traffic enclaving all the negentropic attributes which can be specified within the purview of Shannon's information theory. That is, an optimal cost-function dictates the learning algorithms to inject negentropy into the optimization endeavors, in order to counteract the neural complexity and the stochastical inherency. Should this be evolved in the information-theoretic domain, it must be consistent with learning rules which enable a dynamic search of global minima of such ad hoc cost-function through the space of synaptic couplings. These domain-constrained algorithms can also discover and unveil cohesively the informatic aspects of higher-order features present in the multiple input streams of the neural data. Consistent with the aforesaid biological plausibility of viewing neural learning in the information-theoretic plane, it is also possible to define and formulate a set of cost-functions to train artificial neural networks in the information-theoretic plane. Such functions can also be specified to include some degree of non-locality (non-extensivity) in computational learning procedures so as to mimic the human brain behavior.

The scope of the remaining chapters ramifies the various information-theoretic perspectives of real and artificial neural complex into the relevant generic considerations. The details presented unify the associated concepts for the purpose of achieving better neural network models which represent the learning strategies, memory capabilities, complexity profiles and entropic characteristics in a single analytical framework structured in the information-theoretic plane.

Bibliography

[1.1]   Deco, G. and Obradovic, D.: *An Information-Theoretic Approach to Neural Computing* (Springer-Verlag New York Inc., New York, NY: 1996).

[1.2]   Norwich, K. H. : *Information, Sensation and Perception* (Academic Press Inc., San Diego, CA: 1993).

[1.3]   Shimuzu, H. (Ed.): *Biological Complexity and Information* (World Scientific Publishing Co., Ltd., Singapore: 1990).

[1.4]   Griffith, J. S.: *Mathematical Neurobiology* (Academic Press, London: 1971).

[1.5]   Gatlin, L. L: *Information Theory and Living Systems* (Columbia University Press, New York, NY: 1972).

[1.6]   Sampson, J. R.: *Adaptive Information Processing - An Introductory Survey* (Springer-Verlag New York Inc., New York, NY: 1976).

[1.7]   Sampson, J. R.: *Biological Information Processing - Current Theory and Computer Simulation* (John Wiley & Sons, Inc., New York, NY: 1984).

[1.8]   Haken, H.: *Information and Self-Organization* (Springer-Verlag, Heidelberg: 1988).

[1.9]   Dabrowski, M, Michalewicz, M. and Ras`. Z. (Eds.): *Intelligent Information Systems III* (Instytut Podstaw Informatyki PAN, Warszawa (Poland): 1994).

[1.10] MacGregor, R. J.: *Neural and Brain Modeling* (Academic Press, Inc., San Diego, CA:1987), 138-139.

[1.11] Wiener, N.: *Cybernetics* (The MIT Press, Cambridge, MA: 1961).

[1.12] Bergström, R. M. and Nevanlinna, O.: An entropy model of primitive neural systems. *Intern. J. Neurosci.*, 4, 1972, 171-173.

[1.13] Abu-Mostafa, Y. S. and St. Jacques, S.: Information capacity of the Hopfield model. *IEEE Trans. Inform. Theory*, IT-31, 1985, 461-464.

[1.14]   Caianello, E. R: Outline of thought processes and thinking machines. *J.Theor. Biol.*, 2, 1961, 204-235.

[1.15]   Shannon, C. E. and Weaver, W. : *The Mathematical Theory of Communication*, (University of Illinois Press, Urbana, IL: 1949).

[1.16]   Pfaffelhuber, E.: Learning and information theory. *Int. Rev. Neurosci.,* 3, 1972, 83-88.

[1.17]   Legendy, C. R: On the scheme by which the human brain stores information. *Math. Bioscsi.*, 1, 1967, 555-597.

[1.18]   Little, W. A: The existence of persistent states in the brain. *Math. Biosci.,* 19, 1974, 101-120.

[1.19]   MacGregor, R. J.: *Theoretical Mechanics of Biological Neural Networks,* (Academic Press Inc./Harcourt Brace Jovanovich Publishers, Boston, MA: 1993).

[1.20]   Gardner, E. and Derrida, B.: Optimal storage properties of neural network models. *J. Phys. A,* 21, 1988, 257-270.

[1.21]   MacGregor, R. J. and Gerstein, G. L: Cross-talk theory of memory capacity in neural networks. *Biol. Cybern.*, 65, 1991, 351-355.

[1.22]   Palm, G.: On the storage capacity of an associative memory with randomly distributed storage elements. *Biol. Cybern.*, 39, 1981, 125-127.

[1.23]   Palm, G.: Evidence, information and surprise, *Biol. Cybern.*, 42, 1981, 57-68.

[1.24]   Neelakanta, P. S. and DeGroff, D.: *Neural Network Modeling: Statistical Mechanics and Cybernetic Perspectives* (CRC Press, Boca Raton, FL: 1994).

[1.25]   McEliece, R. J., Posner, E. C., Rodemich, E. R. and Venkatesh, S. S.: The capacity of the Hopfield associative memory. *IEEE Trans. Inform. Theory*, IT-33, 1987, 461-482.

[1.26]   Nakahama, H., Yamamoto, M., Aya, K., Shima, K. and Fujii, H.: Markov dependency based on Shannon's entropy and its applications to neuron spike trains. *IEEE Trans. Syst. Man and Cybernet.*, 13, 1983, 692-701.

[1.27]   Windhorst, V. and Shultens, H.: A measure of transinformation for multiple input/single output neuronal systems. *Biol. Cybern.*, 45, 1982, 57-61.

[1.28]   Eckhorn, R. and Poepel, B.: Rigorous and extended application for information theory to the afferent visual systems of the cat. 1. Basic Concepts. *Biol. Cybern.*, 16, 1974, 191-200.

[1.29]   Tsukada, M., Ishii, N. and Sato, R.: Temporal pattern discrimination of impulse sequence in the computer-simulated nerve cells. *Biol. Cybern.* 17, 1975, 19-28.

[1.30]   Sejnowski, T. J.: On the stochastical dynamics of neural interaction. *Biol. Cybern.*, 22, 1976, 203-211.

[1.31]   Ventriglia, F.: Kinetic theory of neural systems: Analysis of activity of the two dimensional model. *Biol. Cybern.*, 46, 1983, 93-99.

[1.32]   Wilson, H. R. and Cowan, J. D.: A mathematical theory of the functional dynamics of cortical thalmic nervous tissue. *Biol. Cybern.*, 3, 1973, 55-80.

[1.33]   Van der Lubbe, J. C. A., Boxma, Y. and Boekee, D. E.: A generalized class of certainty and information measures. *Info. Sci.*, 32, 1984, 187-215.

[1.34]   Hartley, R. V.: Transmission of information. *Bell Syst. Tech. J.*, 7, 1928, 535-563.

[1.35]   Kullback, S. and Leibler, R. A.: On information and suffency. *Ann. Math. Stat.*, 22, 1951, 79-86.

[1.36]   Csiszár, I.: Information type measures of difference of probability distributions and indirect observations. *Studia. Sci. Math. Hungar.*, 2, 1967, 229-318.

[1.37]   Amari, S.: *Differential Geometrical Methods of Statistics* (Springer-Verlag, New York, NY: 1985).

[1.38]   Kullback, S.: *Information Theory and Statistics* (Wiley Interscience Pub., NY: 1959).

[1.39] Jaynes, E. T.:Information theory and statistical mechanics I. *Phys. Rev.*, 106, 1957, 620-630.

[1.40] Jaynes, E. T.:On the rationale of maximum entropy methods. *Proc. IEEE*, 70, 1982, 939-952.

[1.41] Jaynes, E. T.:*Papers on Probability, Statistics and Statistical Physics* (Reidel, Dordrecht: 1982).

[1.42] Brillouin, L.: *Science and Information Theory* (Academic Press, New York, NY: 1962).

[1.43] Alekseev, G. N: *Energy and Entropy* (Mir Publishers, Moscow: 1986).

[1.44] Bennett, C. H.: Demons, engines and the second law. *Scientific American*, 259 (5), 1987, 108-116.

[1.45] Bennett, C. H. and Landauer, R.:The fundamental physical limits of computations. *Scientific American*, 255E, 1985, 48-56.

[1.46] Abu-Mostafa, Y. S.:Information theory, complexity and neural networks. *IEEE Comm. Magazine*, November 1989, 25-28.

[1.47] Liou, C. Y. and Lin, S. L.: The other variant Boltzmann machine. *Proc. Joint Conf. Neural Networks* (June 18-22, 1990. Washington, D. C.), I 449- I 454.

[1.48] Ackley, D. H., Hinton, G. E. and Sejnowski, T. J.: A learning algorithm for Boltzmann machines. *Cognit. Sci.*, 9, 1985, pp. 147-169.

[1.49] Takatsuji, M.: An information-theoretical approach to a system of interacting elements. *Biol. Cybern.*, 17, 1975, 207-210.

[1.50] Mahalanobis, P. C.: On the generalized distance in statistics. *Proc. Natl. Inst. Sci.*(India), 12, 1936, 49-55.

[1.51] P. C. Mahalanobis: Analysis of race mixture in Bengal. *J. Asiat. Soc.* (India), 23, 1925, 301-310.

[1.52] Fisher, R. A.: The use of multiple measurements in taxonomic problems. *Ann. Eugenics*, 7, 1936, 179-188.

[1.53] Bhattacharyya, A.: On the measure of divergence between two distributions. *Bull. Calcutta Math. Soc.*, 35, 1943, 99-109.

[1.54] Jeffreys, H.: An invariant from for the prior probability in estimation problems. *Proc. Roy. Soc. A.,* 186, 1946, 453-461.

[1.55] Adhikari, B. P. and Joshi, D. D.: Distance discrimination et resume exhaustif. *Publs. Inst. Statist.*, 5, 1956, 57-74.

[1.56] Matusita, K.: Decision rules, based on the distance for problems of, two samples and estimation. *The Annals Math. Stat.*, 26, 1955, 631-641.

[1.57] Bergstrî m, R. M.: Neural micro- and macrostates. In: G. Newton and A. H. Riesen (Eds.): *Advances in Psychobiology*, (Wiley-Interscience, NY: 1972), Vol. 2, Chapter 2.

[1.58] Shore, J. E. and Johnson, R. W.: Axiomatic derivation of the principle of maximum entropy and the principle of minimum cross-entropy. *IEEE Trans. Inform.Theory*, IT-26, 1980, 26-36.

[1.59] Good, I. J.: Maximum entropy for hypothesis formulation, especially for multidimensional contingency tables. *Ann. Math. Stat.*, 34, 1963, 911-934.

[1.60] Hobson, A. and Cheng, B.: A comparison of the Shannon and Kullback information measures. *J. Stat. Phys.* 7, 1973, 301-310.

[1.61] Johnson, R. W.: Axiomatic characterization of the directed divergence and their linear combinations. *IEEE Trans. Inform. Theory*, IT-25, 1979, 709-716.

[1.62] Aczél, J. and Daroczy, Z.: *On Measures of Information and Their Characterizations* (Academic Press, New York, NY: 1975).

[1.63] Jaynes, E. J.: Information theory and statistical mechanics. In *Statistical Physics*, Vol. 3, Brandeis Lectures (Ed.: K. W. Ford) (Benjamin, New York, NY: 1963), 182-218.

[1.64] Jaynes, E. J.: Prior probabilities, *IEEE Trans. Syst. Sci. Cybern.*, SSC-4, 1068, 227-241.

[1.65]   Kapur, J. N. and Kesavan, H. K.: *Entropy Optimization Principles with Applications* (Academic Press/Harcourt Brace Jovanovich Publishers, Boston, MA: 1992).

[1.66]   Kailath, T.: The divergence and Bhattacharyya distance measure in signal selection. *IEEE Trans. Comm. Tech.*, COM-15, 1967, 52-60.

[1.67]   Liu, J.: Divergence measures based on the Shannon entropy. *IEEE Trans. Inform. Theory*, 37(1), 1991, 145-151.

[1.68]   Tildesley, M. L.: A first study of the Burmese skull. *Biometrika*, 13, 1921, 176-262.

[1.69]   Fisher, R. A.: *Contributions to Mathematical Statistics*. (John Wiley, New York, NY: 1950).

[1.70]   Anderson, T. W.: *An Introduction to Multivariate Statistical Analysis*. (John Wiley, New York NY: 1958).

[1.71]   Rao, C. R.: *Advanced Statistical Methods in the Estimation of Statistical Parameters* (John Wiley, New York, NY: 1952).

[1.72]   Chernoff, H.: A merasure of asymptotic efficiency for tests of a hypothesis base on a sum of observations. *Ann. Math Stat.*, 23, 1952, 493-507.

[1.73]   Amari, S., Kurata, K. and Nagoaka, H.: Information geometry of Boltzmann machines. *IEEE Trans. Neural Networks*, 3, 1992, 260-271.

[1.74]   Güttinger, W.: Problems of information processing in the nervous system. *Intl. J. Neurosci.*, 3, 1972, 61-66.

[1.75]   Uttley, A. M.: *Information Transmission in the Neurons Systems* (Academic Press, London: 1979).

[1.76]   Gawronski, R.: *Bionics: The Nervous System as a Control System* (Elsevier Publishing Co., Amsterdam: 1971).

[1.77]   Kjaer, T. W., Hertz, J. A. and Richmond, B. J.: Decoding of cortical neuronal signals: Network models, information estimation and spatial tuning. *J. Comput. Neurosci.*, 1, 1994, 109-139.

[1.78]   Baum, E. B. and Wilczek, F.: Supervised learning of probability distributions by neural netwroks. In D. A. Anderson (Ed.) *Neural Information Processing Systems*.: (American Institute of Physics, New York, NY :1988), 52-61.

[1.79]   Watrous, R. L.: A comparison between squared error and relative entropy metrics using several optimization algorithms. *Complex Syst.*, 6, 1992, 495-505.

[1.80]   Solla, S. A., Levin, E. and Fleisher, M.: Accelerated learning in a layered neural network. *Complex Syst.*, 2, 1988, 625-640.

[1.81]   Park, J. C., Neelakanta, P. S., Abusalah, S., De Groff, D. and Sudhakar, R.: Information-theoretic based Error-Metrics for gradient desecent learning in neural networks. *Complex Syst.*, 9, 1995, 287-304.

[1.82]   Hanson, S. J. and Burr, D. J.: Minkowski-r backpropagation: Learning in connectionist models with non-Euclidian error signals. In: D. Z. Anderson (Ed.): *Neural Information Processing System* (American Institute Physics, New York, NY: 1968), 11, 348-357.

[1.83]   Uttley, A. M.: The informon: A network for adaptive pattern recognition. *J. Theor. Biol.*, 27, 1970, 36-67.

[1.84]   Zak, M.: Introduction to terminal dynamics. *Complex Syst.*, 7, 1993, 59-87.

[1.85]   Zadeh, L. A.: Fuzzy sets, *Inform. Control*, 8, 1965, 238-354.

[1.86]   Kandel, A. and Byatt, W. J.: Fuzzy differential equations. *Proc. Int. Conf. on Cybern. and Soc.* (November 1978), pp. 1213-1216.

[1.87]   Kandel, A.: *Fuzzy Statistics and Its Applications to Fuzzy Differential Equations*. Ph.D. Dissertation, University of New Mexico, Albuquerque, NM (USA), 1977.

[1.88]   Goel, N.S., Maitra, S. and Montroll, E.W.: On the Volterra and other nonlinear models of interacting populations. *Rev. Mod. Phys.*, 43, 1971, 231-276.

[1.89]   Kolmogorov, A.N.: Three approaches to the quantitative definition of information, *Problemy Peredachi Informatsii*, 1, 1965, 3-11.

[1.90] Chaitin, G.T.: Algorithmic information theory. *IBM J. Res. Dev.*, 21, 1977, 350-359.

[1.91] Blahut, R. E.: *Principle and Practice of Information Theory* (Addison Wesley Publishing Company, Reading, MA: 1987).

[1.92] Cover, T. M. and Thomas, J. A.: *Elements of Information Theory* (Wiley Interscience, New York, NY: 1991).

[1.93] Feinstein, A.: *Foundations of Information Theory* (McGraw-Hill Book Co., New York, NY: 1958).

[1.94] Elsasser, W. M.: On quantum measurements and the role of the uncertainty relations in statistical mechanics. *Phys. Rev.*, 52, 1937, 987-999.

[1.95] Wyner, A.D.: Fundamental limits in information theory. *Proc. IEEE*, 69, 1981, 239-251.

[1.96] Ash, R.B.: *Information Theory* (Dover Publications, Inc., New York, NY: 1965)

[1.97] Rényi, A.: On the measures of entropy and information. *Proc. 4th Berkeley Symp. Math. Stat. Prob.*, 1, 1961, 547-561.

[1.98] Kannappan, PL.: On Shannon entropy, directed divergence and inaccuracy. *Z. Wahrschilichkeitstheory Verw. Ge.*, 22, 1972, 95-100.

[1.99] Kannappan, PL.: On generalized directed divergence. *Funkcialay Ekvacioj*, 16, 1973, 71-77.

[1.100] Daróczy, Z.: Generalized information functions. *Info. and Control*, 16, 1970, 36-51.

Appendix 1.1

Concepts and Definitions in Information Theory*

Definition: *Entropy* is a measure of uncertainty of a random variable. The entropy H(X) of a discrete random variable X is defined by:

$$H(X) = - \sum_{x \in c} p(x) \ell og\{p(x)\} \qquad (A\ 1.1.1a)$$

$$= E[\ell og\left\{\frac{1}{p(x)}\right\}] \qquad (A\ 1.1.1b)$$

where E[.] denotes the expectation operator. Further, X is a discrete random variable with symbol x and the probablity mass function: $p(x) = P_r\{X = x\}$, $(x \in X) = p_x(x)$.

Properties of H
1. $H(X) \geq 0$ since $0 < p(x) < 1$ and $\ell og\{p(x)\}$ is always negative.
2. $H_b(X) = (\ell og_b a)H_a(X)$
3. Conditioning reduces entropy: For any two random variables, X and Y:

$$H(X|Y) \leq H(X) \qquad (A1.1.2)$$

with equality *iff* (if and only if) X and Y are independent.
4. $H(X_1, X_2, ..., X_n) \leq \sum_{i=1}^{n} H(X_i)$, with equality *iff* the random variables X_i are independent
5. $H(X) \leq \ell og\ |\mathcal{X}|$ with equality *iff* X is uniformly distributed over **X**.
6. H(p) is concave on p.
7. The units in which entropy is measured depend on the base of logarithm used in the definition. If the base of the logarithm is 2, the unit is bits. If the base is the exponent e, the unit is *nats*.
8. If H(X) = 0, the variable X describes deterministic process and there is an absolute certainty that only one outcome of X is possible.
9. H(x) being maximum refers to p(x) is uniform, meaning that the uncertainty about the random variable X is maximal.

---

* *Note*: The proofs of various theorems and analytical statements indicated in the appendices are available in the standard books on information theory. See, for example [1.91-1.97].

10. H(x) also depicts the average length of the shortest description of the random variable X.

Definition: *Joint entropy* in reference to the pair of random variables X, Y over the discrete sets x and y, respectively, is defined as:

$$H(X,Y) = - \sum_{x \in \chi} \sum_{y \in \eta} p(x,y)\ell og\{x,y\} = E[\ell og\{1/p(x,y)\}]] \qquad (A\ 1.1.3)$$

where p(x,y) denotes the joint probability of occur X and Y.

Definition: *Conditional entropy* in reference to a pair of random variables X, Y over the discrete sets χ and η, respectively, is defined as:

$$H(Y \mid X) = \sum_{x \in \chi} p(x)H(Y \mid X \equiv x)$$

$$= - \sum_{x \in \chi} p(x) \sum_{y \in \eta} p(y/x)\ell og\{p(y/x)\}$$

$$= - \sum_{x \in \chi} \sum_{y \in \eta} p(x,y)\ell og\{p(y/x)\} \qquad (A\ 1.1.4)$$

The conditional represents the average of the *degree of uncertainty* of Y over all concrete outcomes of X.

Definition: The *relative entropy or Kullback-Leibler entropy* D(p∥q) of the probability mass function p with respect to the probability mass function q is defined by:

$$D(p\|q) = \sum_{x} p(x)\ell og\{p(x)/q(x)\} \qquad (A\ 1.1.5)$$

D(p∥q) is also denoted as K(p∥q) and represents the cross-entropy or a measure of "distance" between two distributions p(x) and q(x). That is, it depicts the measure of difference between two distributions.

In the context of differential geometry, D(p∥q) is also the Riemann metric in the space(s) of the distribution considered. It should be noted that relative entropy is not a "true" distance measure, inasmuch as it is not symmetric. That is, D(q∥p) or D(p∥q) denote a "quasi-distance" measure which is always positive and is equal to zero *iff* p(x) ≡ q(x).

Definition: The mutual information between two random variables X and Y is defined as:

$$I(X;Y) = \sum_{x\in\chi}\sum_{y\in\eta} p(x,y)\ell og\left\{\frac{p(x,y)}{p(x)p(y)}\right\} \qquad\text{(A 1.1.5a)}$$

$$= D\{p(x,y), p(x)p(y)\} \qquad\text{(A1.1.5b)}$$

where, D represents the discrete lengths of X, Y adopted.

Definition: Given two pdf's f(x) and g(x), the relative or *Kullback-Leibler entropy* can be defined as:

$$D(f,g) = \int f(x)\ell og\left\{\frac{f(x)}{g(x)}\right\}dx \qquad\text{(A 1.1.6)}$$

pertinent to continuous variable, x.

Definition: If X and Y are two continuous random variables with the associated joint pdfs f(x,y), the *mutual information* between X and Y is defined as:

$$I(X;Y) = \iint_{yx} f(x,y)\ell og\left\{\frac{f(x,y)}{f(x)f(y)}\right\}dxdy \qquad\text{(A 1.1.7)}$$

Note

- Properties of D(f, g) and I(X;Y) are the same as in the respective discrete cases

- Further, when $D \to 0$, the discretized mutual information converges to the corresponding mutual information pertinent to continuous distribution in the Riemann space. That is, $I(X_D;Y_D) \to I(X;Y)$ as $D \to 0$

- Theorems and inequalities on entropy, generally, hold good for both discrete as well as continuous versions of entropy. However, in the continuous case, entropy can be negative

Mutual information measures the statistical independence between two variables X and Y with the corresponding probability distribution p(x) and p(y) respectively. This is also equal to the Kullback-Leibler distance between the joint probabilities and the factorized ones. It is equal to zero, *iff* x and y are considered as independent.

Mutual information is always symmetric. That is,

$$I(X;Y) = I(Y;X) \qquad\text{(A 1.1.8.a)}$$

and

$$I(X;X) \equiv H(X) \qquad\qquad\qquad \text{(A 1.1.8.b)}$$

Mutual information, conceptually, is a measure of the amount of information that Y conveys about X or *vice-versa*. It offers a correlation between X and Y. If X and Y are input and output of a stochastic channel, I(X;Y) refers to the transmitted information through the channel.

Properties of D and I
1. $I(X;Y) = H(X) - H(X|Y) = H(Y) - H(Y|X) = H(X) + H(Y) - H(X,Y)$
2. $D(p\|q) \geq 0$ with equality *iff* $p(x) = q(x)$, for all $x \in \chi$
3. $I(X;Y) = D\big(p(x,y)\|p(x)p(y)\big) \geq 0$, with equality *iff* $p(x,y) = p(x)p(y)$; that is, X and Y are independent.
4. If $|\chi| = m$, and u is of uniform distribution over χ, then $D(p\|u) = \ell og(m) - H(p)$
5. $D(p\|q)$ is convex in the pair (p, q).

Chain rules
Entropy: $H(X_1, X_2, ..., X_n) = \sum_{i=1}^{n} H(X_i|X_{i-1}, ..., X_1)$

Mutual information: $I(X_1, X_2, ..., X_n; Y) = \sum_{i=1}^{n} I(X_i; Y|X_1, X_2, ..., X_{i-1})$

Relative entropy: $D\big(p(x,y)\|q(x,y)\big) = D\big(p(x)\|q(x)\big) + D\big(p(y|x)\|q(y|x)\big)$

Definition: In reference to a continuous random variable X described by a probability density function (pdf), f(x), the definition of entropy can be extended as follows:

$$H_c = - \int_{x \in S} f(x)\ell og\{f(x)\}dx \qquad\qquad \text{(A 1.1.9)}$$

where S is a domain which includes x. If the pdf f(x) is Riemann integrable in S, discretizing the continuous random variable X into X_Δ discrete variable leads to :

$$H(X_\Delta) + \log(\Delta) \to H_c(X) \quad \text{as } \Delta \to 0 \qquad\qquad \text{(A 1.1.10)}$$

The relation given by Eqn. A1.1.1a quantizes implicitly the continuous entropy, $H_c(X)$ into v-bits and equates it approximately to $H_c(X) + v$.

Information Theory and Entropy Inequalities

Jensen's inequality theorem: If f(.) is a convex function, then $E[f(X)] \geq f(E[X])$. This relation can be explicitly written as:

$$\int f(x)p(x)dx \geq f\{xp(x)dx\} \qquad (A\ 1.1.11)$$

where p(x) is the pdf of the random variable X.

In the above relation (Eqn.A1.1.11), f(x) is assumed as a *convex function* meaning that, for every x,y and $0 \leq a \leq 1$, the following inequality is satisfied:

$$f\{ax + (1-a)y\} \leq af(x) + (1-a)f(y) \qquad (A\ 1.1.12)$$

and f(x) is designated as concave, if $-$ f(x) is convex.

Log sum inequality: For a set of n positive numbers, $(a_1, a_2, ..., a_n)$ and $(b_1, b_2, ..., b_n)$,

$$\sum_{i=1}^{n} a_i \log\left[\frac{a_i}{b_i}\right] \geq \left[\sum_{i=1}^{n} a_i\right] \log\left[\frac{\sum_{i=1}^{n} a_i}{\sum_{i=1}^{n} b_i}\right] \qquad (A\ 1.1.13)$$

with equality *iff* (a_i/b_i) = constant.

Data processing inequality: If $X \rightarrow Y \rightarrow Z$ forms a Markov chain, then $I(X;Y) \geq I(X;Z)$.

Information inequality: If p(x) and q(x) are two probability distributions, then,

$$D(p,q) \geq 0 \qquad (A\ 1.1.14)$$

with the equality sign applied, *iff*, p(x) = q(x) for all x.

Mutual information inequality: For any two random variables X,Y.

$$I(X;Y) \geq 0 \qquad (A\ 1.1.15)$$

with the equality sign applied, *iff*, X and Y are independent, that is when the joint and factorized probability are identical.

Conditional information inequality: Specific to any two random variables (X,Y).

$$H(X \mid Y) \le H(X) \qquad\qquad (A\ 1.1.16)$$

implying that conditioning always reduces the entropy. The equality sign always holds good, *iff*, X and Y are independent.

Sufficient statistic: A function T(X) is sufficient *relative* to the family of mass functions $\{f_\theta(x)\}$ for all distributions on $I(\theta; X) = I(\theta : T(X))$. This is the same as the data processing inequality indicated earlier.

This inequality is in reference to an observation that there exists no method of manipulating a data set so as to improve the inferences that can be gathered from the data set.

Specifically, assuming that (X,Y,Z) constitute a markovian chain with the order $(X \rightarrow Y \rightarrow Z)$, no processing of Y, deterministic or random can increase the information that Y contains about X.

Suppose X is a sample from a distribution of a family of probability mass function $\{f_\theta(x)\}$ and T(X) refers to a statistical function of the sample considered, such as sample mean or variance. Then, the inequality relation specified by Eqn.(A.1.1.16) holds and the statistic T(X) is referred to as *sufficient* for θ if it contains all the information in X about θ.

The equality sign holds when no information is lost.

The statistic T(X) is a *minimal sufficient statistic* relative to $\{f_\theta(x)\}$, if it is a function of every other sufficient statistic U. That is, in terms of data processing inequality, $\theta \rightarrow T(X) \rightarrow U(X) \rightarrow X$. Here, the minimal sufficient statistic compresses the information about θ in the sample maximally.

Fano's inequality: Let P_e be the probability of error, $\Pr\{g(Y) \ne X\}$, where g is any function of Y denoting $\hat{X}$ which is an estimation of X. Then

$$H(P_e) = P_e \ell og(|\chi| - 1) \ge H(X \mid Y) \qquad\qquad (A\ 1.1.17)$$

This inequality consideration applies when one makes a "guess" about the value of a random variable X correlated to a known random variable Y. The Eqn.(A 11.17) relates the probability of error in guessing the random variable X to its conditional entropy H(X|Y).

Fano's inequality quantifies the idea that one can estimate X with a low probability of error only if the conditional entropy H(X | Y) is small. It can be noted that $P_e = 0$ implies $H(X \mid Y) = 0$, as intuition suggests.

Eqn.(A 1.1.17) bounds the error probability so that $X \ne \hat{X}$; and, $(X \rightarrow Y \rightarrow \hat{X})$ constitutes a Markov chain.

Fano's inequality can also be "weakened" to:

$$1 + P_e \ell og \mid \chi \mid \; \geq H(X/Y) \tag{A 1.1.18}$$

Entropy and the Second Law of Thermodynamics

The Second Law of Thermodynamics postulates that the entropy of an isolated system is nondecreasing. In a statistical thermodynamics point of view, entropy is the logarithm of the number of microstates in the system.

The above generalization can be focused on heuristical considerations to yield the following attributes *vis-a-vis* entropy of an isolated system wherein transition obeys the markovian chain consistent with the physical laws governing the system:

- Entropy $H(X_n)$ increases, if the stationary distribution is uniform. In reference to the statistical thermodynamics, the uniform distribution implies that the microstate are equally likely.

- Relative entropy $D(\mu_n \| \mu'_n)$ decreases with time.

- Relative entropy $D(\mu_n \| \mu'_n)$ between a distribution and the stationary distribution decreases with time.

- Entropy $H(X_n)$ increases if the stationary distribution is uniform.

- The conditional entropy $H(X_n | X_1)$ increases with time for a stationary Markov chain.

- The conditional entropy $H(X_0 | X_n)$ of the initial condition X_0 increases for any Markov chain. Shuffles increase entropy.

Gibbs Second Theorem

If X_1, X_2, ..., X_k represent a vector random variable according to the density $p(x_1, x_2, ..., x_k)$ with zero mean and covariance matrix **C**, then:

$$H(X_1, X_2, ..., X_k) \leq (1/2)\ell og \left\{ (2\pi e)^k \det(\mathbf{C}) \right\} \tag{A 1.1.19a}$$

where $(1/2)\ell og \left\{ (2\pi e)^k \det(\mathbf{C}) \right\}$ is the entropy of a multivariate normal (Gaussian) distribution with a mean value $\hat{m}$ and covariance matrix [**C**].

That is,

$$p(x_1, x_2, ..., x_k) = \left\{ 1 \Big/ \left[(2\pi)^{k/2} \{\det(\mathbf{C})\}^{1/2} \right] \right\}$$
$$\times \left\{ \exp\left[(-1/2)(x - \hat{m})^T \mathbf{C}^{-1} (x - \hat{m}) \right] \right\} \qquad \text{(A 1.1.19b)}$$

Thus, Gibbs Second Theorem indicates that the normal distribution maximizes the entropy over all distributions with the same covariance matrix. In Eqn.A1.1.19), det($\mathbf{C}$) denotes the determinant of [$\mathbf{C}$] matrix and T refers to the transpose operation. A property of [$\mathbf{C}$] which refers to a non-negative defnition matrix is:

$$\det[\mathbf{C}] \le \prod_{i=1}^{k} C_{ii} \qquad \text{(A 1.1.20)}$$

where the equality sign applies *iff* the matrix is diagonalized. The above condition is known as *Hadamard's inequality*.

APPENDIX 1.2

Functional Equations Related to Information Theory

Kannappan [1.98,1.99] has presented a set of measurable solutions of functional equations connected with Shannon's measure of entropy, directed divergence, or *information gain*, and *inaccuracy*.

Definition : The general, real-valued continuous function that characterizes the Shannon's entropy is given by

$$\sum_{i=1}^{m} \sum_{d=1}^{n} f(x_i, y_j) = \sum_{i=1}^{m} f(x_i) + \sum_{j=1}^{n} f(y_i) \qquad (A1.2.1)$$

where $(x_i, y_j) \geq 0$ and $\left(\sum_{i=1}^{m} x_i = \sum_{j=1}^{n} y_j = 1 \right)$.

Definition : The functional equation that characterizes a *directed divergence* and *inaccuracy*** is given by

$$\sum_{i=1}^{m} \sum_{j=1}^{n} F(x_i, y_j, u_i, v_j) = \sum_{i=1}^{m} F(x_i, u_i) + \sum_{j=1}^{n} F(y_i, v_i) \qquad (A1.2.2)$$

for $(x_i, u_i, y_i, v_i \geq 0)$, $\left(\sum_{i=1}^{m} x_i = \sum_{j=1}^{n} y_j = 1 \right)$ and $\left(\sum_{i=1}^{m} u_i \leq 1, \sum_{j=1}^{n} v_j \leq 1 \right)$.

Theorem: A necessary and sufficient condition that a continuous function f (.) satisfies Eqn.(A 1.2.1) is that

$$f(x) = Ax\log(x) \qquad (A1.2.3)$$

for all $x \in$ Interval [0, 1] and A is a constant decided by the base of the logarithm. The corresponding Shannon's entropy is given by

---

* **Inaccuracy :** This is defined as follows:

$$H_n = \begin{pmatrix} p_1, p_2, \cdots, p_n \\ q_1, q_2, \cdots, q_n \end{pmatrix} = -\sum_{i=1}^{n} p_i \log(q_i)$$

Theorem: The most general continuous solution of the functional Eqn. (A 1.2.2) satisfying further $F(1, 1/2) = 1$ and $F(1/2, 1/2) = 0$ is given by

$$F(x,y) = x[\log(x/y)] \tag{A 1.2.5a}$$

for all $(x,y) \in J$. The corresponding *directed-divergence* or *information gain* is given by

$$I_n\left(\begin{matrix} p_1, p_2, \cdots, p_n \\ q_1, q_2, \cdots, q_n \end{matrix}\right) = \sum_{i=1}^{n} p_i \log(p_i/q_i) \tag{A1.2.5b}$$

Theorem: If a real-valued function F which is continuous in J satisfies the functional Eqn.(A1.2.2) and the continuous $F(1,1/2) = 1$ and $F(1/2,1/2) = 0$, then,

$$F(x,y) = -x\log(y) \tag{A 1.2.6}$$

Jaynes' formalism: This refers to the so-called *maximum entropy formalism* to determine the most unbiased probability distribution, when the given entities are the Shannon's entropy function $-\sum_{i=1}^{n} p_i \log(p_i)$ and some simple linear moment constraints.

Assumptions:
The entropy function is a concave function

Maximization process always leads to a set of probabilities that will never assume negative values

By virtue of being a concave function, the entropy function possesses a global minimum.

Suppose it is required to maximize the entropy function $-\sum_{i=1}^{n} p_i \log(p_i)$ given by Eqn. (A1.2.4) subject to the following constraints:

$$\sum_{i=1}^{n} p_i = 1, \quad \sum_{i=1}^{n} p_i g_r(x_i) = a_r, \quad r = 1,2,\ldots, m. \tag{A1.2.7}$$

The *Lagrangian operation* (L) is specified by:

$$L = -\sum_{i=1}^{n} p_i \log(p_i) - (\lambda_0 - 1)\left[\sum_{i=1}^{n} p_i - 1\right] - \sum_{r=1}^{m} \lambda_r\left[\sum_{i-1}^{n} p_i g_r(x_i) - a_r\right] \tag{A1.2.8}$$

Maximizing L, it follows that

$$\ell og(p_i) + \lambda_0 + \lambda_1 g_1(x_i) + \lambda_2 g_2(x_i) + ... + \lambda_m g_m(x_i) = 0, \quad i = 1, 2,.., n \tag{A1.2.9a}$$

or,

$$p_i = \exp\left[-\left(\lambda_0 + \lambda_1 g_1(x_i) + ... + \lambda_m g_m(x_i)\right)\right], \quad i = 1, 2,..., n \tag{A1.2.9b}$$

Assuming $\ell og(p_i) \Rightarrow \ell n(p_i)$, namely, the natural logarithm. In order to determine the Lagrange multipliers, $\lambda_0, \lambda_1, ..., \lambda_m$, Eqn.(A1.2.7) is substituted into Eqn.(A1.2.9b) to get:

$$\exp(\lambda_0) = \sum_{i=1}^{n} e^{-\lambda_1 g_1(x_i) - \lambda_2 g_2(x_i) - ... \lambda_m g_m(x_i)} \tag{A1.2.10}$$

and

$$a_r \exp(\lambda_0) = \sum_{i=1}^{n} g_r(x_i) \exp\left[\lambda_1 g_1(x_i) - ... - \lambda_m g_m(x_i)\right], \quad r = 1, 2, ..., m \tag{A1.2.11}$$

In the case of a continuous random variant, the maximum entropy principle requires the maximization of the entropy function, H(x):

$$H(x) = -\int_a^b f(x) \ell og[f(x)] dx \tag{A1.2.12}$$

subject to the conditions

$$\int_a^b f(x)dx = 1, \quad \text{and} \quad -\int_a^b f(x)\log[f(x)] \int_a^b f(x)g_r(x)dx = a_r, \quad r = 1, 2,..., m. \tag{A1.2.13}$$

Pertinent results on maximization of the discrete version can be extended to this case without any loss of generality.

Minimum Discrimination Information Principle
Eqn.(A1.2.5b) refers to the Kullback-Leibler measure, in reference to two statistics with distributions $\{p_i\}$ and $\{q_i\}$. That is, $D(p\|q) = \sum_{i=1}^{n} p_i \ell og(p_i/q_i)$ which discriminates the probability P from Q. $D(p\|q)$ is always greater than or equal to zero and has a global minimum value of zero when the two distributions are identical. When referred to uniform distribution U, that is, when $q_i = 1/n$ for each i, then,

63

$$D(P:U) = \sum_{i=1}^{n} p_i \, \ell og(np_i) - \sum_{i=1}^{n} p_i \, \ell og(p_i)$$
$$= H(U) - H(P) = \ell og\ (n) \qquad (A1.2.14)$$

where H(U) is the entropy associated with the uniform distribution, namely, S_{max}, and H(P) is the Shannon's entropy measure for P.

Minimizing D(P:U) implicitly depicts maximizing H(P), which becomes synonymous with the Jaynes' formalism. But this new measure affords another interpretation of the maximum entropy principle: The Jaynes' formalism seeks to determine that distribution P out of those which satisfy the constraints for which D(P:U) is a minimum. In other words, one is seeking that distribution which satisfies the constraints and is close to the uniform distribution.

Kullback's minimum discrimination information principle, however, generalizes this concept in a different perspective. It seeks to minimize the directed divergence D(P:Q) – which means it seeks to determine the distribution P that satisfies all the constraints and is closest to a given distribution Q.

Minimization Procedure

The relevant problem is to minimize $\sum_{i=1}^{n} p_i \, \ell og(p_i/q_i)$ subject to the constraints given in Eqn. A 1.2.7. The corresponding Lagrangian L is given by

$$L = \sum_{i=1}^{n} p_i \, \ell og(p_i/q_i) + (\lambda_0 - 1)\left[\sum_{i=1}^{n} p_i - 1\right] + \sum_{r=1}^{m} \lambda_r \left[\sum_{r=1}^{n} p_i g_r(x_i) - a_r\right]$$

$$(A1.2.15)$$

so that, by assuming natural logarithm,

$$p_i = q_i exp\left[-\lambda_0 - \lambda_1 g_1(x_i) - ... - \lambda_m g_m(x_i)\right] \qquad (A1.2.16a)$$

or,

$$exp[\lambda_0] = \sum_{i=1}^{n} q_i exp\left[-\lambda_1 g_1(x_i) - ... - \lambda_m g_m(x_i)\right] \qquad (A1.2.16b)$$

In the case of a continuous random variable, the minimum Q discrimination information principle requires to minimize the following function:

$$D(f:g) = \int_{a}^{b} f(x)\ell og\left\{\frac{f(x)}{g(x)}\right\}dx \qquad (A1.2.17)$$

subject to the constraints given by

$$\int_a^b f(x)dx = 1, \quad \int_a^b g_r(x)f(x)dx = a_r, \quad r = 1, 2, \ldots, m \qquad (A1.2.18)$$

The result obtained thereof, is given by

$$f(x) = g(x)\exp\left[-\lambda_0 - \lambda_1 g_1(x) - \ldots - \lambda_m g_m(x)\right] \qquad (A1.2.19)$$

where $\lambda_0, \lambda_1, \ldots, \lambda_m$ are determined from the following

$$\int_a^b g(x)\exp\left[-\lambda_0 - \lambda_1 g_1(x) - \ldots - \lambda_m g_m(x)\right] dx = 1 \qquad (A1.2.20a)$$

and

$$\int_a^b g_r(x)g(x)\exp\left[-l_0 - \lambda_1 g_1(x) - \ldots - \lambda_m g_m(x)\right] dx = a_r, \quad r = 1, 2, \ldots, m$$

$$(A\ 1.2.20b)$$

Note:
A. Generalized maximum entropy principle problems
   Suppose the generalization of maximum entropy principle governs a convex entropy function $\phi(.)$ such that $H(P) = -\sum_{i=1}^{n} \phi(p)$, with the constraints $\sum_{i=1}^{n} p_i = 1$ and $\sum_{i=1}^{n} p_i g_r(x_i) = a$, $r=1, 2, \ldots, m$. The methods of Lagrangian multipliers allows maximization of $H(P)$ leading to the first derivative of $\phi(p)$ as:

$$\phi'(p_i) = \lambda_0 + \lambda_1 g_1(x_i) + \ldots + \lambda_m g_m(x_i) \qquad (A1.2.21)$$

Direct problems: Given the entropy measure $\phi(\cdot)$ and the constraint mean values $g_1(x_i), g_2(x_i), \ldots, g_m(x_i)$, find a probability distribution that maximizes the entropy measure:

Solution: Using Eqn. A1.2.21 and the constraints relations lead to (m + 1) Lagrangian multipliers explicitly, yielding the required set of probabilities, $p_i's$.

The first inverse problem — (*Determination of constraints*): Given the entropy measure $\phi(\cdot)$ and the probability distribution for p_i, determine one or more probability constraints that yield the given probability distribution when the entropy measure is maximized subject to these constraints.

Solution: Since $\phi(p_i)$, and $\phi'(p_i)$, are known, the right-hand side of Eqn. A1.2.21 can be determined. This will allow the values for $g_1(x_i), g_2(x_i), ..., g_m(x_i)$ to be identified by matching the terms. Thus, the most unbiased set of constraints required are obtained.

The second inverse problem — (Determination of the entropy measure): Given the constraints $g_1(x_i), g_2(x_i), ..., g_m(x_i)$, and the probability distribution p_i, determine the most unbiased entropy measure that when maximized subject to the given constraints, yields the given probability distribution.

Solution: The given values when substituted in Eqn. A1.2.21, lead to a differential equation which can be solved for $\phi(\cdot)$. Once $\phi(\cdot)$ is obtained, the entropy measure $H(P) = -\sum_{i=1}^{n} \phi(p_i)$ can be determined.

B. *Generalized minimum discrimination information principle problem*
    Let $\phi(\cdot)$ be a convex function and

$$D(P:Q) = \sum_{i=1}^{n} q_i \phi(p_i/q_i) \qquad (A1.2.22)$$

be a measure of directed divergence. Relevant constraints are:

$$\sum_{i=1}^{n} p_i = 1, \ \sum_{i=1}^{n} p_i g_r(x) = a_r, \ r = 1, 2, ..., m \qquad (A1.2.23)$$

Minimizing Eqn.(A1.2.22) subject to the constraints of Eqn.(A1.2.23) gives

$$\phi'(p_i/q_i) = -(\lambda_0 - 1) - \lambda_1 g_1(x) - \lambda_2 g_2(x_i) - ,,, -\lambda_m g_m(x_i) \qquad (A1.2.24a)$$

For the Kullback-Leibler measure,

$$\phi'(p_i/q_i) = \ell og(p_i/q_i) = -\lambda_0 - \lambda_1 g_1(x_i) - \lambda_2 g_2(x_i) - ... - \lambda_m g_m(x_i) \qquad (A1.2.24b)$$

- *The direct problem — (Determination of probability distribution):* If q_i and $g_1(x_i), g_2(x_i), ..., g_m(x_i)$, are known, Eqn.A1.2.24a determines $p_1, p_2, ..., p_n$.

- *First inverse problem — Determination of constraints):* If p_i's, q_i's and $\phi(\cdot)$ are known, Eqn.A1.2.24a determines the constraint function $g_1(\cdot), g_2(\cdot), ..., g_m(\cdot)$.

- *Determination of divergence measure — (Second inverse problem):* If p_i's, q_i's and $g_r(x_i)$ are known, Eqn.A1.2.24 determines $\phi'(p_i/q_i)$ and as such the divergence measure $D(P:Q)$ can be obtained.

- *Determination of a priori distributions — (Third inverse problem):* Finally, if p_i's, q_i's and $\phi(\cdot)$ are known, Eqn.(A1.2.24) determines the q_i's.

Uniqueness of the Solutions

In both maximum entropy principle as well as minimum discrimination information principle problems, the existence of the solution is subject to the following conditions :

- The set of constraints must be consistent; otherwise, there will be no probability distribution satisfying the constraints. Consequently, in such cases, the problem of determining a probability distribution with minimum entropy or minimum directed divergence does not arise.

- The probabilities p_1, p_2, ..., p_n must all be non-negative. The concavity of $H(P)$ and the convexity of $D(P:Q)$ do not automatically ensure the non-negativity condition. The latter can only be ensured by a suitable choice of the measures $H(P)$ and $D(P:Q)$. Otherwise, the non-negativity constraints have to be imposed as additional constraints.

- The series and integrals that are encountered in the solution process must be convergent.

Entropy Measure for Maximum Likelihood Principle

Let $f(x,\theta)$ be the density function for a random variant x. Let x_1, x_2, x_3, Öx$_n$ be a random sample from this population. It is desired to obtain an estimate $\hat{\theta}$ for θ in terms of x_1, x_2, x_3, ..., x_n.

Fisher [1.69] has suggested that θ be chosen such that it maximizes the likelihood function given by:

$$L(x_1, x_2, ..., x_n; \theta) = f(x_1, \theta), f(x_2, \theta), ..., f(x_n, \theta)$$ (A1.2.25)

In terms of the second inverse principle, one can proceed to determine a measure of entropy whose maximization will yield the same result as the maximization of L:

$$H = \sum_{i=1}^{n} \ell og \left[f(x_i, \theta) \right]$$ (A1.2.26)

In general, while using the maximization entropy or minimum discrimination principle, one seeks to determine $(p_1, p_2, ..., p_n)$. In such problems, only the form-function f for the probabilities is known and the unknown quantity is θ. Accordingly, θ is chosen in such a way as to maximize the measure of entropy.

In light of this discussion, it may be noted that the well-established Fisher's principle of maximum likelihood and the second inverse principle are linked to each other.

Appendix 1.3

A Note on Generalized Information Functions

Daróczy [1.100] has furnished the concept of the so-called *information functions* and in terms of these information functions, he has defined *entropies of type β*. It is indicated in [1.100] that these entropies portray a number of interesting algebraic and analytic properties similar to Shannon's entropy. The following is a summary of relevant details.

Definition: A real function f defined in the interval [0,1] is an *information function* if it satisfies the boundary conditions

$$f(0) = f(1) ; \quad f(1/2) = 1 \qquad (A\ 1.3.1)$$

and the functional equations

$$f(x) + (1-x)\ f(y/1-x) = f(y) + (1-y)\ f(x/1-y) \qquad (A\ 1.3.2)$$

for all $(x, y) \in \mathcal{D}$ where $\mathcal{D} = \{(x, y): (0 \le x > 1); (0 \le y < 1); (x + y) \le 1\}$

Definition: If f is an information function and ($p_1, p_2, ..., p_n$; $\Sigma p_i = 1$; $1 \ge p_i \ge 0$) defines a set of discrete probability distribution, then the *entropy* of the distribution with respect to f is given by

$$H_f (p_i)_{i=1,2,...,n} = \sum_{j=2}^{n} s_j f(p_i/s_i) \qquad (A\ 1.3.3)$$

where $s_j = (p_1 + ... + p_j); j = 2, ..., n)$

Definition: A real function f defined on [0, 1] is an *information function of type β* if it satisfies the boundary conditions (A1.3.1) as well as the following functional equation:

$$f(x) + (1 - x)^\beta f(\frac{y}{1 - x}) = f(y) + (1 - y)^\beta f(\frac{x}{1 - y}) \qquad (A1.3.4)$$

where β is a positive number.

Definition: *Entropy of type β* of a probability distribution $\{p_i\}$ is given by

$$H_f^\beta \{p_i\}_{i=1,2,...,n} = \sum_{j=2}^{n} s_j^\beta f(p_j/s_j) \qquad (A1.3.5)$$

where $s_j = (p_1 + ... + p_j); j = 2, ..., n)$ and f is an information function of type β.

Chapter 2

Neural Complex: A Nonlinear C³I System?

P. S. Neelakanta, S. Abusalah and D. De Groff

> *"... to unlock the secrets in
> nonlinear problems..."*
> – H.T. Davis

2.1 Introduction

There is a hierarchy of structure in the nervous system with an inherent C³I protocol[1] stemming from the brain and converging to a cell. Any single physiological action perceived, such as pain or pleasure is the output response of a *collective activity* due to innumerable neurons participating in the decision-making control procedures in the nervous system [2.1]. If the neural complex is "dubbed as a democracy, the neural activity refers to how the votes are collected and how the result of the vote is communicated as a command to all the nerve cells involved". That is, the antithesis must remain that our brain is a democracy of ten thousand million cells, yet it provides us a unified experience [2.2].

Functionally, the neural complex "can be regarded as a three-stage system" as illustrated in Fig. 2.1. Pertinent to this three-stage hierarchy of the neural complex, Griffith [2.2] poses a question: "How is the level of control organized?" It is the control which is required for the C^3I protocol of neurocybernetics. A simple possibility of this is to assume the hierarchy converges to a single cell, a dictator for the whole nervous system. However this is purely hypothetical. Such a dictatorship is overridden by the "democratic" aspect of every cell participating collectively in the decision-process of yielding an ultimate response to the neural complex environment.

The collectiveness of neural participation in the cybernetics of control and communication processes is a direct consequence of the anatomical cohesiveness of the structured neural and brain complexes and the associated *order* in the associated physiological activities.

The cybernetic concepts as applied to the neural complex are applicable at all levels of its anatomy — brain to *cytoblast*, the nucleus of the cell. They refer to the control mechanism conceivable for every neurophysiological activity — at the microscopic cellular level — or at, the gross extensiveness of the entire brain.

In the universe of neural ensemble, the morphological aspects facilitating the neural complex as a self-organizing structure are enumerated by Kohonen [2.3] as follows:

---

[1] C^3I : Command, Communication, Control, and Information — a modern management protocol in strategic operations.

- Synergic (interaction) response of compact neuronal assembly.
- Balance (or unbalance) of inhibitory and excitatory neuronal population.
- Dimensional and configurational aspects of the neural ensemble.
- Probabilistic (or stochastical) aspects of individual neurons in the interaction process culminating as the collective response of the system.

Dependence of ensemble parameters on the type of neural stimuli and upon the functional state of the system.

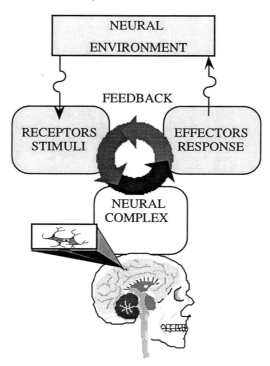

Fig. 2.1: Three-stage model of the neural complex

2.2 Neural Information Processing: C^3I Protocols

The neural complex (real or artificial) is evolved by a progressive multiplication of interneural connections across which the cellular state transitions proliferate randomly in the spatiotemporal domain. The participation of an individual neuron in the collective response of the neural complex is, therefore, largely stochastical in nature. The basis for the analytical modeling of the transfer function pertinent to a neuron relies on the collective behavior of the neural complex and the integrated effect of all the excitatory and inhibitory (post-synaptic) axon potentials which decide the state transitions (or "firing") in the associated neurons.

Conventionally, the network of interconnected neurons is modeled as a set of cellular elements having multi-input state vectors x_i (i =1, 2, ..., M). Corresponding outputs S_i are linked *via* M-component weights states w_{ij} and decided by a nonlinear activation. The relevant input-output relation is specified by (Fig. 1.1a):

$$X_i = \sum_{i=1}^{M} w_{ij}x_i + \theta_i + \zeta_n \qquad (2.1a)$$

and

$$S_i = F_S(X_i) \qquad (2.1b)$$

where θ_i is an external (constant) bias parameter that may prevail at the input and ζ_n is the error (per unit time) resulting from the inevitable presence of intra-and/or extracellular disturbances. The summed-up input signal X_i is processed by a nonlinear activation function F_s to produce the neuron's output signal, S_i. That is, each neuron randomly and synchronously evaluates its inputs and "squashes" its permissible amplitude range of output to some finite value in a nonlinear fashion. In the classical model of a neuron due to McCulloch and Pitts [2.4], the nonlinear activity F_s was specified as a hard-limiting transfer-function given by Eqn. (1.2), where F_s is regarded simply as a bipolar binary function.

However, real neurons have a continuous, monotonic input-output relation with integrative time delays and competing excitatory and inhibitory inputs. This warrants an activation function F_s of finite steepness, rather than the step-like two-state response curve depicted by Eqn. (1.2). In other words, soft-limiting activity could be considered as more realistic and appropriate to represent the function F_s than the hard-limiting logical model *vis-à-vis* the input-output relation of real neurons. Such soft-limiting functions in general, are bounded and S-shaped; they are designated as *sigmoids*.

Ever since the emergence of artificial neural networks developed to mimic the architecture and the spatiotemporal activities of the real neural complex, a number of analytical representations to depict the soft-limiting activity function F_s has been proposed [2.5]. Further, the limits of the activity have been considered either as bipolar $(-1, +1)$ or unipolar $(0, +1)$ characteristics to simplify the numerical scaling. Again, the expressions for F_s have been suitably modified to accommodate these appropriate limiting values.

The analytical representations of F_s have been evolved mostly on the empirical consideration that F_s should be S-shaped, and it should be monotonic and continuous between the squashing limits. Therefore, any function considered to represent F_s should be differentiable everywhere. The function should also accommodate the strategies to minimize the cost-function stipulated by the input-output relation of the neural unit. Such a minimization is specified by the optimal convergence desirability. Consequently, the set of input vectors of the multilayered network of Fig. 1.1b is weighted such that the output vector closely follows a target set. In reference to an isolated neural unit (Fig. 1.1a), a mean-squared error (ε) can be specified by:

$$\varepsilon(w) = \frac{1}{2}\sum_i (X_i - S_i)^2 = \frac{1}{2}\sum_i (X_i - \sum_i w_{ij}x_i)^2 \qquad (2.2)$$

To achieve the minimization of the aforesaid error, a gradient-descent algorithm can be prescribed as follows: Given $\varepsilon(w)$, an improvement on a set of w_{ij}s (by sliding downhill on a surface in the w-space), can be achieved by changing w_{ij} by an amount Δw_{ij} proportional to the gradient of ε at the present location. That is, Δw_{ij} is proportional to $\partial\varepsilon/\partial w_{ij}$. Hence, it follows that:

$$\Delta W_{ij} = \eta \sum_i (X_i - S_i)X_j \qquad (2.3)$$

where η is a constant of proportionality. Denoting the error $(X_i - S_i)$ by δ_i, $\Delta w_{ij} = \eta\delta_i X_j$ refers to the well-known *delta-rule* (or *Widrow-Hoff rule*) that leads to a convergence of the gradient-descent towards "the bottom of the valley" [2.6].

In the above error specification, a linear activity relating the neural input-output entities has been presumed. On the other hand, if the neural activity follows a nonlinear trend quantified by the function F_s, then $\varepsilon(w)$ is given by:

$$\varepsilon(w) = \frac{1}{2}\sum_i \left[X_i - F_s\left(\sum_i w_{ij}x_i\right)\right] \qquad (2.4a)$$

and the corresponding derivative $\partial\varepsilon/\partial w_{ij}$ is as follows:

$$\partial\varepsilon/\partial w_{ij} = -\sum_i \left[X_i - F_s(h_i)\right]F_s'(h_i)X_j \qquad (2.4b)$$

where $h_i = \sum_i w_{ij}x_i$. Hence, it follows that $\Delta w_{ij} = \eta\delta_i X_j$ with $\delta_i = (X_i - S_i) F_s'(h_i)$.

As long as the nonlinear function F_s which depicts the activity across a neural unit is differentiable and monotonic, the quadratic cost-function ε becomes a differentiable function of X_i and S_i and, therefore, could be minimized by letting $S_i \rightarrow X_i$. Alternatively, in terms of learning the correct probabilities of a set of hypotheses represented by output units, using $(1 + S_i)/2$ or the probability that the hypothesis represented by unit i is true (that is, $S_i = -1$ means definitely false, and $S_i = +1$ means definitely true), an information-theoretic cost-function (entropy) can be specified by [2.7, 2.8]:

$$\varepsilon = \sum_i \left(\frac{1}{2}\right)(1 + \tilde{T}_i)\ell og\left\{(1 + \tilde{T}_i)/(1 + \tilde{S}_i)\right\}$$

$$+\left(\frac{1}{2}\right)(1 + \tilde{T}_i)\ell og\left\{(1 - \tilde{T}_i)/(1 - \tilde{S}_i)\right\}$$

(2.5a)

where $\tilde{T}_i$ refers to a target set (or teacher values) and the *tilde* is used to depict the normalization or scaling of the variables limited to values between 0 and 1 so as to represent corresponding probabilities. In Eqn.(2.5a), again $\tilde{S}_i$ can be written as $F_s(h_i)$. The corresponding derivative $d\varepsilon/\partial h$ is given by:

$$\partial\varepsilon/\partial\eta_i = F_S'(h_i)\{S_i - X_i)/[1 - F_s^2(h_i)]$$

(2.5b)

and the error-term $\delta_i = (S_i - X_i) F_S'(h_i)[1 - F_s^2(h_i)]$. In order that this error be independent of the nonlinear activity function F_S, the coefficient $F_S'/(1 - F_S^2)$ can be set equal to a constant, say unity. That is,

$$F_S'(h_i) = 1 - F_S^2(h_i)$$

(2.6a)

or

$$F_S' + F_S^2 - 1 = 0$$

(2.6b)

which is an ordinary, nonlinear (first order) differentiable equation known popularly as the *Riccati differential equation* [2.9, 2.10].

In view of the above considerations, it is attempted in the present chapter to develop a generalized Riccati differentiable equation to describe the nonlinear transfer characteristics of the activities across a neurocellular unit. A set of sigmoidal functions which satisfies this differential equation is elucidated and their physical significance *vis-à-vis* the logistic aspects of input-dictated (output) growth of neural activity are discussed. In general, it is indicated that the sigmoidal functions are either particular solutions or complete primitives of the generalized Riccati equation. Applications of these sigmoidal solutions in the development of a minimizable cost-function in the information-theoretic perspectives are also studied. The conventional sigmoidal functions are shown as particular solutions of the Riccati equation under specific circumstances. The complete heuristics of developing the input-output logistic growth with the pertinent phenomenological considerations associated with the neural complex are discussed. A justifiable sigmoid stemming from the solution of the Riccati differential equation is indicated as the Bernoulli function and its pragmatic aspects in neural network theory are surveyed. Finally, any arbitrarily prescribed nonsigmoidal nonlinearity versus the Riccati equation-based solutions in deciding the *sufficiency condition* to represent a function (or a pattern) at the output of an artificial neural network is analyzed.

2.3 Nonlinear Neuron Activity

The neurobiological basis in the logic of choosing an activation function relies on the fact of that the feature-detecting neurons in the biological sensory system are often idealized as signaling the presence or absence of their preferred features by becoming highly active or inactive respectively. Further, such an activity function formalizes the range of activity notionally. That is, it encodes with finite limits the extent of the preferred features present and delivers it as an output. In other words, the evolution of neural input *versus* output relation depicts a logistic growth of the output as a function of the input.

The nonlinear response of a neuron is decided by how low is the extent of firing activities involved. A neuron being in a persistent state (an attractor) may fire strongly (maximum firing activity) or weakly (close to minimal firing activity). These activities refer to two stable states of the persistent state, exhibiting sigmoidal responses of two distinct slopes.

Let y ($\equiv$ S) depict the output and X refer to the sum of a set of weighted inputs of a neural cell. Then the change (growth) of y with respect to X can be stipulated as [2.11]:

$$(1/y)dy/dX = f(y) \qquad\qquad (2.7a)$$

or

$$dy/dX = yf(y) \qquad\qquad (2.7b)$$

where f(y) is an arbitrary function to be decided.

Under competing inputs, namely, inhibitory and excitatory, the activity of the neuron governs the response towards a preferred input state. That is, a logistic growth of y versus X depicting a fractional change in y, $\Delta y/y$, with respect to change in the input (ΔX) should directly dependent on some function of y denoted by f(y) in Eqn. (2.7). The activity growth (function) is nonlinear in general and also is "squashed", meaning that the function f(y) would decrease as y increases and at a certain stage the growth of y with respect to the input X ceases (saturates). If f(y) is modeled as a simple linear function of y, namely, f(y) = (a – by) where a and b are constants, Eqn.(2.7) reduces to the following nonlinear first order differential equation (known as the *logistic equation*):

$$dy/dX = y(a - by) \qquad (2.8)$$

Here, the constant a can be regarded as a parameter independent of the cellular activity, such as the biochemical action in a real neuron, but it could represent an external neuroenvironmental influence. The factor b in Eqn. (2.8) can be considered as the effect of the squashing influence perceived as a result of an increase in the output entity, y. Both a and b are positive constants. The level of output at which any further increase in the output ceases refers to a state of equilibrium. It corresponds to the state that the neurocelllular influences would offset the input-dictated output growth. This condition is specified by y = a/b in Eqn.(2.8) and can be regarded as the maximum output capacity of the cellular unit.

Sketching dy/dX as a function of y yields Fig. 2.1 and the corresponding y versus X refers to a logistic growth model in the phase-plane with the value of y taken as a constant y_0 for zero-inputs as shown in Fig. 2.1.

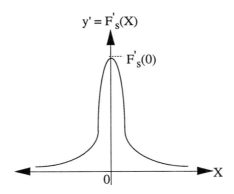

Fig. 2.1: $F_S'(X)$ versus X

(*Note*: The continuous, monotonic characteristic of $F_s(X)$ guarantees dy/dX to be continuous everywhere in the domain **X**)

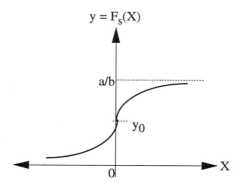

Fig. 2.2: Logistic phase-plane curve of neural input-output relation:
$$y = F_S(X)$$

The logistic phase-plane curve has specific characteristics which can be stated as a set of *lemmas* and proved as presented in the following sub-sections:

Lemma 2.3.1: *The equilibrium value of $y = a/b$ is stable*
Proof: Let $y = (a/b) + \rho y_1$ where ρy_1 is a small perturbation from the equilibrium and $|\rho y_1| \ll a/b$. Substituting this expression for y into the logistic relation of Eqn.(2.8) yields

$$dy_1/d\rho = -by_1[(a/b) + \rho y_1] \tag{2.9}$$

Since ρy_1 is small, the nonlinear (higher order) term in Eqn.(2.9) can be neglected. Hence, it follows that

$$dy_1/dX = -ay_1 \tag{2.10a}$$

which has a solution given by

$$y_1 = c_1 \exp(-aX) \tag{2.10b}$$

where c_1 is a constant of integration. The exponential relation of Eqn.(2.10b) guarantees that the equilibrium value of y would be stable inasmuch as $y_1 \rightarrow 0$ as $X \rightarrow \infty$.

Lemma 2.3.2: *The logistic curve is a sigmoid*
Proof: Writing Eqn. (2.8) as a separation of variables, it follows that

$$dy/[y(a - by)] = dX \tag{2.11}$$

Using partial fractions,

$$1/[y(a - by)] = (1/a)y + (b/a)(a - by) \qquad (2.12)$$

Hence upon integration, Eqn. (2.11) with the substitution of Eqn. (2.12) yields

$$(1/a)\ell n\, |\, y\,| - (1/a)\ell n\, |\, a - by_0\,| \ = X + c_2 \qquad (2.13)$$

where c_2 is a constant of integration and it can be eliminated if the value of y at $X = 0$ is known. Assuming the output value as y_0 for zero inputs during the refractory period, Eqn.(2.13) for positive values of y and y_0 reduces to

$$(1/a)\ell n(y/y_0) + (1/a)\ell n\, |\, (a - by_0)/(a - by)\,| \ = X \qquad (2.14a)$$

which can be solved for y as

$$y = (a/b)[1 + (a - by_0)\exp(-aX)/by_0] \quad y > 0 \qquad (2.14b)$$

The above Eqn. (2.14b) has the following properties:

1. It is defined for all values of $X \in (-\infty, +\infty)$.
2. As $X \to \infty$, $y \to a/b$ for all values of $y_0 > 0$.
3. Substituting $(\alpha = \beta) = (a/2b)$ and with $X_0 = (1/a)\ln[(a - by_0)/by_0]$, Eqn. (2.14b) reduces to:

$$y = \alpha + \beta[1 - \exp\{-a(X - X_0)\}]/[1 + \exp\{-a(X - X_0)\}]$$

$$= \alpha + \beta\tanh\{a(X - x_0)/2\} \qquad (2.14c)$$

The above relation (Eqn. (2.14c)) indicates that y is S-shaped (sigmoidal) inasmuch as tanh $(z) \to 0$ as $z \to 0$ and tanh (z) is an odd function, namely, $\tanh(-z) = -\tanh(z)$.

Specifically, by considering $a = b = 1$ so that $\alpha = \beta = 0.5$ and $X_0 = 0$, Eqn. (2.14c) simplifies to

$$y = (1/2) + (1/2)\tanh(X/2) \qquad (2.14d)$$

which represents a unipolar sigmoid adopted conventionally in the analysis of artificial neural networks. It is bounded (squashed) between the limits of 0 and +1.

Thus, the nonlinear input-output response of a neuron mathematically follows a sigmoidal logistic phase-plane curve. It was assumed in the above considerations that f(y) is a linear function of y with constant-parameters of a and b. From Fig. 2.1, it follows that the nonlinear saturation is dictated by the ratio a/b and the functional attributes of y *versus* X are independent on the refractory, zero-input conditions, or, the state y_0 of the output.

In the following section, it is indicated that the evolution of nonlinear output *versus* input relation in a neuron could be more generally depicted by the Benoulli-Riccati equations. The solution of these equations is a single parameter family of curves representing various sigmoidal functions conventionally adopted in modeling the activity function of a neuron.

2.4 Bernoulli-Riccati Equations

A natural extension of the first-order linear equation, namely, $y' = p_0(X) + q_0(X)$ leads to a nonlinear, first-order equation of the type:

$$y' = p_0(X) + q_0(X)y + r_0(X)y^n \tag{2.15}$$

where $p_0(X)$, $q_0(X)$ and $r_0(X)$ are continuous functions of X and $n \neq 0$. The above equation, is well-known as the *Bernoulli equation* which can be transformed into an integrable form using an appropriately chosen (new) dependent variable [2.12, 2.13].

In Eqn. (2.15), the functional coefficients p_0, q_0 and r_0 specify definite attributes to the nonlinear activity governed by this differential equation. The coefficient p_0 characterizes the external stimulus which enables the nonlinear activity to commence and remain sustained. Depending on the output value y, the prevailing activity is weighted by the extent of that output value leading to the nonlinearity perceived. This is accommodated in Eqn. (2.15) by the coefficient r_0 which decides the output-dictated influence on the nonlinear activity. Should r_0 be equal to zero, the activity degenerates to a simple linear input-output relation. The exponent n stipulates the degree of the output entity in formulating the extent of nonlinearity involved. If $n = 2$, Eqn. (2.15) is known popularly as the *generalized Riccati equation*. The coefficient q_0 in Eqn.(2.15) contributes to the linear input-output relation and could be set equal to zero if the nonlinear activity predominates. It should be noted that in general, p_0, q_0 and r_0 could be constant coefficients as well.

The Bernoulli equation given by Eqn. (2.15), therefore, could be reduced to a simpler Riccati equation pertinent to neural activity, namely,

$$y' + r_0(X)y^2 + p_0(X) = 0 \tag{2.16}$$

Eqn. (2.16) assumes that the neural cellular activity is more likely (or predominantly) nonlinear and the output y that governs such a nonlinear behavior, in the simplest case, offers a second degree influence (with n = 2). Any higher order influences with n > 2 are neglected.

The Riccati equation (Eqn. (2.16)) is specified over an interval I of X and could be transformed such that its general solution can be written in the following form:

$$y(X) = [1/r_0(X)][Au'(X) + Bv'(X)]/[au(X) + bv(X)] \qquad (2.17)$$

where A and B are arbitrary constants, not both zero, and u and v are linearly independent solutions of

$$d/dX\{[1/r_0(X)dz/dX]\} + zp(X) = 0 \qquad (2.18)$$

where z represents either u or v.

Satisfying the above conditions, the following set of functions can be identified as the possible solutions, (either as particular integrals or as complete general primitives of Eqn. (2.16):

Case (I)

$$y(X) = \begin{cases} (1/rp)^{1/2}[K\exp\{2(r_0/p_0)^{1/2}X\} - 1] / [K\exp(2(r_0/p_0)^{1/2}X\} + 1] \\ \qquad\qquad\qquad \text{if } y(0) < (1/r_0 p_0)^{1/2} \\ \\ (1/rp)^{1/2} \qquad\qquad\qquad \text{if } y(0) = (1/r_0 p_0)^{1/2} \\ \\ (1/rp)^{1/2}[K\exp\{2(r_0/p_0)^{1/2}X\} + 1] / K\exp\{2(r_0/p_0)^{1/2}X\} - 1] \\ \qquad\qquad\qquad \text{if } y(0) > (1/r_0 p_0)^{1/2} \end{cases}$$

$$\qquad (2.19a)$$

where K= $[1/(r_0 p_0)^{1/2} + y(0)]/[1/r_0 p_0]$. Further, if y(0) = 0 and $(1/r)^{1/2} = (1/p)^{1/2}$ =1, then

$$y(X) = [\exp(2X) - 1] / [\exp(2X) + 1] = \tanh(X) \qquad (2.19b)$$

Case(ii)

Suppose y(0) = 0, $r_0(X) = 1$ and,

$$p(X) = c^2 + d^2[\coth^2(cX) + \operatorname{cosech}^2(dX) - 2cd[\coth(cX)\operatorname{cosech}(dX)]$$

$$p(X) = c^2 + d^2[\coth^2(cX) + \text{cosech}^2(dX) - 2cd[\coth(cX)\text{cosech}(dX)]$$
(2.19c)

then,

$$y(X) = c[\coth(cX)] - d[\coth(dX)]$$
(2.19d)

In Eqn. (2.19d), setting $c = (1+1/2Q)$ and $d = 1/2Q$,

$$y(X) = L_Q(X)$$
(2.19e)

which is popularly known as the *Bernoulli or modified Langevin function* [2.13]. It is also known as the *Brillouin function* [2.14].

Case(iii)
If $y(0) = 0$, $r_0(X) = 1$ and $p(X) = [1 + C\exp(-CX)] / [1 + \exp(-CX)]^2$, then

$$y(X) = 1 / [1 + \exp(-CX)]$$
(2.19f)

Case(iv)
If $y(0) = 0$, $r_0(X) = 1$ and $p_0(X) = (2/p)[1 + X^2)] + [(2/p)\arctan(X)]^2$, then

$$y(X) = (2/p_0)\arctan(X)$$
(2.19g)

In all the above solutions of $y(X)$ given by Eqns.((2.19a & 2.19b) and (2.19d-2.19g)), the relevant Riccati equation (Eqn. (2.16)), assumes a transformed form, namely:

$$u''(X) - p_0(X)u(X) = 0$$
(2.20)

with $r_0(X) = 1$ and $u(X) \neq 0$ for X on a solution a subinterval I_0 of I; and, the logarithmic derivative $y(X) = u'(X)/u(X)$ constitutes the solution on I_0 of the Riccati equation, (Eqn. 2.16).

Further, if $y(X)$ is a bipolar output squashed between the limits -1 and $+1$, the excitation stimulus function $p_0(x)$ arising from the set of inputs $\{x_i\}$ is bounded by: $y(0) \geq p_0(X) > 1$. In the case of unipolar output squashed by the limiting values of 0 and $+1$, the excitation stimulus function $p_0(X)$ is bounded by $0 < p_0(X) \leq [y'(X) + y^2(X)]$. In both cases the lower limit could be specified only as the greatest lower bound (*infimum*) [2.15].

2.5 Nonlinear Neural Activity: Practical Considerations

From the foregoing discussions, it is clear that the analytical representation of the nonlinear input-output behavior pertinent to a neuronal cell depends on the persisting state output level (y_0) prevailing. In this case, the incoming (summed up) inputs decide the further increments on the output. The Riccati differential equation and the associated solutions, as discussed above, describe the analytical logistic nonlinear (neural) response and practical implications of such representation(s) in studying the real neurons, as well as in modeling artificial neural networks. Relevant details are as follows.

2.5.1 Stochastical response of neurons under activation

By considering the neurons as analogous to a random, statistically isotropic dipole system, the graded response of the neurons under activation has been modeled by Neelakanta et al. [2.12]. Modeling was done by applying the concepts of Langevin's theory of dipole polarization. The continuous graded response of neuron activity corresponding to the stochastical interaction between incoming excitations (that produce true, collective, nonlinear effects) has been elucidated in terms of a sigmoidal function. This function is specified by a gain parameter $\lambda = \Lambda/k_B T$ with Λ being the scaling factor of the cellular inputs and $k_B T$ the pseudo-Boltzmann energy of the system with k_B being the (pseudo) Boltzmann constant and T the (pseudo) temperature. It has been shown by Neelakanta et al. [2.12] that the corresponding sigmoid is the Langevin function L(.), which is the same as Eqn.(2.19e) with $Q \to \infty$. In a statistical mechanics point of view, $Q = 1/2$ refers to a dichotomous state in which the spatial orientation of the dipoles, or the corresponding spatial proliferation of neuronal states across the interconnected cells, depicts two totally discrete aniosotropic (parallel or antiparallel) orientations. Thus, in a statistically isotropic randomly oriented system, the spatial, discrete alignments would approach infinity, (depicted by $Q \to \infty$), and in a totally anisotropic orientation, $Q \to 1/2$. For the intermediate range of orientations, the extent of dipole alignments, or, correspondingly, the output response of a neuron to excitation, could be specified by $1/2 < Q < \infty$. That is, the pertinent sigmoid is given by $L_Q(.)$ as elaborated in [2.16] and by Neelakanta and DeGroff [2.15]. The above consideration of representing a sigmoidal function in reference to a neuronal cell is consistent with the solution (Eqn. 2.19e) of the Riccati equation (Eqn. 2.16). It should also be noted that, as $Q \to 1/2$, $L_Q \to \tanh(.)$. The conventional use of a hyperbolic tangent as a sigmoidal activity function in modeling an artificial neural network presumes implicitly that the spatial proliferation of neuronal state transitions across the interconnected network of cellular units is totally anisotropic. But more realistically, especially in the case of biological neurons, a partial, antisotropic spatial dispersion of nonlinear neural state transitions cannot be per se represented by tanh(x). It warrants a more general solution which is given by Eqn. (2.19e); or its variations and the Langevin function, which constitute only the upper and the lower bounds of the state-vector squashing characteristics of a neural unit, respectively. A summary of the Langevin's theory of dipole polarization and its relevance to the neural complex nonlinearity is presented in Appendix 2.2.

2.5.2 Representation of a neuron as an input-dictated cybernetic regulator with a quadratic cost-function

A neural network represents a complex automaton — a cybernetic system. It represents a self-controlling set of automata, a memory machine, and a homeostat [2.1]. The association of its subsystems *via* feed-forward and/or feedback constitutes an optimal control for neural self-organization. In this endeavor of cybernetically-controlled (self-adapting) activities, the neural system deliberates the minimization of uncertainties arising from the inherent noise or spatiotemporal random characteristics of its activities. A set of learning rules normally facilitate the optimal self-organization envisaged. A cost-function such as the quadratic error given by (Eqn. (2.4a)) can be stipulated on the basis of which the output is regulated as an appropriate nonlinear activity function of the input set. In this perspective, a neuronal cell can be regarded as an input-dictated regulator which can be described *via* a Riccati differential equation as follows:

Let $dy/dX = q_0(X)y + r_0(X)Y + p_0(X)$ with $y(X_0) = y_0$ and X is defined in the interval $X_0 \leq X \leq X_1$. This nonlinear differential equation can be considered to depict the neurocellular activity of self-organization in appropriately relating the output y and the summed up input X. The parameter Y is a control parameter so chosen as to minimize the cost of the performance measure. That is, the output is squashed nonlinearly by introducing a linear deficiency in the neural activity in accordance with a learning rule. The objective of the learning rule is to set the output as close as possible to a target value. Should the output differ from this objective function, an error such as Eqn. (2.4a) is estimated. The error or the cost-function is minimized optimally by proper choice Y(y) which governs the linear deficiency parameters of the system. It can be noted that $Y = y^n$ or y^2, corresponding to Bernoulli and Riccati equations, respectively is the governing relation of the nonlinear regulation in the self-organizing efforts of a neuronal cell.

2.5.3 Generalized information-theoretic entropy measure

As given by Eqn. (2.19e), $L_Q(.)$ refers to general solution of the nonlinear activity which also defines implicitly the isotropic/anisotropic aspects of (stochastical) spatial proliferation of neuronal state transitions. Referring to the information-theoretic cost-function of Eqn. (2.5a) indicated earlier, it is governed by an activity function F_S as decided by the Riccati equation (Eqn.

(2.6)), and the relevant solution is pertinent only to a totally anisotropic system with $Q = 1/2$ as discussed before. Therefore, in view of the more generalized solution of the Riccati equation (Eqn. (2.16)) as given by Eqn. (2.19d), namely $L_Q(.)$, the cost-function or the information-theoretic entropy can be written in a general form as follows:

$$H_\varepsilon = \varepsilon_{1T} = \sum_i (1/2)(1+ X_i)\ell og\{(1+ X_i)/[p_0^{1/2} + F_S(X_i)]\}$$

$$+ (1/2)(1- X_i)\ell og\{(1- X_i)/[p_0^{1/2} - F_S(X_i)]\}$$

(2.21)

so that the minimization of $H_\varepsilon = \varepsilon_{1T}$ would require that the Riccati equation (Eqn. (2.16)) with $r_0 = 1$ and $y = F_S$ is satisfied.

2.5.4 Influence of nonlinear neuronal activity on information

Under information-theoretic considerations pertinent to neural network, the error metric that could be minimized corresponds to the cross-entropy parameters based on the principle of the discrimination-information defined *via* the so-called distance or divergence measures as indicated in Chapter 1. These are conditional (information) discrimination parameters defined by

$$H(P_1, P_2) = \sum_k p_{1k}\ell og(p_{1k}/p_{2k})$$

(2.22)

where P_1 and P_2 are probability distributions of random variables associated with the probabilities p_{1K} and p_{2K}, respectively.

In reference to a neural state transitional activity, p_{2k} could depict the probability of the hypothesis represented by the neural unit k at its output S_k and p_{1k} can be interpreted as the target set of probabilities. Hence, Eqn. (2.22) is a relative (cross) entropy of the probability distributions P_1 and P_2. A typical information-theoretic distance measure (as will be elaborated in Chapter 4) is the Jensen divergence [2.8] which is a symmetric error-metric given by:

$$H(P_1, P_2) = H(P_1, P_2) + H(P_2, P_1)$$
$$= \sum_k (p_{1k} - p_{2k})\ell og(p_{1k}/p_{2k})$$

(2.23a)

In the analysis pertinent to the evaluation and computation of Eqn. (2.23a) with respect to the neural complex, if p_1 of the input probability distribution is given, the corresponding p_2, namely, the output probability distribution, should be known. Since the neural input-output relation is the governed function F_S, p_2 can be elucidated from p_1 *via* the transformation,

$$p_2(X) = \left[p_1(X)/F_S'(X) \right]\Big|_{X=F_S^{-1}(X)} \tag{2.23b}$$

Considering a generalized activity function such as $F_S = L_Q(X)$, the aforesaid transformation could be effected by approximating $L_Q(X)$ as a soft-limiting nonlinear transfer function as follows:

$$L_Q(X) \approx -1 + (2\alpha_0/2\pi j) \int_{C_u} [\{\exp(jXu)\sin(u/\alpha_0)/u^2]du \tag{2.24a}$$

where the contour C_U runs from $-\infty$ to $+\infty$ and is indented downward at the origin [2.18]. Explicitly, Eqn. (2.24a) is written as (see Appendix 2.1)

$$F_S(X) \approx \begin{cases} -1 & X < -(1/\alpha_0) \\ \alpha_0 X & -(1/\alpha_0) < X < +(1/\alpha_0) \\ +1 & X > +(1/\alpha_0) \end{cases} \tag{2.24b}$$

where $\alpha_0 = L_Q'(X)$ at $X = 0$. From Eqn. (2.19d) and Eqn. (2.19e), it follows that, $\alpha_0 = (2c-1)/3 = (2d+1)/3$. Using the above soft-limited approximation of a neural cell, the Jensen measure or any similar measure can be written explicitly. More discussion on the various information-theoretic error-measures which can be adopted in a neural network for optimization purposes are discussed in Chapters 4 and 5. The considerations on the nonlinear operation involved thereof are the same as those discussed above without any loss of generality.

2.5.5 Significance of the parameter Q in the generalized Bernoulli function
   $L_Q(.)$ depicting the neuronal input-output relations
   As indicated in Chapter 1 and illustrated in Figs. 1.2 and Fig. 2.3, $Q \to \infty$ sets the lower bound of the activity function (sigmoid) to signify a totally isotropic, neural spatial configuration with zero long-range order. The magnitude of Q decides the number of orientational states which tends to infinity in the case of an isotropic system. When $Q \to 1/2$, the number of orientational states (given by $2Q + 1$) becomes predominantly, an anisotropic system with a long-range order of total extent. It signifies the upper bound of the sigmoid. In partial anisotropic situations, the number of orientational states is $(2Q + 1)$ with $(1/2 < X < \infty)$. In the event of $Q \to 0$, there is no interaction between the neurons, and the activity of each neuron is decided discretely by an activity function $L_Q(X) \to$ signum function as $Q \to 0$. This correspond to the familiar McCulloch-Pitts' model of a neuron [2.1].

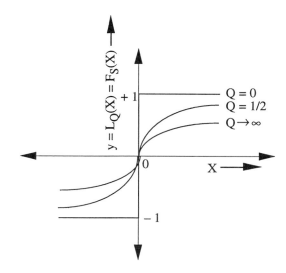

Fig. 2.3: Bernoulli Function $L_Q(X)$ as the neural activation sigmoid

(a)   When $Q \to 0$, $L_Q(X) \to$ signum function (McCulloch-Pitts for limit of noninteracting neurons)

(b)   When $Q \to 1/2$, $L \to Q(X)$ hyperbolic tangent function (for the upper bound of totally anistropic, spatial neural state proliferation)

(c)   When $Q \to \infty$, $L \to Q(X)$ Langevin function (for the lower bound of totally isotropic, spatial neural state proliferation)

Thus, $L_Q(X)$ is a single parametric activity function of a neuron with a physical meaning to the choice of Q which portrays the long-range spatial order associated with the neural complex. In the artificial neural networks, the empirical choice of a sigmoid such as $\tanh(\Gamma X)$ and an arbitrary assignment of a value to the gain factor Γ are presumptuously intuitive — an intuition which stems from rather vague arguments related to the squashing capabilities of the sigmoid and to the fact that the sigmoidal gain (slope) through the axis can be easily manipulated [2.5]. The mathematical ground of justification is choosing a sigmoid or the logic of activation functions as portrayed by Williams [2.6] and Jordan [2.17]. The pragmatic grounds due to Hecht-Nielson [2.18] or Wasserman [2.19] do not model the sigmoids used in the artificial neural networks in the actual (or biological) perspective of nonlinear activity perceived in the real neurons. Further, there is a general attitude in regard to the exact details of the sigmoids being not so critical in a network, such as the one which follows a backpropagation algorithm [2.20, 2.21]. At the most, the modeling strategies pertinent to sigmoids in the artificial neural networks refer to just choosing "an activation function that exhibits the best properties of training" [2.5].

The purpose of such training is to reduce a generic error-function. Kalman and Kwasny [2.2] arrive at a sigmoidal function which provides activations within the range [Low, High] and maintains the paramatic aspects of efficient training. This is given by the function $F_S(X) = [(High - Low)/2]tanh[3X/2] + [(High + Low)/2]$. Again, it can be seen that such a function is an eventual solution of a Riccati equation and the gain/scaling factor (3/2) which appears in the argument of the hyperbolic tangent function is nonetheless the same as the coefficient in order parameter $S^0 = \{3/2[F'_S(0)-1/2\}$ and $F_S(0) = (1/3 + 1/3Q)$, if $F_S(X) = L_Q(X)$. In the range of order parameter, namely $0 < S^0 < 1$, the limiting values $(0,1)$ correspond to the totally isotropic $(Q \to \infty)$ and totally anisotropic $(Q \to 1/2)$ spatial orders of the neural complex [2.1, 2.12, 2.13, 2.22,].

2.6 Entropy / Information Flow across a Neural Nonlinear Process

As discussed in the previous sections, the neural network, in essence operates on the weighted sum of the inputs (nonlinearly) and delivers an output. In this process mediated by a nonlinear operation, it is of interest to study the transformation of entropy, and hence, the associated information content, across the network. Relevant considerations are elaborated in Chapter 4.

2.7 Nonsigmoidal Activation Functions

Sigmoidal functions are generally regarded as candidates to represent the neural nonlinear activity. Various other nonsigmoidal sets of activation functions have also been proposed, which when used in artificial neural networks, have the power of approximating a function to a given extent of accuracy. The following are nonsigmoidal activity functions discussed in the literature [2.5, 2.23-2.30]:

(1) Any rational function which is not a polynomial.
(2) Any root x^α, provided α is not a natural number.
(3) The logarithmic function.
(4) The exponential function.
(5) The gaussian function.
(6) The radial basis functions $(1 + x^2)^\alpha$, $\alpha < 1$, $\alpha \neq 0$.
(7) Bilinear threshold function; and,
(8) Integral representation with priority known kernal

Notable exceptions from the above-listed functions, equivalent to the standard sigmoid, are polynomials, trigonometric polynomials and splines.

Another possible activation function known as the *softmax function* is given by

$$F(x_i) = \exp(x_i)/\{\sum_{j=1}^{m} \exp(x_j)\} \tag{2.25}$$

where, x_i is the total weighted sum input to the ith neuron unit.

Further, an interesting sigmoidal activity function can be constructed from gaussian diffusion theory considerations. It is given by

$$F_s(x) = \frac{\tanh(b\tau) + \tanh[b(x - \tau)]}{\tanh(b\tau) + \tanh[b(1 - \tau)]} \tag{2.26}$$

where, $b(0, \infty)$ is a free-parameter, and $(0 < \tau < 1)$ is a threshold value.

The rationale considerations in prescribing nonsigmoidal activity functions to neural networks rely on the fact that such functions have the comparable power of approximating the functional aspects of a neural network *vis-à-vis* the approximation which could be realized by using a standard sigmoidal function. Comparison of these functions has been made in reference to the computational complexity [2.26], tighter bounds on the network size (number of elements), and the depth of the network depicting the tangent path from any input to the output gate.

The aforesaid considerations indicate how certain nonsigmoidal functions can depict the neural activity at least to an approximate extent. It is further investigated below to see whether these nonsigmoidal functions fall within the logistic-growth functional aspects of the Riccati family of differential equations. For this purpose, the basic mathematical descriptions and *lemmas* attributed to sigmoidal functions, are presented and on the basis of these, nonsigmoidal functions are tested towards their admissibility to the family of the Riccati differential equation.

2.8 Definitions and Lemmas on Certain Classes of Nonlinear Functions

Definition 2.8.1: A function $F_S: R^1 \rightarrow R^1$ is called *a sigmoidal function*, if it satisfies, (*Limit* $X \rightarrow -\infty$, $F_S(x) \rightarrow 0$) and (*Limit* $X \rightarrow +\infty$, $F_S(X) \rightarrow 1$).

Definitions 2.8.2: If F_S is a bounded sigmoidal function, then it is *a Tauber-Wiener* (TW) *function*. A function is a Tauber-Wiener (TW) function [2.26-2.28], if in the finite dimensional space ($R \rightarrow \Re$, it is continuous or discontinuous but satisfies all linear combinations $\sum_{i=1,N} c_i g(\lambda_i X + q_i)$, $\lambda \in \Re$, $\theta \in \Re$, $c_i \in \Re$ (i = 1,2,...., N) are dense in every compact set C[a,b], with the range a to b.

Lemma 2.8.1: A function (continuous or discontinuous) to be qualified as an active function, a sufficient condition is that it belong to TW class [2.23, 2.26-2.28].

Lemma 2.8.2: When a sigmoidal function is used as an activation function in a neural network, the only *necessary* condition imposed on it is its *boundedness* [2.23, 2.26-2.28]. (Note that, invariably in all neural network studies (except in [2.24]), sigmoidal functions are strictly assumed to be monotonic and/or continuous.)

Lemma 2.8.3: Any bounded nonconstant continuous function is qualified to be an activation function [2.23].

Lemma 2.8.4: Under some constraints on the amplitude of a continuous function near infinity, any nonpolynomial function is qualified as an activation function.

Lemma 2.8.5: A continuous, discriminatory sigmoidal function F_S can reconstruct function $f(X)$ approximately by $g(X) = \sum_{j=1,N} \alpha_i F_S(y_j^T X + q_j)$ which are in the compact set $C(I_n)$ with respect to the *suprenum* norm with I_n being the n-dimensional unit cube; and $f \in C(I_n)$ which has *supremum* $\|f\|$.

The above lemma implies that for any given $f \in C(I_n)$ and $e > 0$, there is a sum $g(X)$ of the above form such that $|g(X) - f(X)| < e$ for all $X \in I_n$ [2.23].

Definition 2.8.3: Saturation type of nonlinearity refers to sigmoidal functions which are continuous and bounded within limiting values approached asymptotically as saturation. Typical examples are:

(1) Hyperbolic tangent function: $\tanh(X) = L_{1/2}(X)$
(2) Langevin function: $L_Q(X)$, $(Q = \infty)$
(3) Bernoulli function: $L_Q(X)$, $(Q: 1/2 \text{ to } \infty)$
(4) $1/[1 + \exp(\beta X)]$
(5) $\arcsin(X)$
(6) $\arctan(X)$
(7) Linear soft-limiting function: $sat(X)$
(8) Linear hard-limiting function: $sgn(X)$

Lemma 2.8.6: To be an activation function, the necessary and sufficient condition that it belongs to TW class (as per *lemma 2.8.1*), although not sigmoidal (but bounded and not a polynomial), can still be regarded, at least approximately, as being evolved as a logistic growth nonlinear input-output response.

Proof: Let $F_{ns}(S)$ be a nonsigmoidal or discontinuous function qualified to describe the neural input-output activity, but virtue belonging to the TW class and being bounded.

By *Definition (2.8.2)*,

$$F_{ns}(X) = \Sigma C_i F_S(X) + q_i \qquad (2.25)$$

where F_S represents a sigmoidal function which can be approximately depicted by a linear soft limited function [2.18] by the consideration used to derive Eqn. (2.24). That is

$$\sigma(z) \approx \begin{cases} -1 & z > -(1/\beta_0) \\ \beta_0 z & -(1/\beta_0) < z < +(1/\beta_0) \\ +1 & z > +(1/\beta_0) \end{cases} \qquad (2.26)$$

where β_0 is decided by the slope at $z = 0$. Therefore, it follows that: F_{ns} can be approximated as $F_{ns}{}^a$ given by

$$F_{ns}{}^{(a)}(X) = \Sigma d_i X_i + \theta_i \qquad (2.27)$$

and $| F_{ns}(X) - F_{ns}{}^{(a)}(X) | < (e > 0)$ where e defines the precision of approximation. With the approximation as specified above, the corresponding linearized transformations of the differential equation of the sigmoidal function chosen for the construction of the weighted sum given by Eqn. (2.24) takes the form [2.16]:

$$dF_S(z)/dz \approx \begin{cases} 0 & z > -(1/\beta_0) \\ \beta_0 & -(1/\beta_0) < z < +(1/\beta_0) \\ 0 & z > +(1/\beta_0) \end{cases} \qquad (2.28)$$

which, in the absence of approximation used corresponds to Riccati differential equation as discussed in Section 2.3. Inasmuch as $F_{ns}^{(a)}$ is constructed by linear superposition of approximated F_S-functions, $F_{ns}^{(a)}$ should satisfy the linear version of the differential equation (Eqn. (2.16)) in describing (approximately) the activity of the neural complex. Hence, as $F_{ns}^{(a)} \to F_{ns}$, it also corresponds to the superposition of several logistic growth relations mediating the input and output parameter of the network.

2.9 Concluding Remarks
Consistent with the neurobiological considerations as well as the mathematical and/or pragmatic aspects of training and minimization of the

performance cost-function in an artificial neural network, it is indicated that the nonlinear activity of a neural cell can be represented by the Riccati equation. Among the various possible solutions of the Riccati equation, a generalized solution refers to the Bernoulli function $L_Q(.)$ which is posed here as the justifiable sigmoidal function $F_S(.)$ to depict the input (X) *versus* the output $y = F_S(.)$ of a neurocellular unit.

It is indicated in [2.12, 2.13], that the single parameter Q controls the spatial order of isotropic or anisotropic proliferation of the state transitions in the real neuron. Its choice would enable governing the slope (gain) through the axis required in the efficient training of the artificial neural network as well.

Choice of $L_Q(.)$ as the sigmoid also facilitates the development of generalized error-metrics such as the Jensen divergence measure adopted in the information-theoretics of neural networks. Application of $L_Q(.)$ to the different types of information-theoretic error-measures based learning strategies is indicated in Chapters 4 and 5.

Depiction of the nonlinear activity of a neuron *via* the Riccati differential equation permits modeling the neural complex as a cybernetic self-organizing regulator which is a realistic representation of the adaptive control activities in the neuron.

The nonlinear activity in a neuron stems from the logistic growth of the output as a function of the inputs. It is demonstrated in the present work that such a logistic functional attribute can be described by the Riccati differential equation relating the input and the output entities of the neuronal cell.

Further, the present study generalizes, elaborates, and augments the Langevin machine concepts presented by Neelakanta, et al. [2.13]. Fig 2.3 illustrates the Bernoulli function as a single parameter (Q) neural activation sigmoid. It can be noted that the parameter Q controls the slope of the activation function at the origin $y(0) = (1/3 + 1/3Q)$. It also stipulates the stochastical justification in rendering the activation to correspond to the total anisotropic or isotropic state or an intermediate state with $(1/2 < Q < \infty)$ of neuronal state proliferation across the interconnected cellular space.

The last section of this chapter is concerned with the question of whether sigmoidal nonlinearity is a necessary condition to depict the neural activity. This query is extended further to study whether an arbitrary nonlinearity is sufficient to represent any function or pattern by the neural networks, at least to a given degree of precision. Such intriguing questions have been repeatedly posed in the literature [2.24]. If the arbitrary nonlinearity (other than the sigmoidal versions) is acceptable, the other natural questions which arise are:

- "What concrete type of nonlinear neuron is best to use?"
- "How to implement the (arbitrary nonlinear) function?"

- Is Hecht-Nielson's remark [2.5] that "no constructive method for developing the (arbitrary nonlinear) functions is known" justifiable?
- What constitutes the realistic set of arbitrary (nonsigmoidal) nonlinear functions?
- "Is it possible to approximate arbitrary functions if one uses these realistic nonlinear elements (with the output that is not necessarily the nondecreasing function) of the output?" [2.24].
- With arbitrary (nonsigmoidal) nonlinear functions which can describe the neural input-output relation, at least to a given extent of precision, is it justifiable that the corresponding activity be regarded as being mediated by logistic growth phenomenology?

These questions have been answered at least heuristically, in this chapter through relevant definitions and *lemmas* as available and as proved in the existing literature [2.23-2.28]. Also, they are supplemented and elaborated upon with a commensurate theorem applied to the arbitrary (nonsigmoidal) nonlinear neural activity functions *vis-à-vis* implicit logistic growth superpositions mediated by the input and output parameter of the network.

Bibliography

[2.1]   Neelakanta, P.S. and DeGroff, D: *Neural Network Modeling: Statistical Mechanics and Cybernetic Perspectives* (CRC Press, Boca Raton, FL: 1994).

[2.2]   Griffith, J.S.: *Mathematical Neurobiology* (Academic Press, New York, NY: 1971).

[2.3]   Kohonen, T.: *Self-Organization and Associative Memory* (Springer Verlag, New York, NY: 1988).

[2.4]   McCulloch, W. W. and Pitt, W.: A logical calculus of the ideasimminent in nervous activity. *Bull. Math. Biophys*, 5, 1943, 115-133.

[2.5]   Kalman, B. L. and Kwansny S. C.: Why tanh: Choosing asigmoidal function. *Proc. IEEE Int. Joint Conf. on Neural Networks* (June 1992), pp. IV 578-IV581.

[2.6]   William, R. J.: The logic of activation function. In: Rumelhart, D. E. and McClelland, J. L. (Eds): *Parallel Distributed Processing 1* (MIT Press, Cambridge, MA: 1986), Chapter 10, 423-443.

[2.7]   Baum, E. B. and Wilczek, F.: Supervised learning of probability distributions by neural netwroks. In Anderson, D. A. (Ed.): *Neural Information Processing Systems.*: (American Institute of Physics, New York, NY: 1988), 52-61.

[2.8]   Blahut, R. E.: *Principle and Practice of Information Theory* (Addison Wesley Publishing Company, Reading, MA: 1987).

[2.9]   Davis, H. T.: *Introduction to Nonlinear Differential and Integral Equations* (Dover Publications, Inc., New York, NY: 1962)

[2.10] Bittani, S., Laub,A.J., and Willems, J.C.: *The Riccati Equation.* (Springer Verlag, Berlin: 1991).

[2.11] Haberman, V.: *Mathematical Models: Mechanical Vibration, Populaion Dynamics and Traffic Flow.* (Prentice Hall, Inc., Englewood Cliff, NJ: 1977).

[2.12] Neelakanta, P. S. Sudhakar, R. and DeGroff, D. : Langevin machine: A neural network based on stochastically justifiable sigmoidal function. *Biol. Cybern.*, 65, 1991, 331-338.

[2.13] De Groff, D. Neelakanta, P. S., Sudhakar, R. and F. Medina: Liquid crystal model of neural networks. *Complex Syst.*, 7, 1993, 43-57.

[2.14] Mayer,J.E. and Mayer, M.G.: *Statistical Mechanics* (John Wiley and Sons, Inc., New York, NY: 1946), pp. 346

[2.15] Clark, C.: *The Theoretical Side of Calculus* (Wadsworth Publishing Co., Belmot, CA, 1972), pp. 92-98.

[2.16] Bennet, W.R. and Rice, S. O.: Note on methods of computing modulation products. *Phil. Mag. S. 7.*, 18, 1934, 422-424.

[2.17] Jordan, M. I. An introduction to linear algebra in parallel distributed process, in D. E. Rumelhart and J. L. McClelland, eds., *Parallel Distributed Processing 1* (MIT Press, Cambridge, MA: 1986), Chapter 9, 375-422.

[2.18] Hecht-Nielsen, R.: *Neurocomputing* (Addison-Wesley, Reading, MA: 1990).

[2.19] Wasserman, P. D. *Neural Computing: Theory and Practice* (Van Nostrand Reinhold, New York: 1989).

[2.20] Caudill, M.: *Neural Networks Primer* (Miller Freeman Publications, San Francisco, CA: 1990).

[2.21] Rumelhart, D. E. and McClelland, J. L. (Eds.): *Parallel Distributed Processing*: Vol. I (MIT Press, Cambridge, MA: 1986).

[2.22] Neelakanta, P. S., Abusalah, S. De Groff, D., and Sudhakar: Logistic model of nonlinear activity with neural complex: Representation *via* Riccati differential equation. *Cybernetica*, XXXIX, 1996, 15-30

[2.23] Chen, T. and Chen, H.: Universal approximation of nonlinear operators by neural networks with arbitrary activation functions and its application to dynamical systems, *IEEE Trans. Neural Networks*, 6, 1995, 911-917.

[2.24] Kreinovich, V. Y.: Arbitrary nonlinearity is sufficient to represent all functions by neural networks: A theorem, *Neural Networks*, 4, 1991, 381-383.

[2.25] Ito, Y: Representation of functions by superpositions of a step or sigmoid function.and their applications to neural network theory, *Neural Networks*, 4, 1991, 385-394.

[2.26] Siu, K. Y. Roychowdhury, V. P. and Kailath, T.: Rational approximation techniques for analysis of neural networks, *IEEE Trans. Inform. Theory*, 40, 1994, 455-466.

[2.27] DasGupta, B. and Schnitger, G.: The power of approximating: A comparison activation, In: Hason, S. J., Cowan, J. D. and Giles, C. L. (Eds): *Advances in Neural Information Processing Systems*, 5 (Morgan Kaufman, San Mateo, CA: 1993), 615-622.

[2.28] Cybenko, G.: Approximation by superposition of a sigmoidal function, *Math.Control Signals Syst.*, 2, 1989, 303-314.

Appendix 2.1

Linear Approximation of Sigmoidal Functions

Suppose a sigmoid is approximated as a soft linear limiting function as illustrated in Fig. A2.1 and specified by Eqn. (2.24b):

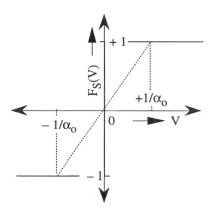

Fig. A2.1: Soft linear-limiting approximation of a sigmoid

That is, by denoting the sigmoid as $F_S(V)$:

$$F_S(V) = \begin{cases} -\alpha D & +V < -D \\ +\alpha V & -D < V < +D \\ +\alpha D & +D < V \end{cases}$$ (A2.1)

it can be represented as [2.18]:

$$F_S(V) = \frac{2\alpha}{p} \int_0^\infty \sin(Vu)\sin(Du)\frac{du}{u^2}$$

$$= -\alpha D + \frac{2\alpha}{2\pi j} \int_{C_U} e^{jVu}\sin(Du)\frac{du}{u^2}$$

$$= -\alpha D + I_{C_u}$$

(A2.2)

where the contour of integration C_U runs from $-\infty$ to $+\infty$ and is indented downward at the origin as shown in Fig. A2.2.

Now, the problem is to evaluate the integral,

$$I = \frac{2\alpha}{\pi} \int_0^\infty \sin(Vu)\sin(Du)\frac{du}{u^2} \qquad (A2.3)$$

Changing the variable in the integration of Eqn. (A2.3),

$$I = \frac{2\alpha}{\pi} \int_0^\infty \sin(Vx)\sin(Dx)\frac{dx}{x^2} \qquad (A2.4)$$

Eqn. (A2.4) may be considered as: $(\alpha/\pi) \times$ imaginary part of

$$I_Z = P\int_{-\infty}^\infty e^{jVz}\frac{\sin(Dz)}{z^2}dz \qquad (A2.5)$$

where P indicates the *Cauchy Principal Value*.

The pole is at $z = 0$. Hence, one can find the residue of Eqn. A2.5 as follows:

$$\lim_{z\to 0}\left[\frac{d}{dz}\left\{z^2\frac{e^{jVz}\sin(Dz)}{z^2}\right\}\right]$$

$$= \lim_{z\to 0}\left\{(jV)e^{jVz}\sin(Dz) + D\cos(Dze^{jVz})\right\}$$

$$= D$$

$$(A2.6)$$

Relevant to the contour chosen, the pole is included as shown in Fig. A2.2.

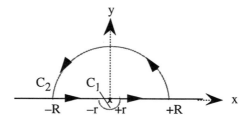

Fig. A2.2: Contour of integration (C_U)
to evaluate the integral of Eqn.(A2.5)

[Note that the integration cannot be done through a singularity before the contour is modified by indenting the path at $z = 0$].

$$\oint e^{jVz} \frac{\sin(Dz)}{z^2} dz = \lim_{R \to \infty} \int_{-R}^{-r} e^{jVx} \frac{\sin(Dx)}{x^2} dx + \int_{C_1} e^{jVz} \frac{\sin(Dz)}{z^2} dz$$

$$+ \lim_{R \to \infty} \int_{r}^{R} e^{jVx} \frac{\sin(Dx)}{x^2} dx + \int_{C_2} e^{jVz} \frac{\sin(Dz)}{z^2} dz$$

$$= 2\pi j \ (\Sigma \ \text{Enclosed residues})$$

(A2.7)

Further, by *Jordan's lemma*,

$$\int_{C_2} e^{jVz} \frac{\sin(Dz)}{z^2} dz \equiv 0$$

(A2.8)

and the integrals (with $\lim_{R \to \infty}$) in Eqn.(A2.7) along the x-axis specified in the ranges (− R to − r) and (+ R to + r) can be combined together as $P \int_{-\infty}^{\infty} e^{jVx} \frac{\sin(Dx)}{x^2} dx$ where P, as indicated earlier, denotes the *Cauchy Principal Value* depicting the limiting process.

Hence, it follows that:

$$\oint e^{jiVz} \frac{\sin(Dz)}{z^2} dz = \int_{C_1} e^{jVz} \frac{\sin(Dz)}{z^2} dz + P \int_{-\infty}^{\infty} e^{jVx} \frac{\sin(Dx)}{x^2} dx$$

$$= 2\pi jD$$

(A2.9)

The integral over the small semicircle (C_1) yields $+ \pi jD$ (positive since the contour is traced counterclockwise).
Therefore,

$$\pi jD + P \int_{-\infty}^{\infty} e^{jVx} \frac{\sin Dx}{x^2} dx = 2\pi jD$$

(A2.10)

Hence,

$$P \int_{-\infty}^{\infty} e^{jVx} \frac{\sin Dx}{x^2} dx = \pi jD$$

(A2.11)

Taking the imaginary part,

$$\int_{-\infty}^{+\infty} \sin(Vu)\sin(Du)\frac{du}{u^2} = \pi D \qquad \text{(A2.12a)}$$

which leads to:

$$\int_{0}^{\infty} \sin(Vu)\sin(Du)\frac{du}{u^2} = \frac{\pi D}{2} \qquad \text{(A2.12b)}$$

Therefore,

$$I = \frac{2\alpha}{\pi}\int_{0}^{\infty} \sin(Vu)\sin(Du)\frac{du}{u^2} = \alpha D \qquad \text{(A2.13)}$$

Similarly, by considering

$$I = \frac{2\alpha}{\pi}\int_{0}^{\infty} \sin(Vx)\sin(Dx)\frac{dx}{x^2} \qquad \text{(A2.14)}$$

which is the imaginary part of I_Z namely, $I_Z = P\int_{-\infty}^{\infty} e^{jDz}\frac{\sin(Vz)}{z^2}dz$, one

obtains $I = \alpha V$.

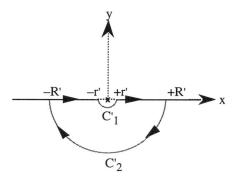

Fig. A2.2: Contour of integration (C'_U)
to evaluate the intergral of Eqn. (A2.15)

Further, the integral: $\displaystyle\int_{-\infty}^{\infty} e^{-iVz}\,\frac{\sin(Dz)}{z^2}\,dz$ has the imaginary part given by

$$I' = -\int_{-\infty}^{\infty}\frac{\sin(Vz)\sin(Dz)}{z^2}dz = -\pi D \qquad (A2.15)$$

Eqn. (A2.15) can be evaluated by the contour integration method as indicated above using the contour C'_U illustrated in Fig. A2.3, $I' = -\pi D$ which leads to $I = -\alpha D$.

Appendix 2.2

Statistical Mechanics Attributes of the Neural Complex: Langevin's Theory of Dipole Polarization

Liquid crystal model of the neural complex

Basically, the analogy between Ising spins system and the neural complex stems from the fact that the organization of neurons is a collective enterprise in which the neuronal activity of interactive cells represents a cooperative process similar to that of spin interactions in a magnetic system. As summarized in [2.1], the strengths of synaptic connections between the cells representing the extent of interactive dynamics in the cellular automata are considered analogous to the strengths of exchange interaction in magnetic spin systems. Further, the synaptic activity, manifesting as the competition between the excitatory and inhibitory processes, is regarded as equitable to the competition between the ferromagnetic and antiferromagnetic exchange interactions in spin-glass systems. Also, the threshold condition stipulated for the neuronal network is considered as the analog of the condition of metastability against single spin flips in the Ising spin-glass model.

While the aforesaid similarities do prevail between the neurons and the magnetic spins, major inconsistencies also persist between these two systems regarding the synaptic coupling *versus* the spin interactions. The inconsistency between neurons with inherent asymmetric synaptic couplings and symmetric spin-glass interactions led Griffith [2.2] to declare the aggregate of neurons *versus* magnetic spin analogy as having "no practical value". Nevertheless, several compromising suggestions have been proposed as discussed earlier showing the usefulness of the analogy between the neurons and the magnetic spins.

The assumption of symmetry and the specific form of the synaptic coupling in a neuronal assembly define what is generally known as the Hopfield model. This model demonstrates the basic concepts and functioning of a neural network and serves as a starting point for a variety of models in which many of the underlying assumptions are relaxed to meet some of the requirements of real systems. For example, the question of W_{ij} being not equal to W_{ji} in a neural system was addressed in a proposal by Little, as detailed in the previous section. Little defined a time-domain long-range order so that the corresponding anisotropy introduces bias terms in the Hamiltonian relation, making it asymmetric to match the neuronal Hamiltonian. That is, Little's long-range order as applied to neurons corresponds to a time-domain based long-time correlation of the states. These persistent states in time of a neuronal network are equated to the long-range spatial order in an Ising spin system.

An alternative method of attributing the long-range order to neurons can be done by following the technique of Little, except that such a long-range order will be applied to the spatial or orientational anisotropy instead of time correlations. To facilitate this approach, the *free-point molecular dipole interactions* can be considered *in lieu* of magnetic spin interactions [2.1]. The free-point molecular dipole interactions with partial anisotropy in spatial arrangement refer to the *nematic phase* in a *liquid crystal*. Hence, the relevant analysis equates the neural statistics to that of a nematic phase system consistent with the known dogma that "the living cell is actually a liquid crystal". The characteristics of molecular patterns, structural and behavioral properties of liquid crystals make them unique model-systems to investigate a variety of biological phenomena. The general physioanatomical state of biological cells depicts neither real crystals nor real liquid phase. This constitutes what is popularly known as the *mesomorphous* state, much akin to several organic compounds which have become known as the "*flussige Kristalle*" or liquid crystals. Both the liquid crystalline materials as well as the biological cells have a common, irregular pattern of side-by-side spatial arrangements in a series of layers known as the *nematic phase*.

The microscopic structural studies of biological cells indicate that they are constituted by very complex systems of macromolecules which are organized into various bodies or "*organelles*" that perform specific functions for the cell. From the structural and functional point of view, Brown and Wolken have drawn an analogy of the description of the living cells to liquid crystals on the basis that a cell has a *structural order*. This in fact is a basic property of liquid crystals as well, for they have a structural order of a solid. Furthermore, in many respects it has been observed that the physical, chemical, structural, and optical properties of biological cells mimic closely those of liquid crystals.

Due to its liquid crystalline nature, a cell through its own structure forms a proto-organ facilitating electrical activity. Further, the anisotropically oriented structure of cellular assembly (analogous to liquid crystals) has been found responsible for the complex catalytic action needed to account for cellular regeneration. In other words, by nature the cells are inherently like liquid crystals with similar functional attributions.

On the basis of these considerations a neural cell can be modeled via liquid-crystal analogy. The squashing action of the neural cells pertinent to the input-output relations depicting the dynamics of the cellular automata, can be described in terms of a stochastically justifiable sigmoidal function and statistical mechanics considerations as presented in the pursuant sections.

Free-Point molecular dipole interactions
Suppose a set of polarizable molecules are anisotropic with a spatial long-range orientational order corresponding to the nematic liquid crystal in the mesomorphic phase. This differs from the *isotropic molecular arrangement,* as in a liquid, in that the molecules are spontaneously oriented with their long axes approximately parallel. The preferred direction or orientational order

may vary from point-to-point in the medium, but in the long-range, a specific orientational parallelism is retained.

In the nematic phase, the statistical aspects of dipole orientation in the presence of an externally applied field can be studied *via Langevin's theory* with the following hypotheses:

1. The molecules are point-dipoles with a prescribed extent of anisotropy.

2. The ensemble average taken at an instant is the same as the time average taken on any element (*ergodicity property*).

3. The characteristic quantum numbers of the problem are so high that the system obeys the classical statistics of Maxwell-Boltzmann, which is the limit of quantum statistics for systems with high quantum numbers. The present characterization of *paraelectricity* differs from spin *paramagnetism*, wherein the quantum levels are restricted to two values only.

4. The dipole molecules in general, when subjected to an external electric field $\mathbf{E}$, experience a moment $\mu_E = \alpha_E \mathbf{E}$, which by definition refers to the *polarizability* of the molecule. The dipole orientation contributing to the polarization of the material is quantified as $\mathbf{P} = N < \mu_E >$ where N is the dipole concentration.

5. In an anisotropic system such as the liquid crystal, there is a permanent dipole moment μ_{PE}, the direction of which is assumed along the long axis of a nonspherical dipole configuration. Consequently, two orthogonal polarizability components exist, namely, α_{E1} along the long axis and α_{E2} perpendicular to this long axis.

The dipole moments in an anisotropic molecule are depicted in Fig. B2.1 Projecting along the applied electric field $\mathbf{E}$, the net-induced electric polarization moment is:

$$\mu_E = \mu_{PE}\cos\theta + [\alpha_{E1}\cos^2(\theta) + \alpha_{E2}\sin^2(\theta)]\mathbf{E}$$

$$= \mu_{PE}\cos\theta + [\Delta\alpha_E\cos^2(\theta) + \alpha_{E2}]\mathbf{E} \qquad (B2.1)$$

where $\Delta\alpha_E$ is a measure of anisotropy.

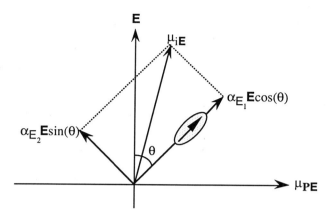

Fig. B2.1 Free-point dipole and its moments

E: Applied electric field; μ_{PE}: Permanent dipole moment; μ_{iE}: Induced dipole moment

The corresponding energy of the polarized molecule in the presence of an applied field **E** is constituted by: (1) The potential energy W_{PE} due to the permanent dipole given by

$$W_{PE} = -\mu_{PE} \bullet \mathbf{E} = -\mu_{PE}\, \mathbf{E} \cos\theta \tag{B2. 2}$$

and (2) the potential energy W_{iE} due to the induced dipole given by

$$W_{iE} = -(1/2)(\alpha_{E1}\cos^2\theta + \alpha_{E2}\sin^2\theta\,)|\,\mathbf{E}|^2 \tag{B2. 3}$$

Hence, the total energy is equal to $W_T = W_{PE} + W_{iE}$. Further, the statistical average of μ_E can be specified by:

$$<\mu_E> = \frac{\int \mu_E \exp[-W_{PE}/k_BT]dW}{\int \exp[-W_T/k_BT]dW} \tag{B2. 4}$$

where $d\Omega$ is the elemental solid angle around the direction of **E**. That is, $d\Omega = 2\pi \sin(\theta)d\theta$. By performing the integration of Eqn.(B2.4) using Eqns.(B2.3) and (B2.4), it follows that

$$<\mu_E> = \mu_{PE}<\cos\theta> + (\Delta\alpha_E<\cos^2\theta> + \alpha_{E2})\mathbf{E} \tag{B2.5}$$

108

where the quantity $<\cos^2(\theta)>$ varies from 1/3 for randomly oriented molecules to 1 for the case where all the molecules are parallel (or antiparallel) to the field **E**. On the basis of the limits specified by $<\cos^2(\theta)>$, the following parameter can be defined:

$$
\begin{aligned}
S^0 &= (3/2) <\cos^2(\theta)> - (1/2) \\
&= 0 \quad \text{(for } <\cos^2(\theta)> = 1/3) \\
&= 1 \quad \text{(for } <\cos^2(\theta)> = 1)
\end{aligned}
\tag{B2.6}
$$

The parameter S^0 which is bounded between 0 and 1 under the above conditions, represents the "order parameter" of the system [2.1]. Appropriate to the nematic phase, S^0 specifies the long-range orientational parameter pertaining to a liquid crystal of rod-like molecules as follows. Assuming the distribution function of the molecules to be cylindrically symmetric about the axis of preferred orientation, S^0 defines the degree of alignment, namely, for perfectly parallel (or antiparallel) alignment $S^0 = 1$, while for random orientations $S^0 = 0$. In the nematic phase S^0 has an intermediate value which is strongly temperature dependent.

For $S^0 = 0$, it refers to an isotropic statistical arrangement of random orientations so that for each dipole pointing in one direction, there is statistically a corresponding molecule in the opposite direction (Fig. 2B.1). In the presence of an external electric field **E**, the dipoles experience a torque and tend to polarize along **E**, so that the system becomes slightly anisotropic. Eventually under a strong field (**E**) the system becomes totally anisotropic with $S^0 = 1$.

Stochastical response of neurons under activation

By considering the neurons as analogous to a random, statistically isotropic dipole system, the graded response of the neurons under activation could be modeled by applying the concepts of Langevin's theory of dipole polarization. The continuous graded response of neuron activity corresponding to the stochastical interaction between incoming excitations that produce true, collective, nonlinear effects can be elucidated in terms of a sigmoidal function specified by a gain parameter $\lambda = \Lambda/k_B T$, with Λ being the scaling factor of σ_i which depicts the neuronal state-vector.

In the pertinent considerations, the neurons are depicted similar to the nematic phase of liquid crystals and are assumed to possess an inherent, long-range spatial order. In other words, it is suggested that $0 < S^0 < 1$ is an appropriate and valid *order function* for the neural complex that $S^0 = 0$.

Specifying in terms of $S^o = (3/2) <\cos^2\theta> - 1/2$, the term $<\cos^2\theta>$ should correspond to a value between 1/3 to 1 justifying the spatial anisotropy.

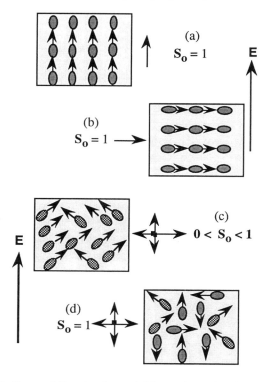

Fig. B2.2: Types of disorders in spatial free-point molecular arrangement subjected to external electric field (**E**)

[Note: (a) & (b) Completely ordered (total anisotropy): Parallel and antiparallel arrangements; (c) Partial long-range order (partial anisotropy): Nematic phase arrangement; (d) Complete absence of long-range order (total isotropy): Random arrangement]

To determine an appropriate squashing function for this range of $<\cos^2\theta>$ between 1/3 to 1 (or for $0 < S^o < 1$), the quantity $<\cos^2\theta>$ can be replaced by $(1/3 + 1/3q)$ in defining the order parameter S^o. Hence:

$$S^o = (3/2) (1/3 + 1/3q) - (1/2) \qquad\qquad (B2.7)$$

where $q \to \infty$ and $q = 1/2$ set the corresponding limits of $S^o = 0$ and $S^o = 1$, respectively.

Again, resorting to statistical mechanics, $q = 1/2$ refers to dichotomous states, if the number of states are specified by $(2q + 1)$. For the dipoles or neuronal alignments, it corresponds to the two totally discrete anisotropic (parallel or antiparallel) orientations. In a statistically isotropic, randomly oriented system, the number of possible discrete alignments would, however, approach infinity, as dictated by $q \to \infty$.

For the intermediate $(2q + 1)$ number of discrete orientations, the extent of dipole alignment to an external field or, correspondingly, the output response of a neuron to excitation, would be decided by the probability of a discrete orientation being realized. It can be specified by [2.1]:

$$L_q(x) = \sum_{m=-q}^{+q} (m/q) \exp(mx/q) \,/ \sum_{m=-q}^{+q} \exp(mx/q)$$

$$= (1 + 1/2q)\coth[(1 + 1/2q)\,x] - (1/2q)\coth(x/2q) \qquad (B2.\,8)$$

The above function, $L_q(x)$, is a *modified Langevin function* and is also known as the *Bernoulli function*. The traditional Langevin function $L(x)$ is the limit of $L_q(x)$ for $q \to \infty$. The other limiting case, namely, $q = 1/2$, which exists for dichotomous states, corresponds to $L_{1/2}(x) = \tanh(x)$.

Thus, the sigmoidal function $F_S(x)$ which decides the neuronal output response to an excitation has two bounds. With $F_S(x) = \tanh(x)$, it corresponds to the assumption that there exists a total orientational long-range order in the neuronal arrangement. Conventionally, $F_S(x) = \tanh(x)$ has been regarded as the *squashing function* (for neuronal nets) purely on empirical considerations of the input-output nonlinear relation being S-shaped. This relation remains bounded between two logistic limits, and follows a continuous monotonic functional form between these limits. In terms of the input variate x_i and the gain/scaling parameter Λ of an ith neuron, the sigmoidal function specified as the hyberbolic tangent function is $\tanh(\Lambda x_i)$. The logistic operation that compresses the range of the input so that the output remains bounded between the logical limits can also be specified alternatively by an exponential form, $F_S(y) = 1/[(1 + \exp(-y)]$ with $y = \Lambda x_i$. Except for being sigmoidal, the adoption of the hyperbolic tangent or the exponential form in the neural network analyses has been purely empirical with no justifiable reasoning attributed to the choice. Pursuant to the earlier discussion, $L(y) = L_{q \to \infty}(y)$ specifies the system in which the randomness is totally isotropic. That is, the anisotropicity being zero is implicit. This, however, refers to rather an extensive situation assuming that the neuronal configuration poses no spatial anisotropicity or long-range order whatsoever. Likewise, considering the intuitive modeling of $F_S(y) = \tanh(y)$, as adopted commonly, it depicts a totally anisotropic system wherein the long-range order attains a value equal to 1. That is, $\tanh(y) = L_{q \to 1/2}(y)$ corresponds to the dichotomous discrete orientations, parallel or antiparallel, specified by $(2q + 1) \to 2$.

In the nematic phase, neither of the above functions, $\tanh(y)$ nor $L(y)$, is commensurable since a partial long-range order depicting a partial anisotropicity is rather imminent in such systems. Thus, with $1/2 < q < \infty$, the true sigmoid of a neuronal arrangement with an inherent nematic, spatial long-range order should be $L_q(y)$.

Therefore, it can be regarded that the conventional sigmoid, namely, the hyperbolic tangent (or its variations) and the Langevin function, constitute the upper and lower bounds of the state-vector squashing characteristics of a neuronal unit, respectively.

Chapter 3

Nonlinear and Informatic Aspects of Fuzzy Neural Activity

P. S. Neelakanta and S. Abusalah

"(Fuzziness)...haziness, vagueness, ambiguity,
cloudiness, unclearness, indistinctiveness, dissonance,
confusion, nonspecificity and sharplessness..."
— G. J. Klir and T.A. Folger
Fuzzy Sets, Uncertainty and
Information

3.1 Introduction

The objective of this chapter is to study and analyze the characteristics of output versus input values in a neural network when the boundaries of the sets constituting the input and output entities are not precise; that is, when they are *fuzzy*. By deducing the fuzzy output activity for fuzzy input values, the associated information-theoretics based fuzzy response of the neural complex are deduced. In order to develop a meaningful functional representation of the fuzzy-output versus fuzzy-input relation pertinent to a neural network, it is precursive to consider the prevailing strategies adopted in modeling the non-fuzzy, neural input-output relation relevant to *crisp* sets of inputs and output values. Modeling of such non-fuzzy, neural network transfer-function has been presented in Chapter 2. The present chapter offers an extended effort on neural transfer-function characteristics relating the input (x) and the output (y) with specified fuzzy attributes. Using the concept of fuzzy non-linearity developed here, the entropy (information) transfer across a fuzzy neuron is elucidated.

3.2 What is Fuzzy Activity?

Fuzzy attribute to an activity refers to the *nonspecificity* of the values and sharplessness of the boundaries of activity variables. When there is a *vagueness* in the activity variables with *indistinctiveness* in their characteristics then, as Kosko [3.1] points out, "Everything is (just) a matter of degree". That is, the attributes of the activity variables specify only fuzzy or gray facts which are true only to some degree, say, between 0 and 1; and, likewise false to some degree.

Fuzzy attributes follow the *multivalue logic* — a logic that builds gray truth into complex schemes of formal reasoning. Such a logic refers to a fuzzy system or mapping from input-to-output that depends on a set of fuzzy rules.

Zadeh [3.2] envisioned the concept of fuzzy logic to study vague linguistic information. He contended that "conventional quantitative techniques of system analysis are intrinsically unsuited for dealing with humanistic systems".

Fuzzy logic and the associated information-theoretics refer to *imprecise* data. It permits the conventional Boolean logic operations to be extended to manipulate information associated with a set having only a *partial* element. Such a partial *membership* of set depicts the fuzzy attributes or vagueness of the universe cast by the set. Therefore, fuzzy set **A** contains x only to some degree, [0,1].

A fuzzy system F is a set of *if-then rules* that maps inputs to outputs. These rules define fuzzy patches in the input-output state-space of $\mathbf{X} \times \mathbf{Y}$. For example referring to Fig 3.1, the fuzzy system (F: X → Y) approximates a function (f: X → Y) by smearing the function f with rule-patches and averaging those patches which overlap along, f.

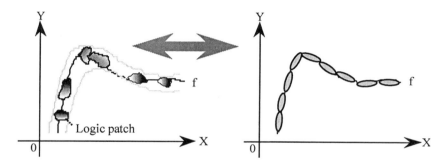

Fig.3.1: Fuzzy-logic based approximation of a function

In the above method of mapping, the associated approximation improves when:

- Fuzzy-rule patches increase in numbers
- Fuzzy-rule patches diminish in size

In general, all fuzzy systems suffer from what is known as the *curse of dimensionality*. That is, they need a stupendously large number of rules; or, *rule-explosion* is encountered while approximating most functions [3.3].

In reference to Fig. 3.1, a simple scalar map corresponding to the approximand (f: $\Re \to \Re$) takes α rule-patches to smear along the graph of f on some rectangle in the plane. But, for (f: $\Re^2 \to \Re$), an order of α^2 rule patches are needed to cover the surface in some three-dimensional enclosure. Likewise, for (f: $\Re^a \to \Re^b$), an order of $\alpha^{(a+b-1)}$ rules are required to represent the graph of f in some hyper-enclosure. Where on the function inflexive points exist (as *extrema* — either as *maxima* or *minima*), *lone optimal rules* may be warranted to "patch the (prevailing) bumps." The fuzzy *if-then rule* refers simply to an if-then statement in which some words are characterized

attributes, say *low* and *high* of a variable x, can be depicted, for example, by continuous membership function (μ) as illustrated in Fig. 3.2.

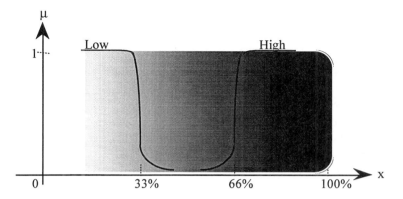

Fig.3.2: Membership characterization of low and high of a variable x

3.3 Crisp Sets *versus* Fuzzy Sets

A crisp set can dichotomize the individuals in some given universe of discourse into two groups, namely, those members that certainly belong to the set and those (nonmembers) who certainly do not. This dichotomous bifurcation is *sharp, unambiguous* and *precise*.

Contrary to the crisp set characterization, a fuzzy set defines a concept pertinent to the classification of its members, as commonly expressed or described in natural languages based on imprecise boundaries of gradual transition from membership to nonmembership or *vice versa*. Accordingly, a fuzzy set can be defined mathematically by assigning to each possible individual in the universe of discourse a value representing its *grade of membership* in the fuzzy set. The grade refers to the extent or degree to which that individual is akin to or compatible with the characteristics or a concept portrayed by the fuzzy set. Greater or lesser extent (or degree) of belongingness of the individual in the fuzzy set defines respectively, the larger or smaller membership grade of the individual concerned.

In essence, fuzzy sets are sets with boundaries that are not precise and the membership in a fuzzy set is not a matter of affirmation or denial, but rather a matter of degree [3.1]. The ability of fuzzy sets to portray gradual transition from memebership to nonmembership (or vice versa) offers a broad utility and a meaningful representation of vague concepts expressed in natural languages and uncertainties associated with physical measurements.

3.3.1 Symbols and notations

The following symbols and notations are used throughout this chapter to represent the appropriate entities:

115

X :   Universal set or universe of discourse

ϕ :   Null set

$x \in A$:   Member or element x of a set A

$X \notin A$:   Non-member of a set A

μ_A :   Membership function of a fuzzy set A

[0, 1] :   Membership interval

$\mu_A : X \rightarrow [0,1]$: Membership interval of a fuzzy set A defined on X.

3.3.2   *Concept of α-cut:*

Given a fuzzy set A defined on X and any member of $\alpha \in$ [0,1], the α-cut, $^\alpha A$ is defined as the crisp set that contains all the elements of the universe X whose membership grades in A are greater than or equal to the specified value of α. That is,

$$^\alpha A = \{ \ x \mid A(x) \mid \geq \alpha \ \} \qquad (3.1)$$

A *strong α-cut*, $^{\alpha+}A$ is a special case in which the membership grades of the elements in A are only greater than the specified value of α. That is,

$$^{\alpha+}A = \{ \ x \mid A(x) \mid \geq \alpha \ \} \qquad (3.2)$$

3.3.3 *Height of a fuzzy set*

The *height*, h(A) of fuzzy set A refers to the maximum membership grade assumed by any element in that set. Or,

$$h(A) = \operatorname*{Sup}_{x \in A} A(x) \qquad (3.3)$$

When the *height*, h(A) = 1, the fuzzy set is said to be *normal*; and, when h(A) < 1, it is *subnormal*. In terms of α, the height h(A) is the *supremum* of α for which $^\alpha A \neq \phi$.

3.4 Membership Attributions to a Fuzzy Set

In reference to the discussions presented above, suppose **A** is a Boolean subset of a universal set of the variable v. Then, **A** is characterized by a membership function $\mu_A(v)$ where {**v**} are generic sets of elements constituting the universal set **V**. The membership function maps the universal set **V** to the real interval [0, 1]. The closer $\mu_A(v)$ is to 1, the more v belongs to A. That is, $\mu_A(v)$ is the *degree of compatibility* of v with the concept represented by A. In terms of qualitative attributes, the membership values indicate a measure of belongings of the variable to each of the linguistic properties, say *low, medium* and *high*. Then, the corresponding {**v**} can be declared as the π-*set* [3.4]. Normally, a π-fuzzy set $\{v\}$, with μ_A in the range [0,1] is specified by the following as an n-dimensional pattern:

$$V = [(\mu_{low}V_1, \mu_{medium}V_1, \mu_{high}V_1),$$
$$(\mu_{low}V_2, \mu_{medium}V_2, \mu_{high}V_2),$$
$$...,$$
$$(\mu_{low}V_n, \mu_{medium}V_n, \mu_{high}V_n)]$$

(3.4)

Conventionally the numerical values of μ specified by the π-function lying in the range [0, 1] in one dimensional form with $v \in V$ and $v \in \Re^n$ is written as:

$$\pi(\mathbf{v};\mathbf{c}^0,\lambda) = \begin{cases} 2\left(1 - \dfrac{\|\mathbf{v}-\mathbf{c}^0\|}{\lambda}\right)^2 & , \quad if \quad \lambda/2 \le \|\mathbf{v}-\mathbf{c}^0\| \le \lambda \\[4mm] 1 - 2\left(1 - \dfrac{\|\mathbf{v}-\mathbf{c}^0\|}{\lambda}\right)^2 & , \quad if \quad 0 \le \|\mathbf{v}-\mathbf{c}^0\| \le \lambda/2e \\[4mm] 0 & , \quad otherwise \end{cases}$$

(3.5)

where $\lambda > 0$ is the radius of the π-function with $\mathbf{c}^0$ as the central point at which $\pi(\mathbf{v}; \mathbf{c}^0, \lambda) = 1$. A fuzzy set with a membership function $\mu = \pi(\mathbf{v}; \mathbf{c}^0, \lambda)$ therefore, represents a set of points clustered around $\mathbf{c}^0$. The choice of $(\mathbf{c}^0, \lambda)$ can be limited to the linguistic properties low, medium and high in the following manner as described by Mitra and Pal [3.25].

Considering the variable v to have a dynamic range with the bounding values v_{max} and v_{min}, the set $\{\mathbf{c}^0, \lambda\}$ with linguistic attributes *low, medium and high* are given by:

$$\lambda_{medium}(v) = (v_{max} - v_{min})/2 \tag{3.6a}$$
$$\lambda_{low}(v) = (c^0_{medium}(v) - v_{min})/|C| \tag{3.6b}$$
$$\lambda_{high}(v) = (v_{max} - c^0_{medium}(v))/|C| \tag{3.6c}$$
$$c^0_{low}(v) = (c^0_{medium}(v) - \lambda_{low}(v))/2) \tag{3.7a}$$
$$c^0_{medium}(v) = v_{min} + \lambda_{medium}(v) \tag{3.7b}$$
$$c^0_{high}(v) = (c^0_{medium}(v) + \lambda_{high}(v))/2 \tag{3.7c}$$

where $0.5 \leq |c| \leq 0.5$ is the *coupling or interacting matrix parameter* controlling the extent of overlapping (interaction).

Apart from the π-set representation, μ has also been depicted by a set of a various parameterized family of functions as given below [3.5]:

$$\mu(v;c^0,\chi_1,\chi_2) = \begin{cases} \chi_1(v - c^0) + 1 & , \quad \chi_1 \in [c^0 - 1/\chi_1, c^0] \\ \chi_1(c^0 - v) + 1 & , \quad \chi_1 \in [c^0, c^0 + 1/\chi_1] \\ 0 & , \quad \text{otherwise} \end{cases}$$

(3.8)

$$\mu(v;c^0,\chi_3) = 1/[1 + \chi_3(v - c^0)^2]$$ (3.9)

$$\mu(v;c^0,\chi_4) = \exp\left[-\left|\chi_4(v - c^0)\right|\right]$$ (3.10)

$$\mu(v;c^0,\chi_5) = \begin{cases} \{1 + \cos[\chi_5 \pi(v - c^0)]\}/2, & \chi_5 \in [c^0 - 1/\chi_5, c^0 + 1/\chi_5] \\ 0 & \text{otherwise} \end{cases}$$

(3.11)

where c^0 is again the center and χ_i ($i = 1, 2, 3, 4, 5$) are parameters which control the width of the μ-function. It can be noted that χ_i decides the rate at which the function decreases with the increasing difference $\left|c^0 - v\right|$ for each value v.

Other variations of the function depicting μ, such as the trapezoidal function, are also popular in fuzzy theory and are evolved from the *B-spline basis functions* as follows. The B-spline basis functions can be judiciously adopted to define the membership function $\mu_A(.)$ pertinent to an invariant fuzzy set [3.6]. This functional representation accommodates implicitly the implementation of interval considerations associated with the fuzzy sets recursively. Defining $N_s^k(.)$ as the sth basis function of order k, $\mu_A^s(.)$ can be identically set equal to $N_s^k(.)$ as follows:

$$\mu_A{}^{(s)}(v) = N_s{}^k(v) = \left[(v - \lambda_{s-k})/(\lambda_{s-1} - \lambda_{s-k})\right]N_{s-1}^{k-1}(v)$$
$$+ \left[(\lambda_s - v)/(\lambda_s - \lambda_{s-k+1})\right]N_s^{k-1}(v)$$

(3.12a)

where,

$$N_s^1(v) = \begin{cases} 1 & \text{if} \quad v \in (I_S = s^{th} \text{ interval } \lambda_{s-1}, \lambda_S) \\ 0 & \text{otherwise} \end{cases}$$

(3.12b)

The knots $\{\lambda_h\}$ of the basis fuzzy membership function determine the size of the intervals and hence the "width" of each fuzzy set $\in \mathfrak{R}$. The order k determines the shape of $\mu_A(.)$. For example, k = 2 leads to the popular triangular shape basis function, $N_s^2(.)$. In general, an order k B-spline basis function has a knot vector of length (k+1).

An sth basis function centered at c_s^0 and variance σ_s is a bell-shaped Gaussian function which also represents a membership function with a positive response. That is,

$$\mu_A^s(v) = \exp\left[-(v - c_s^0)/2\sigma_s^2\right]$$

(3.13a)

Alternatively, to yield a positive as well as a negative response, the following membership function (Gaussian) type has been proposed [3.7]:

$$\mu_A^s(v) = \begin{cases} [1/\exp(-1)]^{-1} \exp[-(\lambda_{s,2} - \lambda_{s,1})/4(\lambda_{s,2} - v)(v - \lambda_{s,1})] \\ \qquad\qquad\qquad\qquad\qquad\qquad \text{if } v \in (\lambda_{s,1}, \lambda_{s,2}) \\ 0 \\ \qquad\qquad\qquad\qquad\qquad\qquad \text{otherwise} \end{cases}$$

(3.13b)

3.5 Fuzzy Neural Activity

The discussions in [3.8-3.24] and in Chapter 2 refer to neural nonlinear activity in respect of crisp sets $\{x\}$ and $\{y\}$. In these sets, certain definitive values can be assigned to each element of the respective universal set so as to discriminate between members and nonmembers of the crisp set under consideration. The uncertainty features of real world problems, however, warrant that the values so assigned to the elements of these universal sets fall within a specified range and qualify the associated uncertainty by a membership grade given to each of the elements in the set in question. That is, the input set $\{x\}$ and the output set $\{y\}$ should be regarded as fuzzy sets with corresponding membership functions assigned appropriately. Denoting the relevant fuzzy variables as x^f and y^f (with the superscript f denoting explicitly the fuzzy considerations), the membership gradation should allow the representation of the range concepts attributed to the variables involved to be expressed in natural language within the scope of the context pertinent to $\{x^f, y^f\}$.

The fuzzy simulation of the nonlinear neural complex integrates a fuzzy, qualitative approach based on a linguistic description of the range of variables involved and applies the traditional methods of calculus in order to prescribe quantitative attributes to these variables. That is, the qualitative aspects considered may refer to a fuzzy description of the parameters involved and the quantitative method being prescribed can follow the mathematics of logistic functions and the associated calculus of differential equations. Hence, the resulting algorithm can be regarded as a *fuzzy differential equation* based on conventional differential calculus but characterized *via* fuzzy numbers. The solution of this fuzzy differential equation is shown in the following sections as the *sigmoidal evolution* of a bounded region of uncertainty representing the fuzzy set $\{\mathbf{y}^{\mathrm{f}}\}$ *versus* the fuzzy set $\{\mathbf{x}^{\mathrm{f}}\}$. This is similar to the Riccati differential equation based logistic functional description of crisp input and output variables discussed in Chapter 2.

3.6 Fuzzy Differential Equations
The calculus of fuzzy variables has been elaborated in the literature [3.25-3.29] from two perspectives. The first one refers to the fuzzy differential calculus due to Dubois and Prade [3.25-3.27] who used the "extension principle" of Zadeh [3.2] to evaluate the integral of fuzzy mappings in the probabilistic framework of fuzzy events (as conceived by Zadeh). Also, in [3.27], they have dealt with the calculus of ordinary functions at a fuzzy point and have indicated a method of differentiation applied to fuzzy-valued mappings.

The other aspect of fuzzy differential equations refers to the notions of Kandel and Byatt [3.28] who formulated such equations with the relevant solutions being probabilities of fuzzy events (in the sense of Zadeh. [3.2]).

In contrast to the conceptualization by Kandel and Byatt [3.28], the spirit of differential calculus indicated in [3.25-3.27] by Dubois and Prade is completely different. Their analytical framework does not involve probabilistic solutions, but is rather concerned with fuzzy interval-valued mappings and related definitions of "fuzzy slope" etc.

Relevant to fuzzy variables and fuzzy parameters, Bonarini and Botempi [3.29] have also considered an interval-value calculus-based approach. They have evolved a method of fusing the quantitative calculus of interval-values and the qualitative aspects of fuzzy attributions in order to simulate fuzzy dynamical models. Essentially, the work addressed in [3.29] refers to the time-evolution of a fuzzy variable *via* differential equation formulation.

In the present study, the general considerations (algebraic notions not involving probabilistic attributes) of ordinary differential equation modeling using interval-value calculus [3.29] is envisaged with a qualitative fusing of fuzzy concepts, in order to simulate the fuzzy neural nonlinearity under discussion. That is, the "extension principle" [3.2] due to Zadeh is adopted notionally which enables a general method of extending the standard mathematical concepts in order to deal with fuzzy quantities. Further, the procedure[1] adopted presently is similar to that due to Banarini and Botempti [3.29] except that the independent variable considered here is the summed input X and not the time variable (t) used in [3.29, 3.30].

In view of the aforesaid approach contemplated, first the general algebraic description of the nonlinear differential equation relating the output set $\{y_i\}$ and the input set $\{X_i\}$ bearing crisp values in describing the nonlinearity of a neural cell (as developed in [3.22] and elaborated in Chapter 2) is considered with relevant details vis-à-vis the interval-value calculus pertinent to the Bernoulli-Riccati equation, namely,

$$dy/dX + R(X) - P(X) = 0 \qquad (3.14)$$

with the solution given by

$$L_Q(X) = y(X) = g_Q \coth(g_Q X) - h_Q \coth(h_Q X) \qquad (3.15)$$

where $g_Q = (1 + 1/2Q)$ and $h_Q = 1/2Q$ with details as follows.

The sigmoidal function given by Eqn. (3.15) is bounded within two limits (corresponding to $Q = 1/2$ and $Q = \infty$) as shown in Fig. 3.3a and its functional representation varies between the squashing limits $[-1, +1]$. As stated in the last chapter, it has been demonstrated in [3.19-3.21], on the basis of statistical mechanics considerations that $L_Q(.)$ (Bernoulli function) sigmoid with $Q = 1/2$ represents the neural state proliferation with a totally anisotropic spatial long-range order; whereas, $Q = \infty$ in $L_Q(.)$ denotes, the totally isotropic random spatial order of the interconnected neurons. Any neuronal activity, either due to crisp or fuzzy variables, should, therefore, be confined within these two bounding physical states of the order regardless of the extent of fuzziness involved. Further, it should be noted at this stage that when $Q \rightarrow 0$, $L_{Q \rightarrow 0}(.) = \text{signum}(.)$. This represents the McCulloch-Pitts regime of neuronal activity [3.23], wherein the hypothetical considerations of no interaction prevailing between neurons is presumed implicitly.

Having prescribed an algebraic/conventional calculus format of a differential equation to the neural activity function *via* Eqn. (3.14), its extension to a qualitative fuzzy calculus can be addressed as follows: Rewriting Eqn. (3.14) with $R(X) = +1$, it follows that

---

[1] *Note*: The fuzzy differential equation developed in this chapter does not bear any relevance to the probabilistic solution-based differential equation portrayed by Kandel and Byatt in [3.28].

$$dy_i(X_i)/dx + y_i^2(X_i) - P_i(X_i) = 0 \qquad (i = 1, 2, ..., M)$$

$$(3.16)$$

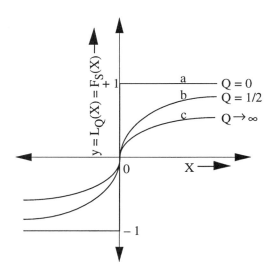

Fig. 3.3a: Bernoulli function $L_Q(x)$ as the neural activation sigmoid.

a.   When $Q \to 0$, $L_Q(x) \to$ signum function (McCulloch-Pitts limit for a noninteracting neuron)
b.   When $Q \to 1/2$, $L_Q(x) \to$ hyperbolic tangent function (for the upper bound of totally anisotropic, spatial neural state proliferation)
c.   When $Q \to \infty$, $L_Q(x) \to$ Langevin function (for the lower bound of totally isotropic, spatial neural state proliferation)

By denoting $dy_i(X_i)/dX_i$ as a vector function $\Phi_{i\ell}[y(X_i)]$ with ($\ell = 1, 2,$ ..., n) on an euclidean n-dimensional vector space $\Re^n$, Eqn. (3.16) can be cast into an autonomous system of differential equations of finite orders. Inasmuch as the solution of Eqn. (3.16), namely $y_i(X)$ is a sigmoid $F_S(.)$, it is continuous, and monotonic. That is, continuous derivatives of all orders for an interval-value $X \in [X_L, X_H]$, are presumed to exist. Therefore, the Taylor's theorem with remainder asserts that for $X \in [X_L, X_H]$ and $k > 1$, the following expansion of $y_{i\ell}(X)$ ($\ell = 1, 2, ..., n$) is valid for some $X_a \in (X_L, X)$:

$$y_{i\ell}(X) = y_{i\ell}(X_L) + \sum_{j=1}^{k-1} [\Phi_{i\ell}^{(j-1)}(y_1(X_L),...,y_n(X_L)]X^j/j!$$

$$+ [\Phi_{i\ell}^{(k-1)}(y_1(X_a),...,y_n(X_a)]X^k/k!$$

$$(3.17)$$

It should be noted that the coefficients of the Taylor's formula representation of the numerical differential system as above are strictly numerical coefficients.

Suppose the set of input values of $\{X_i\}$ are uncertain or fuzzy. Extending the Taylor's formula to intervals, the output $y_i(X)$ has a fuzzy value $Y_i^f(X)$ for a given generic input (summed) set $\{X_i\}$. That is, $\{Y_i^f\}$ refers to the fuzzy transformation of the crisp output set $\{y_i\}$ whose generic elements are $\{X_i\}$. The ith component of $\{Y_i^f\}$ can, therefore, be written in respect of the uncertain (fuzzy) boundary value of an interval vector $Y_i^f(X_L)$ in the domain $\Re^n$ as

$$Y_{i\ell}^f(X) \cong Y_{i\ell}^f(X_L) + \sum_{j=1}^{(k-1)} {}^f\Phi_{i\ell}^{(j-1)}(Y_{i\ell}^f(X_L)X^j/j! \qquad (3.18)$$

where ${}^f\Phi_{i\ell}(.)$ with ($\ell = 1, 2, ..., n$) is the corresponding function of $\Phi_{i\ell}(.)$ defined in respect of the fuzzy set $Y_i^f(X)$.

Relevant to the above relation, inasmuch as the parameters, each of which commences at a certain boundary value (such as $X = X_L$), refers to an interval, the Taylor's formula given by Eqn. (3.18) becomes an algebraic sum of *addenda*, each one of which taking an interval value. Summation so computed *via* interval arithmetic leads to the width of the result being the superposition (sum) of the widths of the two *addenda*. Hence, in using the interval arithmetic, any interval-value of Y_i^f becomes wider with increasing value of X; or the interval-value of y may become narrower with decreasing value of $-X$.

The Riccati equation (Eqn. (3.16)), can be written in a vectorial form where the variables and parameter values can be expressed as intervals with k being the number of interval-valued parameters and n is the order of the system. That is,

$$dY^f(X,I)/dX + (Y^f)^2(X,I) - P(X,I) = 0 \qquad (3.19)$$

where the vector boundary values of Y_i^f are specified by $Y^f(X_{L,H}) = Y_{L,H}^f$

Further, $\{I\}$ represents the vector set of parameters whose value is an interval and $I = [I_1, I_2, ..., I_k]$. Since Eqn. (3.16) is a first order (nonlinear) equation, its order is n = 1. The explicit solution of this fuzzy differential equation (Eqn. (3.19)), namely, $Y^f(X) = {}^fF_S(X)$ is indicated in the next section.

Pertinent to the aforesaid vectorial representation of the differential equation written in terms of interval-valued parameters, Bonarini and Bontempi [3.22] have stated a theorem and proved it. The modified version of it written appropriately in terms of the variables (X, Y^f) is as follows:

3.6.1 A theorem on the ordinary differential equation mapping a region of uncertainty

An ordinary differential equation (ODE) maps the external surface of its region of uncertainty at the value of the independent variable X, into the external surface of its region of uncertainty at the value of the independent variable (X + dX). That is, $S(X + dX) = S'(X + dX)$ where S(X) is the extreme surface delimiting an N-cube at the value of the independent variable X.

Here, the N-cube represents the surface generated by the interval-values of N = (n + k) entities at the boundary value(s) of X; and, every point of this hypercube is the boundary value(s) of a solution of the differential equation system under discussion. For a crisp case, the variable occupies a corner of this hypercube. Further, it is stated in [3.29], that this N-cube represents the "region of uncertainty" of the system at the boundary value of X. Considering an interval [X, X + ΔX], suppose S'(X) is obtained by mapping the surface S(X) in accordance with the test differential equation written as the Nth order ODE in terms of an N-dimensional state vector **Z**, $dZ_{i\ell}(X)/dX = g[Z_{i\ell}(X)]$ with g being a functional and $Z_{i\ell} = Y_{i\ell}^f$ for $\ell = 1$, 2, ..., n; and, $Z_{i\ell} = I_{i,(\ell-n)}$ for $\ell = (n+1)$, ..., N. The evolution of the external surface of the region of uncertainty with the discourse of independent variable X is depicted in Fig. 3.3b. Suppose $Z(X_{L,H})$ is the N-cube representation at the boundary value(s) fuzzified in the interval limit $X_L \le (X \to X^f) \le X_H$, then $S(X_{L,H})$ corresponds to the external surface of $Z(X_{L,H})$.

In [3.29], the above theorem is proved on *ab absurdo* principle and it is concluded that it is sufficient to calculate the trajectories of the points belonging to the external surface of the region of uncertainty to know the evolution in X of the region itself.

In simulating a fuzzy system which can be described by a differential equation as discussed above, implementation of the evolution of a surface constituted by a fuzzy dependent variable versus the independent fuzzy variable (as per the interval-value algorithm) is as follows: Suppose the boundary conditions of X and y are specified by the fuzzy sets $[\alpha_A^f, \alpha_B^f]$ and $[\beta_B^f, \beta_C^f]$, respectively. The corresponding uncertainty region can be illustrated by a rectangular domain ABCD. The differential relation maps this boundary uncertainty region to A'B'C'D' when X is incremented to (X + ΔX) as shown in Fig. 3.4a. A noninteracting algorithm (in which the variables are presumed to be noninteracting) would transform A'B'C'D' to $A^TB^TC^TD^T$. Alternatively, an interacting algorithm would transform ABCD to $A^IB^IC^ID^I$ as shown in Fig. (3.4b).

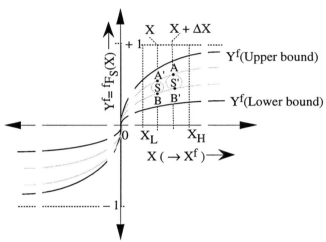

Fig. 3.3b: Discourse of the region of uncertainty S → S' versus
X → (X + ΔX) over an interval X_L to X_H

In Fig. 3.4b, ABCD and $A^IB^IC^ID^I$ are interacting via a coupling matrix, $[C] = \begin{bmatrix} C_{11} & C_{12} \\ C_{21} & C_{22} \end{bmatrix}$. In [3.30], a method of computing this coupling matrix in terms of the Jacobian matrix coefficient associated with the differential equation relation, is indicated. Alternatively, a correlation coefficient matrix can be also be assigned for [C], assuming that all of the uncertainty sources are correlated [3.29].

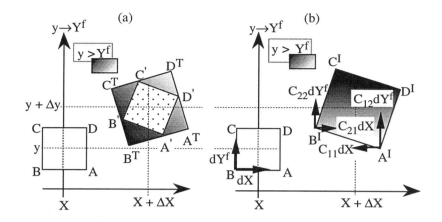

Fig. 3.4: Transformations of regions of uncertainty due to nonlinear neural transfer (activity) function

a : Transformation *via* noninteracting algorithm

b: Transformation *via* interacting algorithm

3.7 Membership Attributions to Fuzzy Sets via Activation Function

As discussed earlier, in concordance with the logistic growth of neural output $\{y\}$ in response to a set of summed inputs $\{X\}$ in the fuzzy domain, the input-output relation can be specified by a fuzzy Riccati differential equation given by Eqn. (3.14).

Now, the question is whether the membership attributes can also be incorporated in the aforesaid fuzzy activity function described by Eqn. (3.14) corresponding to the order of the system $n = 1$. Again, the specifications of membership functions as discussed earlier are $\mu_A(.) = [0, 1]$ and its sth interval I_s is specified by $(\lambda_{s-1}, \lambda_s)$. Suppose a basis function is derivable from the fuzzy activity function represented by the Eqn. (3.14). It should then meet the specifications of $\mu_A(.)$ as stipulated above.

Consider the neural activity represented by the fuzzy sigmoid and illustrated in Fig. 3.5 which corresponds to the solution of Eqn. (3.14).

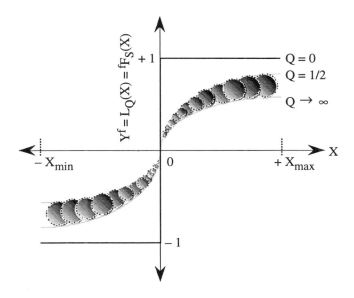

Fig. 3.5: Fuzzy nonlinear neural transfer (activity) function

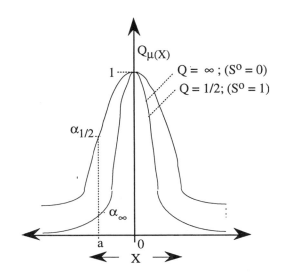

Fig. 3.6: Membership function controlled via Q or S^0

The variables X and Y are bounded by: $X \in (- X_{min} \to -\infty, + X_{max} \to +\infty)$ and $Y = [L_{1/2}(X), L_\infty(X)]$ where $L_Q(.)$ is the Bernoulli function mentioned earlier. For a crisp variable X, differentiating $L_Q(X)$ with respect to X yields $L'_Q(X)$ given by

$$L'_Q(X)/L'_Q(0) = \{-(1 + 1/2Q)\text{cosech}^2[(1 + 1/2Q)X]$$

$$+ (1/2Q)\text{cosech}^2(X/2Q)]/[(1 + Q)/3Q]\}$$

(3.20)

where $L'_Q(0) = [(1 + Q)/3Q]$ is the value of $L'_Q(X)$ with $X \to 0$ and is used here for normalization. Denoting $L'_Q(X)/L'_Q(0) = {}^Q\mu(X)$, it is plotted as a function of X in Fig. 3.6 for Q = 1/2 and Q = ∞.

As discussed earlier and indicated in [3.21], (1/Q) represents the number of orientational states of the interconnected neurons. When $Q \to \infty$, it refers to a neural spatial configuration which has a totally isotropic randomness with zero long-range order. Another extreme condition pertinent to a totally anisotropic neural spatial configuration with the long-range order realized to a total extent corresponds to Q = 1/2. For a partial long-range order of the neural spatial configuration (nemetic phase) with partial anistropic randomness, the number of orientational states is equal to (2Q + 1). The factor Q can be specified alternatively in terms of an order parameter S^0 of the system given by $S^0 = (3/2)[(1 + Q)/3Q] - (1/2)$, so that: $Q \to \infty$, $S^0 = 0$ (zero anisotropicity) and $Q \to 1/2$, $S^0 = 1$ (total anisotropicity). Hence, ${}^Q\mu(X)$ can be represented (equivalently) by ${}^{S^0}\mu(X) = L'_{S^0}(X)/L'_{S^0}(0)$ and $S^0 \in [1, 0]$.

Fig. 3.6 characterizes a membership function with specified upper and lower bounds. It is, therefore, compatible as a membership function for the interval-valued fuzzy sets defined formally by the functions of the form $\mu : \mathbf{X} \to [1, 0]$ where [1, 0] denotes the family of all closed intervals of real number in [0, 1]. For each value of X, $\mu(X)$ is represented by the segment between two curves which denote the identified lower bound and upper bounds decided by S^0 (or by Q). Thus in Fig. (3.6), ${}^Q\mu(a) = [\alpha_\infty, \alpha_{1/2}]$.

The extent of overlaps (or interactions) of contiguous fuzzy subsets can be accomplished *via* proper choice of the parameter S^0 associated with each membership function ${}^{S_0}\mu(.)$. That is, S^0 decides the choice of the interval-value of each fuzzy subset implicitly. Depicted in Fig. 3.5, are the three possible overlapping (interacting) subsets of a universe of discourse $\mathbf{X} \supset \{X_n\}$ each having a specific extent of overlap or width of the membership function, or the interval-value, being distinctly decided by the parameter S^0; and, the centers of the locations of each subset are denoted by c_a^0, c_b^0 and c_c^0.

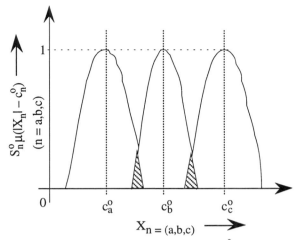

Fig. 3.7: Membership functions controlled *via* Q or S^0 and their overlap characteristics

Further, the order parameters S_a^0, S_b^0 and S_c^0, in general, need not be equal, but $S_{a,b,c}^0 \in [0,1]$. Inasmuch as S^0 decides the interval-value of $\{X\}$, at any s^{th} interval, $S_s^0 ® S_{s-1}^0$ should decide the extent of the overlap of the subsets; or $S_s^0 ® S_{s-1}^0 \equiv |C|$, the determinant of the connection matrix discussed earlier and ® denotes a *relativeness measure* (such as difference or ratio) between the value of S_s^0 and S_{s-1}^0 compared. That is, $® \in (-, /, ...)$.

Since the crisp form of the differential equation whose solution is $L_Q(X)$ is given by Eqn. (3.15), the corresponding nonfuzzy differential equation for $^{S_0}\mu(X)$ can be written in the Riccati form as

$$U''(X) - P(X)U'(X) = 0 \qquad (3.21)$$

where,

$$U'(X)/U(X) = \int_{z=-\infty}^{X} \left[(1 + 2S^0)/3\right] \left[^{S_0}\mu(z)\right] dz \quad \Re \in \{X\}$$

$$(3.22)$$

The solution of the above Eqn.(3.21) yields an explicit expression for $^{S_0}\mu(X)$ given by Eqn.(3.18), which can be written in terms of S^0 as follows:

$$^{S_0}\mu(X) = \left[3/(1 + 2S^0)\right]\left[-(1 + S^0)\text{cosech}^2[(1 + S^0)X] + S^0\text{cosech}^2(S^0X)\right]$$

$$(3.23)$$

In summary, in the above deliberations, it is hypothesized that the functioning of a neural network in fuzzy domain can be specified by a fuzzy nonlinear input-output relation mediated by a fuzzy differential equation to which relevant considerations of interval calculus applies justifiably. The corresponding solution yields a fuzzy set governed by a sigmoidal function which includes implicitly the membership characterizations as well.

It can also be inferred that such a membership function can be deduced from the solution of the Riccati differential equation, namely $L_Q(.)$. Fuzzification of $L_Q(.)$ yielding a fuzzy sigmoid $^f F_S(.)$ is either controlled by Q or by an equivalent parameter S^0 through which the membership function $^{S_0}\mu(.)$ is constructed using $L'_Q(X)$. Further, the parameter $S^0 \in [0, 1]$ offers a method of incorporating a width or interval-value for the sth interval of the fuzzy variable specified by λ_{s-1} and λ_s using corresponding values of S^0_{s-1} and S^0_s, respectively. Hence, the overlaps of the subsets of the discourse of the fuzzy variable can be quantified via $S^0_{s-1} \circledR S^0_s$.

3.8 Neural Architecture with a Fuzzy Sigmoid

There are two categories of fuzzy-neural networks generally adopted in practice. The first group of neural networks uses a set of fuzzy weights in the network for fuzzy reasoning. For example, Hayashi et al. [3.31] and Ishibushi et al. [3.32, 3.33] have considered fuzzy neural networks with fuzzy signals processed by fuzzy weights and biases. They have also discussed the relevant "if-then" rule-based learning characteristics. In the second category of fuzzy neural networks [3.34, 3.35], the input data are fuzzified at the first or the second layer, but the network weights themselves are not fuzzy [3.33]. A typical version of the second category is due to Lin and Cunningham III [3.36] who have introduced the concept of "fuzzy curves" to identify the significant input variables, estimate the number of rules needed in the fuzzy model and determine the initial weights for the neural network. The fuzzy curve is generated *via* centroid defuzzification for each input variable.

The approach indicated in this section follows general protocols of the second category of fuzzy-neural networks indicated above, but membership grading is done at an intermediate layer using the fuzzy-sigmoid parameter (Q or S^o) as an implicit part of the associated nonlinear transformation. The architecture of the proposed fuzzy neural network is depicted in Figs. 3.8a and 3.8b.

In terms of the nonlinear modeling of a typical neural cell exposed to fuzzy environment as discussed in the earlier sections, a multiple input-output fuzzy neural architecture can be represented as shown in Figs. 3.8a and 3.8b. For comparison, in Fig. (3.9) a conventional method of representing such a fuzzy neural network [3.35] is presented. It is based on fuzzifying directly the classical (crisp) version of a layered neural network architecture.

The differences in the architectural aspects and functional protocols between the proposed network (Figs. 3.8a and 3.8b) and the conventional type (Fig. 3.9) can be summarized as follows:

* Conventionally, real numbers characterizing a (crisp) neuron are translated to fuzzified counter parts in the fuzzy neuron (Fig. 3.9)

* Corresponding to the fuzzy input $\{x_i^f\}$ and fuzzy interconnecting weights $\{x_i^f\}$, a fuzzy arithmetic operation would yield $A_k = \sum_i^n w_{ki}^f x_{ki}$

* Further, in the conventional, if-then rule-based fuzzy neural structure (Fig. 3.8b) using a crisp sigmoid $F_S(.)$, the fuzzy output is obtained via the $extension\ principle$. That is, $Y_k^f = F_S(A_k)$

* Relevant to the present considerations, where the fuzzy neuron deploys a fuzzy sigmoid $^f F_S(.)$ as illustrated in Fig. 3.8a, the transformation $A_k \rightarrow Y_k$ is mediated by a membership function $^{S_0}\mu(.)$ deduced via a fuzzy sigmoid $^f F_S(.)$, as discussed

* From the fuzzy output Y_k, the corresponding crisp output can be extracted by a defuzzification procedure such as the center of area method [3.36]

* In contrast to the conventional procedure (Fig.3.9), the essence of deliberations presented in this chapter, refer to the feasibility of using a fuzzified, nonlinear sigmoid depicting the neural input-to-output transformation.

* Based on the notions of logistic growth of output versus input in a crisp neuron, as discussed in [3.22], the fuzzy neural sigmoid is evaluated as $^f F_S(.)$ which represents quantitatively the same functional trend as the crisp sigmoid, namely $F_S(.) = L_Q(.)$. It also bears certain qualitative or linguistic attributes in terms of the associated membership function(s), $^{S_0}\mu(.)$, where $S^0 \in [0, 1]$ is an order function (related to Q of the Bernoulli function $L_Q(.)$. This order function decides the interval-value and, hence, the extent of overlaps (connection matrix) of the membership functions

* As conceived in [3.22], the nonlinear activity in a crisp neuron can be represented by a logistic input-output relation mediated by a nonlinear Riccati differential equation. In extending the relevant notions to a fuzzy neuron, the corresponding nonlinear neural activity specified in the present study by a Riccati differential equation is subjected to fuzzification via interval-value

calculus [3.29]. The resulting solution implies that X^f *versus* Y^f evolution in a fuzzy domain depicted in terms of the evolution of the fuzzy surface constituted by the set $\{X^f, Y^f\}$ versus X^f is, the same as the trajectory surface $S(X^f, Y^f)$. Therefore, the fuzzy sigmoid $^fF_S(.)$ has the same functional trend as the crisp sigmoid $F_S(.) = L_Q(.)$ quantitatively

* The uncertainty spread of $^fF_S(.)$ is bounded between the function $L_{Q \to 1/2}(.)$ and $L_{Q \to \infty}(.)$ as depicted in Fig. (3.3). For the fuzzy output set $\{Y^f\}$, $L_{1/2}(.)$ and $L_\infty(.)$ constitute the upper and lower bounds, respectively.

The statistical mechanics considerations attributed in depicting the crisp sigmoid as the Bernoulli function $L_Q(.)$ of order Q can be extended to explain the fuzzy sigmoid $^fF_S(.)=^fL_Q(.)$ presented in Fig. 3.6. Fuzzification of the variable X_i leading to the fuzzy set $\{X_i^f\}$ corresponds to a fuzzifying effort in constituting the directed anisotropic neuronal state proliferation between the input and the output sections. Relevant fuzzified number of neural orientational states namely, fuzzified value of $(2Q +1)$ or equivalently, the fuzzified extent of the order parameter S^0, assign a corresponding value of membership grading, $^f\mu(.)$ to each value of $\{X_i^f\}$ with a specified width or overlapping interval.

Such assignments of $^f\mu(.)$ also facilitate the degree of belongingness expressed in terms of linguistic norms to $\{X_i^f\}$ as done in the conventional fuzzy neurons.

Further, the shape of the membership function $^f\mu(.)$ deduced *via* the Bernoulli function is similar to the bell-shaped (Gaussian) function, as can be seen from Fig. 3.6.

A fuzzy neuronal architecture employing a fuzzy sigmoid as depicted in Fig. 3.8a is functionally identical to the conventional fuzzy neural architecture (Fig. 3.9) which uses the crisp sigmoidal nonlinearity, except for the stage of protocol in assigning the membership function.

(a)

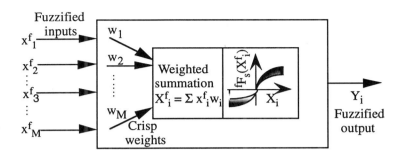

(b)

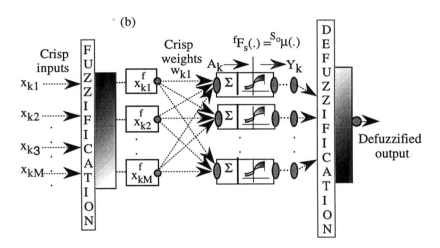

Fig. 3.8: Fuzzy neural architecture employing fuzzy nonlinear activation function

a. Fuzzy neuron: Cellular structure with nonlinear activation using a fuzzified sigmoid $^{f}F(X_i) = L_Q(X_i)$ specified by a corresponding membership assignment *via* $^{f}\mu(X_i) = L'_Q(X_i)/L'_Q(0)$

b. Fuzzy neural network using the fuzzy sigmoidal function

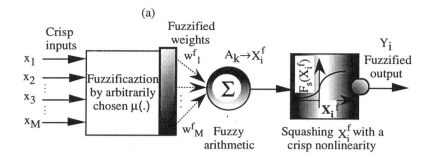

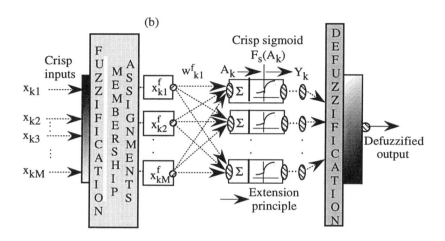

Fig. 3.9: Fuzzy neural architecture employing crisp nonlinear activation function

a. Fuzzy neuron: Cellular structure with nonlinear activation using a crisp sigmoid
b. Fuzzy neural network using the crisp sigmoidal function

3.9 Fuzzy Considerations, Uncertainty and Information

Random attributes of events and processes prescribe an imprecise vagueness and portray an associated uncertainty at all levels of such processes whether at a microscopic stratum or at a macroscopic plane. The conflict of order and disorder existing in a system leads to an organized simplicity or a disorganized complexity, respectively. Most of the time, the system is, however, neither totally disorganized nor completely ordered.

Probabilistic concepts which stemmed form Aristotelian two-level logic, namely, the truth and falsity of the associated proposition, constitute a base for the classical description of uncertainty. Reduction of such uncertainty (or entropy) refers to gaining information.

The entropic attributes of a system which quantitatively measure the information content in terms of uncertainty reduction are described in Chapter 1 and constitute Hartley-Shannon's perspective of uncertainty or entropy-based information. Formally, the associated notion has the root proliferating into the axiomatic concepts of probability. Distinct from this entropy-based conceptualization of information, it may be noted that *theory of computability* (in the Turing sense) specifies the amount of information represented by an object as the length of the shortest possible program written in some standard language, so as to describe the object in computable terms. This type of information is designated as the *descriptive* or *algorithmic information*.

In reference to fuzzy sets or events, the information associated with its member elements is "fuzzy" with three types of attributable uncertainty [3.38]:

- *Non-specificity* or *imprecision vis-á-vis* the sizes
  (cardinalities) of relevant sets of alternatives.
- *Vagueness* or *fuzziness* of the imprecise boundaries of fuzzy sets.
- *Strife* or *discord* which expresses the conflicts among the various sets of alternatives.

In order to measure the information content associated with the three aforementioned fuzzy uncertainties, the scope of classical entropy-based information theory can be considered in an expanded framework. Following this strategy, a side-by-side description of information-theroetic considerations pertinent to crisp and fuzzy sets are presented in the following section.

3.10 Information-Theoretics of Crisp and Fuzzy Sets

3.10.1 Hartley-Shannon function: A nonspecificity measure

In reference to a finite set A of non-zero cardinality $|A|$, the Hartley-Shannon class of functions which measure the amount of uncertainty or entropy associated with a finite set of possible alternatives are given by

$$H(A) = k\ell og|A| \qquad (3.24)$$

where k is a constant of proportionality. When $k = 1$ and the base of the logarithm is taken as 2, H(A) is measured in bits. When H refers to non-empty finite subsets of the universe, the elements of which are nonnegative real numbers, it follows that

$$O \le H(A) \le \ell og_2 |X| \qquad (3.25)$$

H(A) denotes a measure of *predictive uncertainty* when A is a set of possible values ascertained from a relevant sample space of predicted states of variables. H(A) may indicate a *diagnostic uncertainty* when A represents a set of statistical evidences. Further, when a stochastical framework of unsettled (historical) question is depicted by A, then H(A) is a *retrodictive uncertainty*. Lastly, A may represent a set of prescribed possible values, in which case H(A) becomes a *prescriptive uncertainty*.

In all the above cases, the non-specificity is inherent in each set. That is, by virtue of the associated uncertainty or less specific attributes of predictions, diagnoses, retrodictions and prescriptive values, the Hartley-Shannon measure is synonymous as far as the non-specificity of the sets concerned.

Having considered the non-specificities of crisp sets, it could be of interest to generalize the information-theoretic concepts (based on the Hartley-Shannon principle) to fuzzy sets. In early 1980, a proposal in that direction was presented [3.39] under the paradigm of *U-uncertainty*. For a universal set X, a generalized Hartley-Shannon function representing the nonspecificity of fuzzy sets was hence defined as follows:

$$U(A) = H^f(A) = \frac{1}{h(A)} \int_0^{h(A)} \ell og_2 |^\alpha A| d\alpha \qquad (3.26)$$

where as indicated earlier, $|^\alpha A|$ denotes cardinality of the α-cut of A and h(A) is the height of A. The superscript f in H(.) again denotes explicitly the fuzzy considerations under discussion. Here the height represents simply the validity or credibility of the information expressed by the fuzzy sets.

3.10.2 Uncertainty: A measure of fuzziness

Exclusive to fuzzy sets, the associated vagueness represents a measure of fuzziness denoted by a function f such that

$$(f : \mathcal{F}(X) \to \mathfrak{R}^+) \qquad (3.27)$$

where the characteristic function f maps all the fuzzy subsets of X, namely $\mathcal{F}(X)$, to elements of the set of all nonnegative real numbers, $\mathfrak{R}^+$. That is, for each fuzzy set A, this function f assigns a nonnegative real number f(A) which depicts the degree to which the boundary of A is not crisp in that it is not defined precisely. It may be noted that f(A) = 0, if and only if (*iff*), A is a crisp set. In general, f(A) can be formalized as follows:

$$f(A) = \sum_{x \in X} [1 - |2A(x) - 1|] \qquad (3.28)$$

It measures the lack of distinction between the set and its complement. The less a set differs from its complement, the fuzzier it is. The lack of distinction between the set and its complement can also be specified by a distance measure. Among several options for such feasible specifications, the concept of Hamming distance offers a simple direction. That is, f(A) can be specified by the sum of absolute values of the differences between A(x) and its complement [1-A(x)]. Hence, the corresponding distance measure is given by

$$|A(x) - [1 - A(x)]| = |2A(x) - 1| \qquad (3.29)$$

The sum total of such local measures pertinent to all x ε X yields f(A) of Eqn.(3.28).

3.10.3 Shannon entropy: An uncertainty measure of discord or conflict

Disagreement in choosing among several alternatives leads to a uncertainty based on the associated aspects of conflict, discord or strife. Suppose m = { m_1, m_2, m_3,.....m_n} represents the n-tuple set of basic probabilities assignment corresponding to x for some finite n $\in$ N, where m(X) = 1. Then , the Shannon entropy is given by

$$H(m) = - \sum_{x \in X} m(\{x\}) \log_2 m(\{x\})$$
$$(3.30)$$

as discussed in Chapter 1.

Eqn.(3.30) measures the average uncertainty (in bits) associated with the prediction of outcomes in a random experiment in the sample space of X.

Shannon's entropy can be regarded also as the mean expected value of the conflict among evidential claims within a given probabilistic body of evidence portrayed in the sample space.

The summary of uncertainty measures discussed above is illustrated in Fig. 3.10. There are also other uncertainty theories such as *evidence theory, possibility theory* and *fuzzified evidence theory*, the details of which can be gathered from [3.40] .

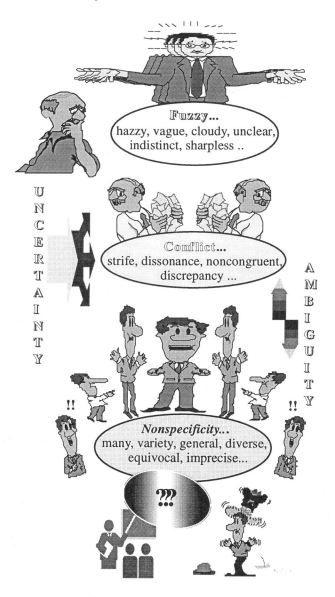

Fig. 3.10: Types of uncertainties

3.10.4 Theorems on fuzzy entropy

The information-theroetics of a fuzzy activity can be viewed essentially in Hartley-Shannon perspectives so as to relate fuzziness and entropy. Kosko [3.1] extends the theory of discrete fuzzy sets to define a "fuzzy cube" and explores the information-theoretics of the continuum of fuzzy sets which fill this cube. Relevant definitions and details are summarized below:

Definitions

Fuzzy cube: This is the unit hypercube which contains all fuzzy subsets of a set X of n objects. The 2^n bivalents subsets of X lie at the 2^n corners of the n-cube $[0,1]^n$.

Mutual entropy of a fuzzy set: This, (in the Kullback-Leibler sense), refers to a distance measure between a fuzzy set F and its set complement F^c. Implicitly, it measures the logarithm of a unique measure of fuzziness of the set F.

The first theorem of fuzzy mutual entropy: This refers to the basic operation of addition of real numbers. The corresponding logistic map equates the sum of n-components of a real vector with the mutual entropy of some fuzzy set F and its complement F^c. That is, $x_1, x_2, ..., x_n$ is equal to fuzzy mutual entropy of fuzzy set F in the unit hypercube: $I^n = [0,1]^n$

$$\sum_{i=1}^{n} X_i = H\left(F/F^c\right) - H\left(F^c/F\right) \qquad (3.31)$$

where $H[(.)/(.)]$ depicts the mutual entropy function.

3.10.5 Fuzzy mutual entropy versus Shannon entropy

Fuzzy mutual entropy is equal to the negative of the divergence of Shannon entropy. That is,

$$H\left(F/F^c\right) - H\left(F^c/F\right) = \nabla \bullet H(f) + \nabla \bullet H(f^c) \qquad (3.32)$$

where ∇ is the del operator. Inasmuch as divergence specifies a net average out- or in-flow of a flux, Eqn. (3.32) suggests the existence of an information flux or field in the fuzzy cube I^n. Every fuzzy set point has a flux of information (or entropy) across it and this flux emanates from a source point, say, at the mid-point or a vertex of the fuzzy cube.

3.10.6 Real space and fuzzy cubes

Fuzzy cubes map smoothly onto extended real spaces of the same dimension and *vice versa*. For example, Fig. 3.11 shows how each real point x maps to a unique fuzzy set F. That is, $\Re^n \rightarrow I^n$ is a one-to-one ratio and maps onto the differentiable map f with a differentiable inverse f^1. It corresponds to a fuzzy system (f:X) in I^n that approximates a scalar function (f:X) in $\Re^n$. For example, an unbounded real variable x_i (in the real space $\Re^n$) can be diffeomapped onto a fuzzy space I^n by converting the unbounded real input x_i to a bound value or *a fit value*, f_i as follows:

139

$$f_i = 1/[1 + \exp(-x_i)] \qquad (3.33)$$

so that *iff* $x_i \to \infty$, $f_i \to 1$ and *iff* $x_i \to -\infty$, $f_i \to 0$.

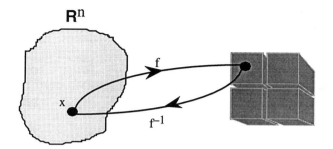

Fig.3.11: Real-space to fuzzy space diffeomapping

3.10.7 Fuzzy chain: Definition and concept
The concept of Shannon entropy is governed by the following chain of probabilities:

$$p_i \to (1/p_i) \to \log(1/p_i) \to \sum p_i \log(1/p_i) \qquad (3.34)$$

Likewise, a *fuzzy chain* can be formulated in terms of f_i:

$$f_i \to (1/1 - f_i) \to \log[f_i/1 - f_i] \to \sum f_i \log[f_i/1 - f_i] = H(F/F^c) \qquad (3.35a)$$

Similarly,

$$\sum (1 - f_i)\log[(1 - f_i)/f_i] = H(F^c/F) \qquad (3.35b)$$

Hence, it follows that, by defining a fuzzy information unit as $\log[f_i/(1 - f_i)]$ the following relation can be written:

$$\sum_{i=1}^{n} \log[f_i/(1 - f_i)] = H(F/F^c) - H(F^c/F) \qquad (3.36)$$

The relevant features of fuzzy information flux can be summarized as follows:

- Fuzzy mutual entropy is flux-like and is specified by the information field in a fuzzy cube

140

- Points in the fuzzy-cube parameter cube are defined by the fuzzy uncertainty descriptions
- Fuzzy mutual entropy is equal to the negative of the divergence of Shannon entropy
- Shannon entropy and fuzzy mutual entropy define vector fields on the fuzzy cube
- Shannon entropy is similar to a potential of the conservative mutual entropy vector field. Thus, the dynamical aspects of information flux flows on the fuzzy cube correspond to a governance by the second law of thermodynamics.

3.11 Fuzzy Neuroinformatics

This section considers the relevant aspects of applying fuzzy information-theoretic considerations to the neural complex.The behavior of the neural complex in mapping the fuzzy inputs-to-output relation in a non-linear fashion was discussed in Section 3.4. Suppose such a mapping refers to fuzzy trans-information-theoretics. In other words, it is of interest to study transfer of fuzzy information across the neural complex governed by the associated nonlinear considerations. Again referring to the fuzzy input and fuzzy output sets as $\{X_i^f\}$ and $\{y_i^f\}$, respectively, the analysis presented in Section 3.4 suggests that

$$y_i^f = F_s^f(x_i) \tag{3.37}$$

where $\{x_i\}$ refers to the crisp counterpart of $\{X_i^f\}$ and $\{y_i^f\}$ is the neural fuzzy sigmoid which has the same functional trend as the crisp sigmoid $F_s(.) = L_Q(.)$. The corresponding extent of overlaps of contiguous fuzzy subsets can be described by the membership function given by

$$\mu^Q(.) = L'_Q(.)/L'_Q(0) \tag{3.38}$$

The bounds on fuzziness is set by $Q = 1/2$ to $Q = \infty$.

To express the transformation of entropy (or information) across a fuzzy neuron, that is, to establish a relation between $H(X_i^f)$ and $H(y_i^f)$ it is convenient to linearize the fuzzy sigmoid $F_s^f(.)$. This can be accomplished by considering the linear version of the crisp sigmoid $F_s(.)$ indicated in Chapter 2. Reproducing Eqn.(2.24b) here for reference:

$$F_s(X) \approx \begin{cases} -1 \\ \alpha_0 X \\ +1 \end{cases} \quad \begin{array}{l} X < (-1/\alpha_0) \\ (-1/\alpha_0) < X < (+1/\alpha_0) \\ X < (+1/\alpha_0) \end{array} \tag{3.39}$$

141

where $\alpha_0 = L_Q'(X)$ at $X=0$. The fuzzy sigmoid of Fig 3.12 is redrawn in its (approximate) linear form in Fig. 3.12.

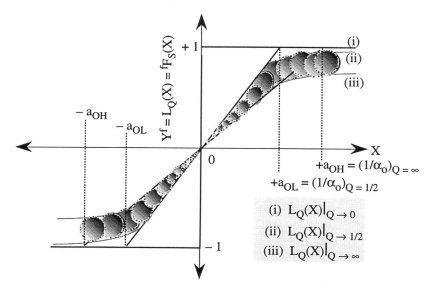

Fig.3.12: Linearized fuzzy sigmoid

In terms of $L_Q(.)$, the unbounded real variable X_i (in real space $\mathfrak{R}^n$) when diffeomapped onto fuzzy a space $\mathbf{I}^n$, by transforming the unbounded real input X_i *via* ${}^f F_s(.)$, a bound set of values g_i is realized. That is,

$$g_i = \frac{1}{2}\left[1 + L_Q(X_i)\right]_{\substack{Q=1/2 \\ Q=\infty}} \qquad (3.40)$$

so that *iff* $X_i \to \infty$, $g_i = 1$, and $X_i \to -\infty$, $g_i = 0$.

In terms of this output (fit) value, as indicated earlier, the fuzzy information can be specified as :

$$\sum \ell og\left[y_i/(1-y_i)\right] = H({}^y F/{}^y F_c) - H({}^y F_c/{}^y F) \qquad (3.41)$$

where ${}^y F$ and ${}^y F_c$ refer to the fuzzy set $\{y_i\}$ and its complement, respectively; and, again $H[(.)/(.)]$ denotes the mutual entropy functions. Further, fuzzy mutual entropy equals the negative of the divergence of the Shannon entropy.

Thus, for a crisp set of X_i at the input of a fuzzy neuron, the transformed entropy (information) of the output fuzzy set $\{y_i\}$ can be specified in terms of fuzzy mutual entropy functions as in Eqn.(3.41).

For a crisp case involving variable X_i and y_i, the transformation of entropy pertinent to $y = G(X)$ is straightforward. Assuming x as a continuous random variable with an entropy $H(X)$, the corresponding transformed entropy $H(y)$ is given by

$$H(y) = - \int_{-\infty}^{+\infty} f_X(x)\ell og\left[f_X(x)\right]dx + \int_{-\infty}^{+\infty} f_X(x)\ell og\left[G'(x)\right]dx$$

$$= H(x) + E\left[\ell og|G'(x)|\right]$$

(3.42)

where E[.] denotes the expected value and it is assumed $y = G(X)$ has a unique inverse, namely, $X = G^{-1}(y)$. Further, $f_X(x)$ in Eqn. (3.42) represents the probability density function of x. In general, if the random variable x is continuous, then

$$H(y) \leq H(x) + E[\ell og|G'(x)|]$$

(3.43a)

with equality *iff* the transformation $y = G(X)$ has a unique inverse.

With the linearized $F_s(.)$, the input-output relation with respect to a crisp neuron is given by Eq. 3.35. Hence in the range $(-1/\alpha_0)$ to $(+1/\alpha_0)$ of x,

$$\begin{aligned} H(y) &= H(x) + E[\ell og|\alpha_o|] \\ &= H(x) + \ell og|\alpha_o| && -1/\alpha_o \leq x \leq +1/\alpha_o \\ &= 0 && \text{otherwise} \end{aligned}$$

(3.43b)

In the case of fuzzy variables, the membership function $^Q\mu(x)$ given by Eqn. 3.20 (and shown in Fig.3.5) can be approximated using the linearized $F_s(.)$ as follows:

$$^Q\mu(x) = \begin{cases} 0 & -\infty < x < -a_{OH} \\ 0 \text{ or } 1 & -a_{OH} < x < -a_{OL} \\ 1 & -a_{OL} < x < +a_{OL} \\ 0 \text{ or } 1 & +a_{OL} < x < +a_{OH} \\ 0 & +a_{OH} < x < +\infty \end{cases}$$

(3.44)

where $\alpha_{oL} = (1/\alpha_o)_{Q=1/2}$ and $\alpha_{oH} = (1/\alpha_o)_{Q=\infty}$, as shown in Fig. 3.12.

From the considerations of the above cited membership aspects of "belongingness" of the fuzzy output $^yF:\{y_i\}$ vis-à-vis the fuzzy input set $\{X_i\}$, the following entropy relation can written:

$$H(^yF) = \begin{cases} H(^xF^c) & -\infty < x < -a_{OH} \\ [H(\mu_L\,^xF^c), H(\mu_H\,^xF)] & -a_{OH} < x < -a_{OL} \\ H(^xF) & -a_{OL} < x < +a_{OL} \\ [H(\mu_L\,^xF^c), H(\mu_H\,^xF)] & +a_{OL} < x < +a_{OH} \\ H(^xF^c) & +a_{OH} < x < +\infty \end{cases}$$

$$(3.45)$$

where the superscript c on F denotes the complement of F. Thus, the entropy of the fuzzy neural output set is decided by the entropy of the fuzzy neural input set or its complement. The overlap regions are decided by the bounding constraints $a_{OL} = (1/\alpha_o)_{Q\,=\,1/2}$ and $a_{OH} = (1/\alpha_o)_{Q\,=\,\infty}$ pertinent to the slopes of the bounding sigmoids. In these overlap regions, the linguistic concept of a range $[H(^xF), H(^xF^c)]$ applies to the entropy of the fuzzy neural output set. In general $[\mu_L H(^xF^c), H(\mu_H\,^xF)]$ define the interval values for example, $\mu(X_R)$ = $[\mu_{LR}, \mu_{HR}]$ in Fig 3.13. Thus, in the overlap regions, $H(^yF)$ corresponds to an interval-valued fuzzy set. A graphical illustration relating $H(^yF)$ *versus* $H(^xF)$ and $H(^xF^c)$ is shown in Fig 3.14.

To summarize, the neural informatics assume fuzzy attributes in specifying the entropy of fuzzy neural output in respect to the entropy of fuzzy neural input over a pair of overlap regions as decided by a set of bounded values of X, namely $\{-a_{OL}, -a_{OH}\}$ and $\{+a_{OL}, +a_{OH}\}$. These values are, in turn, governed by the sigmoidal nonlinearity of the neural complex.

3.12 Concluding Remarks

To conclude, this chapter offers a new perspective for viewing the nonlinear activity in a fuzzy neuron. It is conceived in this work that without any loss of generality, the underlying aspects of logistic growth of output *versus* input viewed in the perspectives of statistical mechanics (and spin-glass theory) pertinent to a crisp neuron, apply even to the fuzzified state of the neuron in deliberating its transfer functional characteristics. The concept of fuzzy sigmoidal nonlinearity specifies an order function (denoted by Q or a related value S^0) which also decides the extent of overlaps or the membership gradation attributed to the neural input-output fuzzy variables.

The associated nonlinearity of a fuzzy neuron enables characterizing the entropy relations between the input and output sets of a fuzzy neuron. Relevant entropy considerations and information transfer characteristics are discussed.

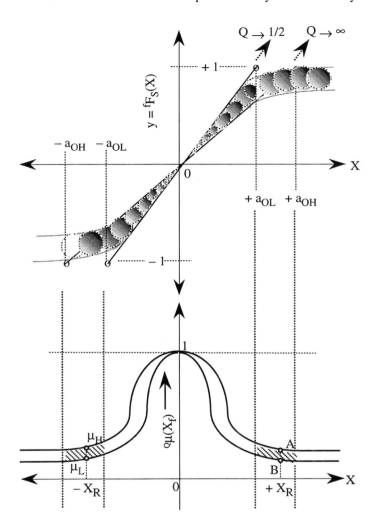

Fig.3.13: Linearized bounds of fuzzy sigmoid and the corresponding
membership designations

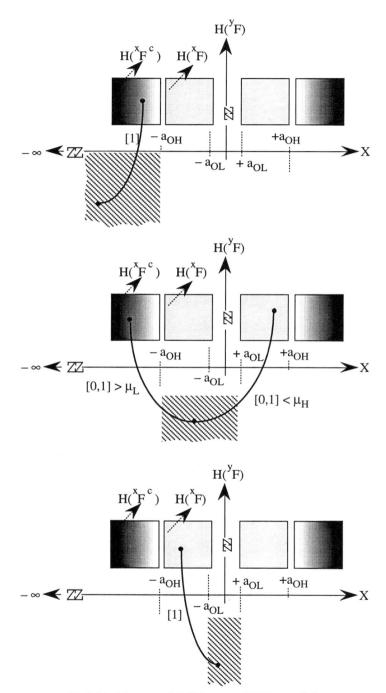

Fig 3.14: Mapping of $H(^yF)$ *versus* $H(^xF)$ & $H(^xF^c)$

146

Bibliography

[3.1]   Kosko, B.: Neural Networks and Fuzzy Systems (Prentice Hall, Englewood Cliff, NJ: 1992).

[3.2]   Zadeh, L. A.: Fuzzy sets, *Inform. Control*, 8, 1965, 338-354.

[3.3]   Pearl, J.: *Heuristics: Intelligent Search Strategies for Computer Problem Solving* (Addison Wesley, Reading, MA: 1984)

[3.4]   Mitra, S. and Pal, S. K. : Self-organizing neural network as a fuzzy classifier, IEEE Trans. Syst. Man and Cybernetics, 24, 1994, 385-399.

[3.5]   Klir, G. J. and Yuan, B.: *Fuzzy Set and Fuzzy Logic: Theory and Applications* (Prentice Hall PCR, Upper Saddle River, NJ: 1995).

[3.6]   Brown, M. and Harris, C.: *Neurofuzzy Adaptive Modelling and Control* (Prentice Hall, New York: 1994).

[3.7]   Werntges, H. W.: Partitions of unity improve neural function approximation, *Proc. IEEE Int. Conf: Neural Networks*, Vol 2 1993).

[3.8]   William, R. J.: The logic of activation functions. In: Rumelhart, D. E. and McClelland, J.L.(Eds.): *Parallel Distributed Processing 1* (MIT Press, Cambridge, MA:1986), Chapter 10, 423-443.

[3.9]   Jordan, M. I.: An introduction to linear algebra in parallel distributed process. In: Rumelhart, D.E. and McClelland, J. L. (Eds):*Parallel Distributed Processing 1* (MIT Press, Cambridge, MA: 1986), Chapter 9, 375-422.

[3.10]   Hecht-Nielsen, R.: *Neurocomputing* (Addison-Wesley, Reading, MA: 1990).

[3.11]   Wasserman, P. D.: *Neural Computing: Theory and Practice* (Van Nostrand Reinhold, New York, NY: 1989).

[3.12]   Caudill, M.: *Neural Networks* Primer (Miller Freeman Publications, San Francisco, CA: 1990).

[3.13]   Kalman, B. L. and Kwansny, S. C.: Why tanh: Choosing a sigmoidal function, *Proc. IEEE Int. Joint. Conf. on Neural Networks* (June, 1992), pp. IV 578 - IV 581.

[3.14]   DasGupta, B. and Schnitger, G.: The power of approximating: a comparison of activation functions, in: Hason, S. J. Cowan, J. D. and Giles, C. L.(Eds.): *Advances in Neural Information Processing Systems*, 5 (Morgan Kaufman, San Mateo, CA: 1993), 615-622.

[3.15]   Ito, Y.: Representation of functions by superpositions of a step or sigmoid function and their applications to neural network theory, *Neural Networks*, 4, 1991, 385- 394.

[3.16]   Siuand, K. Y., Roychowdhury, V. P. and Kailath, T.: Rational approximation techniques for analysis of neural networks, *IEEE Trans. Inform. Theory*, 40, 1994, 455-466.

[3.17]   Kreinovich, V. Y.: Arbitrary nonlinearity is sufficient to represent all functions by neural networks: A theorem, *Neural Networks* 4, 1991, 381-383.

[3.18]   Chen, T. and Chen, H.: Universal approximation of nonlinear operators by neural networks with arbitrary activation functions and its application to dynamical systems*, IEEE Trans. Neural Networks* 6, 1995, 911-917.

[3.19]   Neelakanta, P. S., Sudhakar, R. and DeGroff, D.: Langevin machine: A neural network based on stochastically justifiable sigmoidal function, *Biol. Cybern.* 65, 1991, 331-338.

[3.20]   DeGroff, D., Neelakanta, P. S., Sudhakar, R. and Medina, F.: Liquid crystal model of neural networks, *Complex Syst.*, 7, 1993, 43-57.

[3.21]   Neelakanta, P. S. and DeGroff, D.:*Neural Network Modeling: Statistical Mechanics and Cybernetic Perspectives* (CRC Press, Boca Raton, FL: 1994).

[3.22]   Neelakanta, P. S. Abusalah, S., DeGroff, D. and Sudhakar, R.: Logistic model of nonlinear activity in a neural complex: Representation via Riccati differential equation, *Cybernetica,* XXXIX, 1996, 15-30.

[3.23]   McCulloch, W. W. and Pitts, W.: A logical calculus of the ideas imminent in nervous activity, *Bull. Math. Biophys.*, 5, 1943, 115-133.

[3.24] Cybenko, G.: Approximation by superposition of a sigmoidal function, Math.*Control Signals Systems*, 2, 1989, 303-314.

[3.25] Dubois, D. and Prade, H.: Towards fuzzy-differential calculus Part 1: Integration of fuzzy mappings, *Fuzzy Sets and Syst.*, 8, 1982, 1-17.

[3.26] Dubois, D. and Prade, H.: Towards fuzzy-differential calculus Part 2: Integration of fuzzy mappings, *Fuzzy Sets and Systems*, 8, 1982, 105-116.

[3.27] Dubois, D. and Prade, H.: Towards fuzzy-differential calculus Part 3: Integration of fuzzy mappings, *Fuzzy Sets and Systems*, 8, 1982, 225-233.

[3.28] Kandel, A and Byatt, W. J.: Fuzzy differential equations, *Proc. International Conf. on Cybernetics and Society*, (November 1978), pp.1213-1216.

[3.29] Bonarini, A. and Botempi, G.: A qualitative simulation approach for fuzzy dynamical models, *ACM Trans. Modeling and Computer Simulation*, 4, 1994, 285-313.

[3.30] Moore, R. E.: *Interval Analysis* (Prentice Hall, Englewood Cliffs, NJ: 1966).

[3.31] Hayashi, Y., Buckley, J. and Czogal, E.: Fuzzy neural network with fuzzy signals and weights, *Int. J. Intell. Syst.*, 8, 1993, 527-537.

[3.32] Ishibuchi, H., Fujioka., R. and Tanaka, H.: Neural networks that learn from fuzzy if- then rules, *IEEE Trans. Fuzzy Systems*,1, 1993, 85-97.

[3.33] Ishibuchi, H., Twanaka, H. and Okada, H.: Fuzzy neural networks with fuzzy weights and fuzzy biases. *Proc. Int. Conf. Neural Networks*, 1993, pp. l650-1655.

[3.34] Horikawa, S., Furuhashi, T., and Uchikawa, Y.: On fuzzy modeling using fuzzy neural networks with the backpropagation algorithm. *IEEE Trans. Neural Networks*, 3, 1992, 801-814.

[3.35] Takagi, H., Suzuki. N., Koda, T. and Kojiwa, Y.: Neural networks designed on approximate reasoning architecture and their applications. *IEEE Trans. Neural Networks*, 3, 1992, 752-760.

[3.36]   Lin, Y. and Cunningham III, G. A.: New approach to fuzzy-neural system modeling, *IEEE Trans. Fuzzy Systems*, 3, 1995, 190-198.

[3.37]   Fu, L.: *Neural Networks in Computer Intelligence* (McGraw-Hill, Inc., New York, NY: 1994).

[3.38]   Klir, G.J. and Folger, T.A.: *Fuzzy Sets Uncertainty and Information* (Prentice-Hall, Englewood Cliff, NJ: 1988)

[3.39]   Higashi, M., and Klir, G.J.: Measures of uncertainty and information based on possibility distribution, *Int. J. Gen. Systs.*, 9, 1983, 43-58.

[3.40]   Pal, N.R. and Bezdek, J.C.: Measuring fuzzy uncertainty. *IEEE Trans. Fuzzy Syst.,* 2.1994, 107-118.

Chapter 4

Csiszár's Information-Theoretic Error-Measures for Neural Network Optimizations

P.S. Neelakanta, S. Abusalah and J.C. Park

"I do not know what I may appear to the world, but to myself I seem to have been only like a boy playing on the sea shore, diverting myself in now and then finding a smoother pebble or a prettier shell than ordinary, while the great ocean of truth lay all undiscovered before me"

– Isaac Newton

4.1 Introduction

The neural network is essentially an information-processing system in which randomness coexists due to the inevitable presence of noise and, therefore, represents an entropy-dictated system. In general, as indicated in earlier chapters, a neural network is amenable for training to yield an output close to a true (or teacher) value *via* training optimization methods. Typically, in learning algorithms pertinent to neural networks, an error-measure which is minimized towards optimization of the network's performance so as to yield a desired output. Also, as mentioned in earlier chapters, such a cost-function can be determined directly in terms of the network parameters by evaluating the difference or mean-square of the difference between the network output and the desired (target) value. Alternatively, the cost-function can be defined in terms of relative entropy associated with the network parameters. Traditionally, the quadratic or square-error (SE) based error-measure have been used to train the neural networks with conventional protocols such as a back-propagation algorithm. However, by considering the information-theoretic aspects of the network described in Chapter 1, it is also possible to elucidate a cross-entropy error-measure based on the principle of discrimination-information (or mutual/conditional information) associated with the output and target (teacher) values. Such error-measures are known as the *divergence* or *distance* error-functions and they refer to the divergence (or distance) between the statistical attributes of the output and target parameters as elaborated in Chapter 1.

In a typical neural cell (Fig. 4.1a), the integrated effect of all the excitatory and inhibitory postsynaptic action potentials decide the state transition (or firing). The corresponding state vectors $\{S_i \ (i = 1,2,....N)\}$ of the ith cell can be specified in terms of the multi-inputs x_j ($j = 1, 2, ...,N$) linked to the ith cell *via* the N-component weight states w_{ij}, so that,

$$X_i = \sum_{j=1}^{N} w_{ij}x_j + \Theta_i + e_n; \quad S_i = F_S(X_i) \tag{4.1}$$

where Θ_i is a bias parameter and e_n is the error due to random noise. Further $F_S(.)$ represents an activity function which is invariably nonlinear and bounded ("squashed"). Similar to the real neural complex, an artificial multilayer assembly (as depicted in Fig. 4.1b) can be constructed, by randomly interconnecting a set of "neuronal cells". The network shown in Fig. 4.1b represents a typical multilayer perceptron [4.1, 4.2] trainable *via* backpropagation algorithm [4.3]. It consists of a set of input units (sensory receptors) and one or more hidden layers (associative arrays) of trainable neurons. That is, the weights on interconnecting links between the neuronal units can be modified dynamically by an error signal derived by comparing the output of the network with a desired (target) value; and, the resulting error is backpropagated layer-by-layer from the output unit. The method of obtaining an error-measure by comparing the network output *versus* a target (teacher) value and using it to modify the interconnecting weights is known as *supervised learning*.

In reference to an artificial neural network (ANN), the essence of backpropagation algorithm, also known as "generalized delta rule", is that the synaptic weights of the trainable neurons are adjusted to minimize the local error (or the cost-function) of the network with a given knowledge of the "correct" or target output T_i. The fundamental quantity which serves as the cost-function in adjusting the weight states is, thus, the error (ε_i) or the deviation of the network output (O_i) of the ith unit from the target value T_i. The adjustment of synaptic weights *via* backpropagation is done iteratively until the error function ε_i is minimized. The set of iterative epochs leading to a convergence ($\varepsilon_i \rightarrow$ a minimum value) constitutes one training run or cycle.

The basic prescription to adjust the weights at the rth training step or iteration as facilitated by backpropagation is given by:

$$w_{ij}(r) = w_{ij}(r-1) + \Delta w_{ij} \tag{4.2}$$

where $\Delta w_{ij} = \eta \delta_{ij} O_i$ and $w_{ij}(r) = w_{ij}(r-1) + \Delta w_{ij} w_{ij}$ is the connecting weight from unit i to j, η is the learning rate, and δ_{ij} is the effective gradient. The effective gradient has two distinct definitions depending on whether a target value is available or not for a particular unit.

In the case of a network output unit for which a target is known, δ_{ij} is defined as: $\delta_{ij} = (\partial O_i / \partial X_j)\varepsilon_j$ where ε_j is the error-measure between the output and the teacher value and X_j represents the summed input to the activation function, or $O_j = F_s(X_j)$. When the unit resides in a hidden layer or at the input layer, a target value is not, however, explicit in order to determine the effective gradient. Therefore, the definition of δ_{ij} is modified in which the products of cumulative effective gradients with the interconnection weights are backpropagated to these units. It is given by the relation:

152

$$\delta_{ij} = (\partial O_i / \partial X_j) \sum_k \delta_{kl} w_{kl} \qquad (4.3)$$

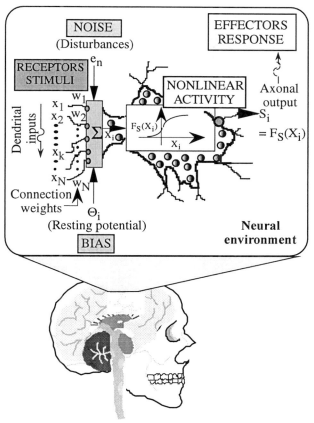

Fig. 4.1 (a): A biological neuron and the associated input-output activity

In regions where larger gradients exist over the error-surface, the effective gradient terms may become inordinately large. The resulting weight modifications will, therefore, be extensive, leading to large oscillations of the output bypassing the true error minimum sought. The learning coefficient can then be set to a very small value to counteract this tendency; however, this may appreciably increase the training time. To avoid this problem, the weight modification can be given a 'memory' so that it will no longer be subjected to abrupt changes. That is, the weight-change algorithm can be specified by

$$\Delta w_{ij}(r) = \eta \, \delta_{ij} \, O_i + \lambda \, [\, \Delta w_{ij}(r-1) \,] \qquad (4.4)$$

153

where λ is known as the *momentum parameter* [4.3]. If λ is set close to 1, the search in the parameter space will be determined by the gradient accumulated over several epochs instead of from a single iteration, thereby improving the stability of the network towards convergence.

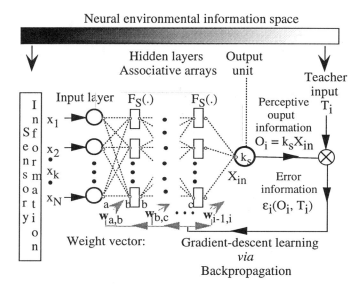

Fig. 4.1(b): An artificial multilayer neural network ("Perceptron")

4.2 Disorganization and Entropy Considerations in Neural Networks

The information content [4.4] associated with a neural network refers to the negative entropy (in Shannon's sense) pertinent to the neural activity or state-transitional events across the interconnected neurons. These include the excitatory and inhibitory considerations as well as the inherent disturbances (noise) prevalent in the system. The processing of neural information in general, depicts all the relevancies to yield a converged output value close to the objective (or target) function as a result of the supervised learning process improvised. It includes the controlling (or error) information which is looped into the system (through backpropagation as indicated in Fig. 4.1b) in order to refine the output towards the goal or target specified.

Fig. 4.2 is a spatial representation of the disorganization in the neural network. That is, the extent of randomness due to noise prevailing in the network causing a state of disorganization, which offsets the network's performance by yielding an output veering away from the target value, is specified by a measure of deviation of a selected variable, say, output y_k, with reference to a target standard $\mathbf{y}_\ell$.

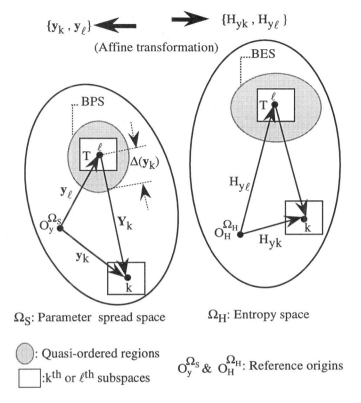

Fig. 4.2: Parameter spread space (Ω_S) and entropy space (Ω_H) representation of the neural network

[T: Locale of the objective or target function presumed to be in the ℓth subspace. BPS: Boundary of the quasi-ordered region of parameter space. BES: Boundary of the quasi-ordered region of entropy space]

The entity, $\mathbf{y}_\ell$ is denoted by a vector corresponding to the center of a region of orderliness (indicated as the target subspace **T**) wherein a stipulated (stochastic) extent of orderliness prevails and is dictated within certain bounds. Here, the orderliness refers to the extent of nonrandomness being present in the subregion concerned. The disorderliness at the kth realization in the spread space of the variable parameter **y** can be written as [4.5]:

$$\mathbf{Y}_k = (\mathbf{y}_k - \mathbf{y}_\ell) - \Delta\mathbf{y}_k \qquad (4.5)$$

where $(\mathbf{y}_k - \mathbf{y}_\ell)$ refers to the error vector and $\Delta\mathbf{y}_k$ is the vector from the center of the target subspace **T** to the boundary of a quasi-ordered region in the vicinity of the target subspace (T) as shown in Fig. 4.2.

Within the quasi-ordered region, it is presumed that the randomness becomes vanishingly small on approaching the center of the target subspace (**T**). Further, a positional entropy, namely, the uncertainty associated with the variable y_k by virtue of being at (or realized at) the kth position in the parameter spread space Ω_S, can be specified by H_{yk} at the corresponding k^{th} subspace in an entropy space Ω_H (as indicated in Fig. 4.2). Likewise, corresponding to the ℓth subspace of the target standard y_ℓ in Ω_S, H_{y_ℓ} represents the goal-associated (or target-specified) positional entropy in Ω_H. Now, one can seek information in the sample space Ω_S of y for discrimination in favor of $\mathbf{y}_\ell$ against $\mathbf{y}_k$, or symmetrically for discrimination in favor of $\mathbf{y}_k$ against $\mathbf{y}_\ell$. This discrimination information can be deduced from H_{yk} and H_{y_ℓ}, namely, the entropy functions specified in the (entropy) space, Ω_H. It offers implicitly a cost-function which can be minimized towards reducing the offset that prevails between k^{th} and ℓ^{th} realizations in Ω_S.

The entropy space Ω_H is affinely similar to the parameter spread space Ω_S. This is in the sense that each value of H_y in the entropy space (Ω_H) could be mapped onto the parameter spread space (Ω_S) on a one-to-one basis by a transformation which permits both rotation (a linear transformation) as well as translation. In Fig. 4.2, the probability of encountering the kth cell, wherein some randomness or disorganization is observed, can be denoted by p_k. Similarly, a probability p_ℓ can be assigned to the ℓth subspace to characterize the randomness (or the state of disorganization) existing at this space. By virtue of these probabilistic attributes, the kth and ℓ^{th} subspaces of Ω_S represent corresponding entropies H_{y_k} and H_{y_ℓ} respectively in Ω_H. These entropies depict the uncertainty measures which can be described analytically in the framework of information theory using the relevant concepts due to Shannon [4.4].

The geometrical representation of $\{y_k\}$ in the domain Ω_S (Fig. 4.2) can be regarded equivalently to depict the output vector set $\{\mathbf{O}_i\}$ of the neural network (Fig 4.1b). Hence, it follows that $p_i \equiv p_k$ refers to the probability associated with $\mathbf{O}_i$. Likewise, $\{\mathbf{y}_\ell\}$ in Fig. 4.2 corresponds to the vector set of the teacher values $\{\mathbf{T}_i\}$ in the network of Fig. 4.1b. Therefore, the probability attributes of $\{\mathbf{y}_\ell\}$ namely, $\{q_\ell\}$ correspond identically to the statistics of the teacher values in the network topology (Fig. 4.1b). That is, $q_i \equiv q_\ell$ is the probability associated with $\mathbf{T}_i$. Thus, the parametric set namely, $\{\mathbf{O}_i, \mathbf{T}_i\}$ of a neural network with the corresponding probability values $\{p_i, q_i\}$ can be described in the entropy-dictated information-theoretic plane, such as Ω_H in Fig. 4.2. That is, $\{O_i, T_i\}$ is identical to the set $\{y_k, y_\ell\}$ represented by entropy functions H_{y_k} and H_{y_ℓ} (respectively) in the Ω_H plane.

Consistent with the above description of neural network parameters and the associated entropies, the development of an optimization algorithm towards the minimization of the error-measure, estimated in respect of the objective or target value, can be sought in the information-theoretic (IT) plane. Adjunct tasks also include determining the convergence rate and the performance aspects of the network during the training cycle, as well as ascertaining the stability of the network, in terms of the error-measures deduced *via* entropy or information-theoretic considerations.

In the relevant pursuits, a relative entropy (RE) measure *in lieu* of the conventional square-error (SE) metric has been indicated in the literature to decide the training performance of a neural network by Solla et al. [4.6]. Further, Watrous [4.7] has presented a comparison between SE and RE error measures using several optimization algorithms pertinent to a specific classification problem presented to a multilayer perceptron. The studies addressed by Solla et al. [4.6] and Watrous [4.7], refer to a RE measure which is deduced directly from the parametric values of O_i and T_i in terms of the scaled (or normalized) values $\tilde{O}_i \in [0,1]$ and $\tilde{T}_i \in [0,1]$ (where tilde denotes normalization). This RE error-measure is then presented as an entropy-dictated interpretation of the output and target activities for the entire training set. It represents an algorithm in the spread-space of the scaled parameters. It is given explicitly by [4.7, 4.8]:

$$\varepsilon_{RE} = (1/2)(1 + \tilde{T}_i)\ell og\left[(1 + \tilde{T}_i)/(1 + \tilde{O}_i)\right]$$

$$+ (1/2)(1 - \tilde{T}_i)\ell og\left[(1 - \tilde{T}_i)/(1 + \tilde{O}_i)\right]$$

(4.6)

Studies performed by Solla et al. [4.6] and Watrous [4.7] essentially suggest the use of relative entropy as an error-measure. This error-measure describes the relative statistical attributes of the network output and target activities and indicates certain merits of RE error-function over the traditional quadratic (SE) function when adopted in the gradient-descent based backpropagation algorithm. Improvements towards convergence specified in terms of iterations required to reach an acceptable performance (that is, the output being close to a target value) in a neural network have been observed with the RE measure. Such improvements have been regarded as the realization of an "accelerated learning" [4.6] resulting from the use of RE measure in lieu of the SE measure. Specifically, the simulation experiments due to Watrous [4.7] indicate that the RE measure converges to a "good solution slightly more often than the SE metric", given the same distribution of initial weights for training. However, Watrous [4.7] has also observed that the network performance is dictated by the choice of the step-size in the training iteration; and, the optimal step-size is both "problem- and metric-specific".

While implementing gradient descent techniques, "smooth solutions" refer to guaranteed stability under a given set of constraints. In Eqn.(4.6), the first and second terms of the sum, each with a coefficient 1/2, try to balance the entropy so as to force the solution to converge rapidly towards the attractor basin. An alternative cost-function akin to Eqn. (4.6) but based on mean-field approach would be that which guarantees a constraint preservation and permits a convergence towards a smooth solution.

Apart from Solla et al. [4.6] and Watrous [4.7], in reference to a layered network architecture (Fig. 4.1b), Ackley et al. [4.9] have proposed a *distance function* to depict the deviation of neural statistics in the presence of environmental inputs from that in the absence of such inputs using the concepts of *reverse cross-entropy* (RCE). Minimization of RCE has been referred to as a network's performance in delivering an output close to the target function. Alternatively, *cross-entropy* (CE) concepts using Bayesian considerations have also been proposed [4.10] as feasible error-measures for neural network training strategies. Both RCE and CE based error-measure considerations describe implicitly an entropy-based, information-theoretic approach of training schemes, in reference to a layered framework of neural units.

The present study addresses the feasibility of developing a new class of explicit measures on the basis of cross-entropy principles and applies them to train a multilayer perceptron comprehensively. Designating these error-functions as information-theoretic error-measures, relevant theoretical considerations are addressed in the following section.

4.3 Information-Theoretic Error-Measures

Pertinent to the multilayer perceptron architecture of an artificial neural network illustrated in Fig. 4.1b, minimization of the cost-function towards the optimization in network (learning) performance refers to minimizing an energy functional (*Hamiltonian*). Such minimization of energy function can also be viewed alternatively as minimizing the state of disorganization promoted as a result of inevitable random noise being present in the network, or introduced by the inputs themselves manifesting as a deviatory response of the network output from the target.

For the purpose of comparison and discussion, the various error-measures which can be backpropagated towards optimization of gradient-descent learning in neural networks are categorized and detailed below. A specific class of information-theoretic error-measures are deduced *via* the so-called *Csiszár's family* [4.11-4.13] of cross-entropy functionals.

4.3.1 *Square-error (SE) measure in the parametric space*

This is the most conventional specification of the neural network error-metric. It represents the mean-square value of the difference between the network output and the teacher (target) value to be minimized so as to achieve optimal network training. Inasmuch as this error-measure is obtained directly from the parametric values of the output and the target, it is referred to here as a cost-function enabling the goal-seeking endeavor (that is, enabling the output to optimally converge towards the target value) in the parametric spread space. If the output and target parameters are specified by the O_i and T_i respectively, the square error-measure is given by

$$\varepsilon_{SE} = 1/2\Sigma(T_i - O_i)^2 \tag{4.7}$$

4.3.2 *Relative entropy (RE) error measure*

Alternative to the quadratic error-measure indicated above in training the neural network via a backpropagation learning algorithm, another convenient error-measure deduced directly from the parameters O_i and T_i is the relative entropy (RE) measure (Eqn. (4.6)) described in the previous section. The RE error-measure given by Eqn. (4.6), can also rewritten modified as

$$\begin{aligned}
\varepsilon_{RE} &= (1/2)(1 + \tilde{T}_i)\ell og\big[(1 + \tilde{T}_i)/(P_i + \tilde{O}_i)\big] \\
&+ (1/2)(1 - \tilde{T}_i)\ell og\big[(1 - \tilde{T}_i)/(P_i + \tilde{O}_i)\big]
\end{aligned} \tag{4.8a}$$

where $\tilde{T}_i$ and $\tilde{O}_i$ are the scaled variables of O_i and T_i, respectively; and,

$$\begin{aligned}
P_i^2 &= g_Q^2 + h_Q^2[\coth^2(h_Q X_i) + \operatorname{cosech}^2(h_Q X_i)] \\
&- 2g_Q h_Q \coth(h_Q X_i)\coth(g_Q X_i)
\end{aligned} \tag{4.8b}$$

when the Bernoulli (nonlinear) activation function, namely, $L_Q(X_i) = g_Q\coth(g_Q X_i) - h_Q\coth(h_Q X_i)$, where $g_Q = (1+1/2Q)$ and $h_Q = (1/2Q)$, is used [4.14]. Otherwise, $P_i = 1$, as in Eqn. (4.6) when the conventional, hyperbolic tangent, namely, $\tanh(\Gamma X_i)$ with Γ being the gain factor, is presumed to describe the nonlinear activity associated with the neural units.

4.3.3 *Kullback-Leibler family of error-measures*

A third class of error-measures which can be defined over a training set for the purpose of minimizing the output error in a layered neural network refers to cross-entropy measures. They are conceived to establish a distance measure of dissimilarity between the probability distribution of the output and the target [4.15, 4.16]. Such measures of dissimilarity between the mapping of the probability density function (pdf) of the input patterns and that of the target activities can be specified in the domain of information-theoretics. In contrast, relevant to the measures of parametric spread space, the dissimilarity mapping is elucidated directly from of the output and target parameters themselves. Hence, the domain of optimization of the error so inferred is designated as the parametric spread-space. In the present case, however, the domain of error representation is referred to as the information-theoretic plane. The reason for this designation is as follows.

In the context of certain real world problems using neural networks to which backpropagation can be applied wherein noisy set of examples are encountered, the output vector is amenable for interpretation in probabilistic terms. In these instances, a relative information (mutual information) measure of divergence between the pdfs of the output values and target parameters can be elucidated [4.15]. Relevant considerations can be elaborated in reference to Fig. 4.2 as follows:

Suppose a basic set of elements y_k's which deviate from the target value y_ℓ (Fig. 4.2) have probability measures p_k's; and the probabilities associated with y_ℓ's are q_ℓ's. These probabilities can be specified by generalized probability density functions (pdf's) $f(y_k)$ and $g(y_\ell)$ respectively such that $-\infty < f(y_k) < \infty$ and $-\infty < g(y_\ell) < \infty$. In terms of these pdfs, $p_k = \sum_{r=-\infty}^{k} f(y_r)$ and $q_\ell = \sum_{s=-\infty}^{\ell} g(y_s)$. The average information for discrimination in favor of H_{y_k} against H_{y_ℓ}, where H_{y_k} and H_{y_ℓ} refer to the hypotheses governing the random variables y_k and y_ℓ, respectively, in Fig. 4.2, can be written as

$$D(k : \ell, y) = \sum_k p_k \, log(p_k/q_\ell) \qquad (4.9)$$

As indicated earlier, in reference to the network depicted in Fig. 4.1b, $p_k \equiv p_i(O_i)$ and $q_\ell \equiv q_i(T_i)$, where O_i and T_i represent the output and teacher values respectively. Hence, consistent with the definition of Eqn. (4.9), the relative distance measure between O_i and T_i can be specified by

$$D_{KL}(p : q) = \varepsilon_{KL} = \sum_i p_i \, log(p_i/q_i) \qquad (4.10a)$$

or,

$$D_{KL}(q : p) = \varepsilon_{KL} = \sum_i q_i \, log(q_i/p_i) \qquad (4.10b)$$

Here, the subscript KL depicts the conventional designation of D_{KL} being the Kullback-Leibler measure [4.17-4.19] which constitutes an error ε_{KL} measuring the distance between the patterns of O_i and T_i. Also, considering the average information for discrimination in favor of H_{y_k} against H_{y_ℓ}, namely,

$D(\ell : k, y)$, a symmetrical measure of divergence can be written as follows [4.15, 4.20, 4.21]

$$D_J(k : \ell; \ell : k, y) = D(k : \ell, y) + D(\ell : k, y) = \sum_k (p_k - p_\ell) \ell og(p_k/p_\ell)$$
(4.11)

As defined earlier, the disorganization in reference to a system refers implicitly to the extent of randomness associated with a subspace in the system relative to an organized reference or target subspace. The entropy functionals of Eqns. (4.10) and (4.11) are dictated by the relative probabilistic attributes, or randomness, of the kth realization y_k against the ℓth realization y_ℓ or vice versa, summed over all values of $k = 1, 2, 3, ..., n$. They can be regarded as the global disorganization measures in the perspectives of entities presented in Fig. 4.2.

D_J of Eqn. (4.11) (known as *the Jensen/Jeffreys measure* or *J-divergence*[*] [4.21]), can be defined in reference to O_i and T_i, of Fig. 4.1b as an error measure ε_J as decided by the set $\{p_i, q_i\}$ given by

$$D_J(p_i : q_i) = \varepsilon_J = \sum p_i \ell og(p_i/q_i) + \sum q_i \ell og(q_i/p_i)$$
(4.12)

Additionally, each of the random realizations pertinent to a network can be weighted in respect to its corresponding probability distribution in order to specify its individual strength in the goal-seeking endeavor of the network. Suppose p_k and p_ℓ ($p_k, p_\ell) \geq 0$ and $p_k + p_\ell = 1$) are the weights of the two probabilities p_k and p_ℓ respectively. Then, a generalized divergence measure (known as the *Jensen-Shannon measure*) can be stipulated in terms of the

---

[*]*Divergence* concept: In general, $L(x) = p(x)/q(x)$ is defined as a *likelihood function* in reference to the probability $p(x)dx$ of observation x, when a certain hypothesis h_P is true. And, $q(x)dx$ refers to the probability of observation x, when the hypothesis h_Q is true. Jeffreys [4.20] indicates the *divergence* (J) as the difference in the mean values of the *log-likelihood ratio* under the two hypotheses h_P and h_Q . J is given explicitly by:

$$J = E_p[\ell og\{L(x)\}] - E_q[\ell og\{L(x)\}]$$

where $E_{p,q}[.]$ depicts expectations. That is,

$$E_p[\ell og\{L(x)\}] = \int [\ell og\{L(x)\}]p(x)dx \text{ and } E_q[\ell og\{L(x)\}] = \int [\ell og\{L(x)\}]q(x)dx$$

entropies (H) associated with the random or probabilistic attributes of p_k and p_ℓ as follows, [4.21]:

$$D_{JS}(p_k : p_\ell) = H(\pi_k p_\ell + \pi_j p_\ell) - \pi_k H(p_\ell) - \pi_j H(p_\ell)$$

(4.13a)

Here, the entropy function H(z) refers to Shannon's information measure specified by the (negative) entropy $-k\sum_i [p(z_i)]\ell og[p(z_i)]$ corresponding to a set of random variables $\{z_i\}$ with probabilities $\{p(z_i)\}$ [4.4] and k is a constant as decided by the base of the logarithm.

The measure given by Eqn. (4.13a) is nonnegative and equal to zero only when $p_k = p_\ell$. It also provides the upper and lower bounds for the Bayes' probability of error. The Jensen-Shannon (JS) divergence is ideal to describe the variations between the subspaces or the goal-seeking realizations as in the entropy space of the neural network. It measures the distances between the random-graph depictions of such realizations pertinent to the entropy plane Ω_H in Fig. 4.2. Written in terms of p_i and q_i pertinent to O_i and T_i values of Fig. 4.1b, the error measure ε_{JS} corresponding to D_{JS} is given by:

$$D_{JS}(p : q) = \varepsilon_{JS} = \pi_1 \sum p_i \ell og(p_i/q_i) + \pi_2 \sum q_i \ell og(q_i/p_i)$$

(4.13b)

with $\pi_1, \pi_2 < 1$, and $(\pi_1 + \pi_2) = 1$.

A special case of Kullback-Leibler measure of directed divergence with a weighting parameter $0 \leq \mu < 1$ can be written in a weighted symmetrized form as [4.21]:

$$D_{WKL}(p : q) = \varepsilon_{WKL} = \sum p_i \ell og\{[p_i /[\mu p_i + (1-\mu)q_i]\}$$
$$+ \sum q_i \ell og\{[q_i /[\mu q_i + (1-\mu)p_i]\}$$

(4.14)

where $0 \leq \mu < 1$. This error-measure can be designated *as weighted and symmetrized Kullback-Leibler measure* (ε_{WKL}) constituted by the weighted fractions of the probabilities.

4.3.4 *Generalized directed divergence of degree 1*
Suppose outcomes of a certain experiment are specified by a set of complete probability distribution (P = $p_1, p_2, ..., p_n$) with $(1 \geq p_i \geq 0)$ and $\sum_i p_i = 1$. Let the corresponding probability distributions estimated by two independent observers be: (Q = $q_1, q_2, ..., q_n$) and (R = $r_1, r_2, ..., r_n$) with $(1 \geq q_i \geq 0)$ and $(1 \geq r_i \geq 0)$. Further, $\sum_i q_i = 1$ and $\sum_i r_i = 1$.

The generalized directed divergence of degree 1 is, then, given by [4.16]:

162

$$I_n(P|Q|R) = \sum_{i=1}^{n} p_i \ell og(q_i/r_i)$$ (4.15)

The characteristics of the aforesaid divergence or distance measures, in general are as follows:

1. $D(p_k: p_\ell)$ continuous of p_k's and p_ℓ's

2. When $D(p_k: p_\ell) = D(p_\ell:p_k)$, the divergence refers to the symmetric property of the error-measure

3. $D(p_k: p_\ell) \geq 0$. This is, the nonnegativity property of the error-measure and the equal sign applies if and only if $p_k = p_\ell$, in which case, the corresponding property is known as the *identity property*

4. Triangle inequality property: $D(p_k: p_\ell) + D(p_\ell:p_m) \geq D(p_k:p_m)$

5. $D(p_k: p_\ell)$ is a convex function of $(p_1, p_2, ..., p_n)$

6. When $D(p_k:p_\ell)$ is minimized with respect to p_k or p_ℓ subject to known linear (moment) constraints, none of the resulting minimized probabilities should be negative.

In reference to the aforesaid properties, the relative-information (or cross-entropy based error) should satisfy the conditions (1) and (3) absolutely in order to be classified as a divergence measure. Further, conditions (2) and (4) are not essential while (5) and (6) are highly desirable for mathematical convenience towards a minimization procedure such as *Lagrange's method* [4.16]. Within the framework of these stipulations, the Kullback-Leibler measures can be used in a neural network as error-measures for gradient-descent learning *via* backpropagation.

4.3.4 *Csiszár's family of minimum directed divergence measures*
The concept of minimum directed divergence measure (as proposed by Kullback-Leibler and described in the Subsection 4.33) can be further generalized to represent a family of measures given by [4.11- 4.13, 4.16]:

$$D(p:q) = \sum_{i=1}^{n} q_i \Phi(p_i/q_i)$$ (4.15a)

or,

$$D(q:p) = \sum_{i=1}^{n} p_i \Phi(q_i/p_i) \qquad (4.15b)$$

where Φ is a twice differentiable convex function for which $\Phi(1) = 0$ and D satisfies the following essential and desirable conditions indicated earlier:

- D(p:q) or D(q:p) is a convex function of both $\{p_i\}$ and $\{q_i\}$
- D(p:q) = 0 or D(q:p) = 0, if and only (*iff*), p = q
- D(p:q) $\geq$ 0; D(q:p) $\geq$ 0
- When D(p:q) or D(q:p) is minimized (using Lagrange's method) subject to given a set of linear (moment) constraints, the minimizing probabilities should be greater than or equal to zero. (Optional condition.)

Eqn. (4.15) is known as the *f-divergence*[1] (f representing the function Φ) introduced and characterized by Csiszár [4.11, 4.12]. These measures can be regarded as a generalized set of information measures, which when applied to a specific class of problems may indicate certain interesting properties akin to Shannon-type information measures [4.4]. As Aczél [4.22] points out specific to those problems, "if the properties are indeed intuitive and significant, then there is a good chance the measures thus obtained may have future application".

It is attempted in this present work to consider the Csiszár's family of error-measures as possible candidates for neural network optimization. In the existing literature, except for the relative entropy (RE) error-measure [4.6, 4.7, 4.8] deduced in terms of parameters O_i and T_i themselves and not in terms of their probabilities as given by Eqns. (4.6) and (4.8), the other measures which could be derived *via* Csiszár's formulation have not been applied *per se* as error-measures (cost-functions) in neural network optimization problems.

The f-divergence measures are numerical measures of relative "informativity" of two stochastical experiments specified in terms of the associated probability distributions. Conventionally, in the perspectives of Shannon's definition of information indicated earlier, the cross-entropy based relative information content (I) between two random experiments X and Y with pdf's $p_X(X)$ and $p_Y(Y)$ is determined by the relative entropy between the

---

[1] *Csiszár's f-divergence concept* is indicated in [4.12] as a measure of *informativity*. In general, f-divergence of two probability distribution P and Q on Y is defined as:

$$I_f(p, q) \underset{=}{\Delta} \int q(y) f\{p(y)/q(y)\} \lambda dy$$

where f(u), $u \in (0, \infty)$ is an arbitrary convex function which is strictly convex at u = 1; λ is some (finite or σ-finite) dominating measure; and p(y), q(y) are densities of P and Q, respectively, with respect to λ.

joint distribution $p_{XY}(X,Y)$ and the product distribution $p_X(X) \times p_Y(Y)$. That is, $I(X;Y) = k\sum_X \sum_Y p_{XY}(X,Y)\log[p_{XY}(X,Y)/p_X(X)p_Y(Y)]$ where k again decides the choice of the base of the logarithm. In Csiszár's definition of f-divergence, the logarithmic function in defining the information content, I is replaced by Φ as indicated in Eqn. (4.15). Correspondingly, in order to distinguish the entity defined by Eqn. (4.15) from the definition of information content in Shannon's sense which uses the logarithmic function, the term "informativity" is used here as well as by Csiszár [4.12]. This term signifies its relevance as a measure analogous to Shannon's information measure, but distinguishes itself from the Shannon's measure by virtue of using the function Φ.

The objective of present discussions is to elucidate the possibilities of applying Csiszár's family of relative information (or "informativity") formulations as error-measures in neural network learning schemes mediated through backpropagation and gradient-descent methods. The relative "informativity" specified by Csiszár's formulations refers, in essence, to comparing the of negative entropies (elucidated in terms of Φ) associated with the network output and the target value(s). In the present study, to use the existing Csiszár measures for neural network applications, it is also assumed a priori that only a limited subset of such measures could possibly be applicable for relevant optimization efforts inasmuch as "measures depend not just upon the probabilities but also upon the object matters; (hence, there could exist some) promising and not so promising generalizations" of f-divergence measures [4.22].

In the following paragraphs, the divergence measures of Csiszár's family as reported in the literature are listed. Their characteristics or properties are enumerated with the scope of using such measures as information-theoretic error-measures in neural network training strategies. These measures, in general, are representations of Eqn. (4.15) with the convex function Φ being expressed explicitly.

The conditions on D(p:q) or D(q:p) relevant to Eqn. (4.15) can be elaborated more as follows: Suppose p_i refers to the set of probabilities in the entropy space of the observed network output, namely $\{p_1, p_2, p_3, ..., p_n\}$, and q_i represents the set of *a priori* probabilities, $\{q_1, q_2, q_3, ..., q_n\}$ relevant to target values with $\sum_i^n p_i = \sum_i^n q_i = 1$. Further, a linear moment constraint attribute can be prescribed to p_i or q_i, in general, of the form:

$$\sum_{i=1}^n p_i g_r(O_i) = a_r; \quad \sum_{i=1}^n q_i h_r(T_i) = b_r \qquad (4.16)$$

where $r = 0, 1, 2, ..., m;$ $g_0(O_i) \equiv 1;$, $a_0 \equiv 1;$ $h_r(T_i) \equiv 1;$, $b_0 \equiv 1$.

In Eqn. (4.16), $g_r(.)$ and $h_r(.)$ are called the *moment functions*, and $\sum_{i=1}^{n} p_i g_r(O_i) = a_r$ and $\sum_{i=1}^{n} q_i h_r(T_i) = b_r$ are moments with a_r and b_r ($r = 0, 1, 2, ..., m$) representing the values of the respective moments. The pdfs of p_i and q_i satisfy, in general, a set of independent moment constraints as decided by the nature of the respective random processes associated with them.

The relative information (or cross-entropy) error-measure $D(p:q)$ or $D(q:p)$ with the aforesaid characteristics when applied to neural network training, is a minimization postulate which guarantees the most unbiased delivery of the network. That is, the output pattern being different from the teacher pattern to a minimum extent on a posteriori basis. Such a relative information measure can be specified uniquely by an explicit family of functions representing Φ which concurrently assure the nonnegativity of the probabilities, namely, p_i's= 1, 2, 3, ..., n $\geq$ 0. Such specific functions are typically as follows [4.16]:

Case (1)
With $\Phi(x) = x\log(x)$, Eqn. (4.15) leads to

$$D_{KL}(p:q) = \varepsilon_{KL} = \sum_i (q_i)(p_i/q_i)\log(p_i/q_i)$$
$$= \sum_i (p_i)\log(p_i/q_i) \tag{4.17a}$$

or;

$$D_{KL}(q:p) = \varepsilon_{KL} = \sum_i (p_i)(q_i/p_i)\log(q_i/p_i)$$
$$= \sum_i (q_i)\log(q_i/p_i) \tag{4.17b}$$

The above measure is same as the Kullback-Leibler measure indicated by Eqn. (4.10) previously.

Case(2)
Suppose $\Phi(x) = x\log(x) - (1/a)(1 + ax)\log(1 + ax) + (1/a)(1 + a)\log(1 + a)$, $a \geq 1$. Then,

$$D_{KP(1)}(p:q) = \varepsilon_{KP(1)} = \sum_i p_i\log(p_i/q_i)$$
$$- (1/a)\sum_i (q_i + ap_i)\log[(q_i + ap_i)/q_i]$$
$$+ (1/a)(1 + a)\log(1 + a)$$

$$\tag{4.18}$$

The above result is due to Kapur [4.23] and it can be seen that, as a → 0, $\varepsilon_{KP(1)} \to \varepsilon_{KL}$ Further, $\varepsilon_{KP(1)}$ is designated here as *Kapur's Type 1 error-measure*.

Case (3)

Havrda and Charvát [4.24], while elucidating the quantificatory theory of classificatory processes, proposed the concept of structural *α-entropy* leading to the function $\Phi(x) = (x^\alpha - x)/(\alpha - 1)$ with $\alpha > 0$ and $\alpha \neq 1$. That is, the concept of structural α-entropy refers to introducing an order parameter α (α > 0 and α ≠ 1) deliberately in the function defining the negative entropy or information content such that, as α → 1. This function reduces to Shannon's definition of negative entropy or information content.

The relative information of order α due to Havrda and Charvát [4.24] is given by

$$D_{HC}(p:q) = \varepsilon_{HC} = \left[1/(\alpha - 1)\right] \sum_i q_i \left[(p_i/q_i)^\alpha - (p_i/q_i)\right]$$
$$= \left[1/(\alpha - 1)\right] \sum_i \left[p_i^\alpha q_i^{(1-\alpha)} - p_i\right]$$

(4.19)

By applying *L'Hospital's rule* to Eqn. (4.19), it can be shown that, when α → 1, $D_{HC}(p:q) \to \sum_i p_i \ell og(p_i/q_i) = D_{KL}(p_i : q_i)$. That is, when α → 1,

$\varepsilon_{HC} \to \varepsilon_{KL}$

Case (4)

Specific to the entropy measure of information (Shannon's information) [4.4] pertinent to a discrete set of probabilities, namely $P_S = \{p_i\}$ and $Q_S = \{q_j\}$, an additivity property can be stated as follows:

$$H(P_S{}^*Q_S) = H(P_S) + H(Q_S)$$ (4.20)

with $\sum_{i=1}^{n} p_i \leq 1$ and $\sum_{j=1}^{m} q_j \leq 1$. Further, $P_S{}^*Q_S \equiv \{(p_1q_j), ..., (p_nq_j)\}_{j=1,2,...,m}$, and H(z) as stated earlier denotes the Shannon's entropy, namely, $H(z) = -k \sum_i z_i \ell og(z_i)$. Also, there is a mean value property which is valid subject to the condition $W(P_S) + W(Q_S) \leq 1$ (where $W(P_S) = \sum_i^n p_i$ and $W(Q_S) = \sum_j^m q_j$) given by

$$H(P_S \cup Q_S) = \mathfrak{S}^{-1}\left[\frac{W(P_S)\mathfrak{S}\,[H(P_S)] + W(Q_S)\mathfrak{S}\,[H(Q_S)]}{W(P_S) + W(Q_S)}\right]$$

$$(4.21)$$

where $\mathfrak{S}$ is a strictly monotonic and continuous function [4.25].

For stochastical outcomes which belong to non-additive sum generalizations (in contradiction to the above properties), Sharma and Mittal [4.26] have developed an expression for $\Phi(x)$ given by: $\Phi(x) = (x^{\alpha} - x^{\beta})/(\alpha - \beta)$, $\alpha > 1$, $\beta \leq 1$; or $\alpha < 1$, $\beta \geq 1$; and, α and β are positive numbers. The corresponding $D_{SM}(p{:}q)$ is obtained as

$$D_{SM}(p:q) = \varepsilon_{SM} = \left[1/(\alpha - \beta)\right]\left[\sum p_i{}^{\alpha}q_i{}^{(1-\alpha)} - \sum p_i{}^{\beta}q_i{}^{(1-\beta)}\right]$$

$$(4.22)$$

As special cases, by applying L'Hospital's rule, it can be seen that if $\beta = 1$, $\varepsilon_{SM} \to \varepsilon_{HC}$; and, if $\alpha \to 1$ and $\beta = 1$, $\varepsilon_{SM} \to \varepsilon_{KL}$.

These nonadditive measures due to Sharma and Mittal address the mean value of nonadditive self-information contents associated with the happenings of various single events.

Case (5)

In an attempt to evolve a mutual (relative) information measure, Rényi [4.25] developed "an information gain of order α" which characterizes the informativities of one set of distributions $\{p_i\}$ against the other $\{q_i\}$ with an entropy order α. The *Rényi's measure of directed divergence*, $D_{RY}(p{:}q)$, is given by

$$D_{RY}(p:q) = \varepsilon_{RY} = \left[1/(\alpha - 1)\right]\left[\log \sum p_i{}^{\alpha}q_i{}^{(1-\alpha)}\right]$$

$$(4.23a)$$

where $\alpha \neq 1$ and $\alpha > 0$.

This Rényi measure has the additive entropy and converges to the Kullback-Leibler measure ε_{KL} as $\alpha \to 1$. Physically, the Rényi measure specifies the information of order α obtained if the distribution set $\{p_i\}$ is replaced by the distribution set $\{q_i\}$; and, $D_{RY}(p{:}q) \to D_{RY}(q{:}p)$ as $\alpha \to 1$.

Generalizations of Rényi's entropy have been presented by Varma [4.27]. They are as follows

$$H^n{}_{A\alpha} = [1/n{-}\alpha]\,\log\{\sum p_i{}^{\alpha{-}n{+}1})/(\sum p_i)\}$$

$$(4.23b)$$

$$H^n{}_{B\alpha} = [n/n{-}\alpha]\,\log\{\sum p_i{}^{\alpha/n})/(\sum p_i)\}$$

$$(4.23c)$$

Case (6)

Kapur [4.28] extended the Rényi's measure to a weighted sum of two such measures and obtained a new directed divergence $D_{KP(2)}$ designated here as *Kapur's type 2 error-measure* $\varepsilon_{KP(2)}$ given by

$$D_{KP(2)} = \varepsilon_{KP(2)} = \left[\beta/(\alpha + \beta)\right]\left[1/(\alpha - 1)\right]\sum \ell og p_i^{\alpha} q_i^{(1-\alpha)}$$
$$+ \left[\alpha/(\alpha + \beta)\right]\left[1/(\beta - 1)\right]\sum \ell og p_i^{\beta} q_i^{(1-\beta)}$$

(4.24)

where, $\alpha < 1$, $\beta < 1$; and $(\alpha + \beta) = 1$. Further, $\alpha \neq 1$ and $\beta \neq 1$.

Apart from the six cases of Csiszár's measures as discussed above, another set of generalized Csiszár's measures can also be obtained with the modification of Eqn. (4.15) to represent the following two general forms [4.16]:

$$D(p:q)\big|_{\text{modified}(1)} = \sum (q_i + c)\Phi[(p_i + c)/(q_i + c)] \qquad c > 0$$

(4.25a)

and

$$D(p:q)\big|_{\text{modified}(2)} = \sum (1 + aq_i)\Phi[(1 + ap_i)/(1 + aq_i)], \qquad a > 0$$

(4.25b)

Relevant to the above expressions, the following cases of generalized *Csiszár's measures* are obtained similar to the Kullback-Leibler format of the function $\Phi(x) = \left[x\ell og(x)\right]$

Case (7)

$$D_{GCZ(1)}(p,q) = \varepsilon_{GCZ(1)} = \sum (p_i + c)\ell og\left(\frac{p_i + c}{q_i + c}\right), \qquad c > 0$$

(4.26)

Case (8)

$$D_{GCZ(2)}(p,q) = \varepsilon_{GCZ(2)} = \sum (1 + ap_i)\ell og\left(\frac{1 + ap_i}{1 + aq_i}\right), \qquad a > 0$$

(4.27)

Case (9)

$$D_{GCZ(3)}(p,q) = \varepsilon_{GCZ(3)} = \sum p_i \ell og\left(\frac{p_i}{q_i}\right)$$
$$-(1/a)\sum (1 + ap_i)\ell og\left(\frac{1 + ap_i}{1 + aq_i}\right), a > -1$$

(4.28)

4.3.5 *Other possible cases of Csiszár's function Φ(x)*

Considering a functional relation such as $(\Phi')^{-1}(x) = \exp[v(x)]$, it is possible to arrive at a set of explicit functions for $\Phi(x)$ such that $\Phi(x)$ is convex and $v(x)$ is a positive monotonic increasing function of x. The resulting directed divergence presents nonnegativities of minimizing probabilities. The following set of functions and the associated D(p:q) are relevant to the these considerations [4.16]:

Case (10)
If $v(x) = ax + b$, $a > 0$,
$$\Phi(x) = (1/a)\ell og(ax + b) - x - (1/a)(a + b)\ell og(a + b) + 1$$

(4.29a)

Hence,

$$D_{GCZ(4)}(p,q) = \varepsilon_{GCZ(4)} = (1/a)\sum (ap_i + bq_i)\ell og\left[\frac{ap_i + bq_i}{q_i}\right]$$

$$- (1/a)(a + b)\ell og(a + b)$$

(4.29b)

In the above measure, if $a = 1$ and $b = 0$, $\varepsilon_{GCZ(4)} \to \varepsilon_{KL}$; and, if $b = 1$, $\varepsilon_{GCZ(4)} \to \varepsilon_{KL} - \varepsilon_{KP(1)}$.

Case (11)
If $v(x) = (ax+b)/(cx+d)$ with $(a, b, c, and d) > 0$, then,

$$\Phi(x) = (1/a)(ax + b)\ell og(ax + b) - (1/a)(a + b)\ell og(a + b)$$
$$- (1/c)(cx + d)\ell og(cx + d) - (1/c)(c + d)\ell og(c + d)$$

(4.30a)

The corresponding D(p:q) is given by:

$$D_{GCZ(5)}(p,q) = \varepsilon_{GCZ(5)} = (1/a)\sum (ap_i + bq_i)\ell og[(ap_i + bq_i)/q_i]$$
$$- (1/a)(a + b)\ell og(a + b)$$
$$- (1/c)\sum (cp_i + dq_i)\ell og[(cp_i + dq_i)/q_i]$$
$$- (1/c)(c + d)\ell og(c + d)$$

(4.30b)

with $(ad - bc) < 0$. If $a = 1$, $b = 0$, $c = 0$, and $d = 1$, $\varepsilon_{GCZ(5)} \to \varepsilon_{KP(1)}$.

4.4 Neural Nonlinear Response *vs.* Optimization Algorithms

Considering the neural network illustrated in Fig. 4.1b, as indicated earlier, O_i represents the output at the ith cell and the corresponding target sought is specified as T_i. Here, $(i = 1,2,....,N)$ represents the number of cells (an index for the output units); and, p_i is the probability of O_i which complies with the following hypothesis. In terms of learning, the correct probabilities of a set of hypotheses represented by the output limits using p_i for the probability that the hypothesis represented by the ith units is true; that is, $p_i = 0$ means definitely false, and $p_i = +1$ means definitely true. Similarly, q_i refers to a target set of probabilities such that $q_i = 0$ and $q_i = +1$ indicate the false-true limits of the target value.

As indicated in Fig. 4.1a, the weighted sum (X_i) of the multi-inputs $\{x_j\}$, with w_{ij} being the weighting factor across the interconnection between the ith and the jth cells, the summed input is processed by a nonlinear activation function F_s to produce the neuron's output signal S_i. That is, each neuron evaluates its inputs and "squashes" its permissible amplitude range of output to some finite value $S_i = F_s(X_i)$ in a nonlinear fashion. Conventionally, F_S is taken as the hyperbolic tangent function to represent the sigmoidal nonlinearity. However, as indicated by Neelakanta et al. [4.14], the Bernoulli function $L_Q(z) = g_Q \coth(g_Q z) - h_Q \coth(h_Q z)$ (with $g_Q = (1 + 1/2Q)$ and $h_Q = 1/2Q$) can be regarded as a stochastically justifiable sigmoidal function in representing the nonlinear activation F_S *in lieu* of the conventional sigmoids, such as the hyperbolic tangent function.

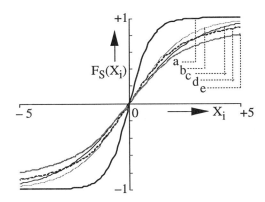

Fig. 4.3: Sigmoidal activity functions: $F_S(X_i)$ versus X_i

a: Bernoulli function $L_Q(X_i)$ with $Q = 1/2$. (Same as hyperbolic tangent function, $\tanh(\Gamma X_i)$ with $\Gamma = 1$)
b: $L_Q(X_i)$ with $Q = 5$
c: $\tanh(\Gamma X_i)$ with $\Gamma \approx 0.4$
d: $\tanh(\Gamma X_i)$ with $\Gamma = 1/3$
e: $L_Q(X_i)$ with $Q = \infty$ (Langevin function)
    (*Note*: Slope of $L_Q(X_i)$ at $X_i = 0$ is $(1/3 +1/3Q)$ and slope of $\tanh(\Gamma X_i)$ at $X_i = 0$ is Γ)

The function $L_Q(.)$ is monotonic and differentiable everywhere. It also, is squashed between the limits -1 and $+1$. When $Q = 1/2$, the Bernoulli function becomes $\tanh(.)$, as shown in Fig. 4.3, and $1/2 \leq Q < \infty$ determines the slope of the sigmoid or the steepness parameter. Further, as indicated in Chapter 2, $L_Q(.)$ can also be approximated as

$$L_Q(z) = \begin{cases} +1; & z > + (1/\alpha_0) \\ \alpha_0 z; & -(1/\alpha_0) < z + (1/\alpha_0) \\ -1; & z < - (1/\alpha_0) \end{cases}$$

(4.31)

where $\alpha_0 = (2g_Q - 1)/3 = (2h_Q + 1)/3$. It may be noted that the slope of $L_Q(z)$ is decided by α_0, that is, implicitly by the order parameter Q.

In terms of the Bernoulli function chosen as the sigmoid, $p_i(S_i)$ can be deduced as follows. Let the probability density function ($f_{X_i}(X_i)$) of the weighted sum of the inputs, namely X_i, be uniform (of constant value $\alpha_0/2$) over the interval $- 1/\alpha_0$ to $+ 1/\alpha_0$. Here, the pdf of X_i is taken as uniform on the considerations of *Laplace's principle of insufficient reason* which states that in the absence of any reason to the contrary, the outcomes may be taken to be equally likely. This assumption however, does not preclude the relevant studies to use other types of pdf's without any loss of generality. For example, the pdf of X_i can be Gaussian when it refers to the sum of weighted random variables with a bias towards a central tendency. Hence, X_i could be rendered as a Gaussian variable by the central limit theorem.

Consistent with the uniform pdf as assumed presently, $f_{X_i}(X_i)$ is equal to $\alpha_0/2$ in the interval $-1/\alpha_0 \leq X_i \leq +1/\alpha_0$ or zero otherwise. Hence, the probability density function of the output of a neuron, namely, $f_{S_i}(S_i)$ can be obtained by the following transformation:

$$f_{S_i}(S_i) = [f_{X_i}(X_i)/F_S'(X_i)]_{X_i = F_s^{-1}(S_i)}$$

(4.32a)

where the prime denotes the differentiation with respect to the argument. With the assumed uniform distribution of X_i, the above relation (Eqn. (4.32a)) reduces to the following uniform distribution:

$$f_{S_i}(S_i) = \begin{cases} 1 & 0 \leq S_i \leq 1 \\ 0 & \text{otherwise} \end{cases}$$

(4.32b)

Hence, $p_i(S_i)$ can be deduced as:

172

$$p_i(S_i) = \int_0^{S_i} f_{S_i}(v)\,dv \quad = (1 + \alpha_0 x_i)/2 \tag{4.33}$$

which guarantees that, at $x_i = -1/\alpha_0$, $p_i(S_i) = 0$ and at $x_i = +1/\alpha_0$, $p_i(S_i) = 1$.

Since the teacher T_i is a known function of the input set $\{x_i\}$, that is, $T_i = G(x_i)$, the probability density function of T_i can be obtained from the following transformation:

$$f_{T_i}(T_i) = K[g_{X_i}(x_i)/G'(x_i)]_{x_i = G^{-1}(T_i)} \tag{4.34a}$$

where $g_{X_i}(x_i)$ denotes the pdf of the input variables $\{x_i\}$ and K is a normalization constant required to set the following identity on total probability:

$$K \int_{-\infty}^{T_i} f_{T_i}(z)\,dz \equiv 1 \tag{4.34b}$$

Denoting $\left. G'(x_i) \right|_{x_i = G^{-1}(T_i)} = R_1(T_i)$ and $R_2(T_i) = \int_0^{T_i} d\tau_i / R_1(\tau_i)$, the probability of T_i is obtained as

$$q_i(T_i) = [R_2(T_i) - R_2(0)]/[R_1(1) - R_2(0)] \tag{4.35}$$

Now, since as $p_i(O_i)$ and $q_i(T_i)$ are known explicitly, various distance measures defined in the information-theoretic plane earlier can be evaluated. (*Note*: In the above expressions, the integrals used correspond to the cases where continuous random variables are used. If the random variables are discrete, the integrals should be replaced by summations.)

4.5. A Multilayer Perceptron Training with Information-Theoretic Cost-Functions

4.5.1 *Description of the network and on-line experiments*
A multilayer perceptron is implemented to evaluate the training effectiveness of backpropagation in the prediction of a function $|\sin(m\alpha_0\pi x_i)|$ where the various distance measures specified in the previous sections are used in the backpropagation algorithm. Appropriate value of m in the argument of the function allows the choice of the frequency of the waveform being predicted.

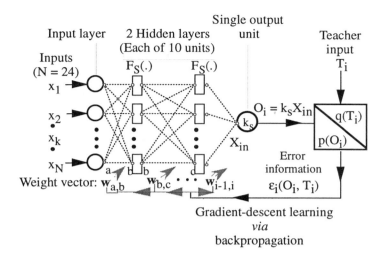

Fig. 4.4: Neural network used in the simulations
(A multilayered perceptron with 24 input units; 2 hidden layers each with 10 trainable
neurons and 1 output unit)

A practical version of the multilayered perceptron of Fig. 4.1b is illustrated in Fig. 4.4. It consists of 24 inputs, 2 hidden layers each with 10 neural units and a single output unit. The activation functions considered are: (i) At the two hidden layers, it is a Bernoulli function and (ii) at the output unit, it is a linear function, $k_S X_i$ with k_S being a scaling constant. The experimental task in predicting the said function using the perceptron of Fig. 4.4 consists of the following phases:

1. Training (learning) phase
2. Testing (prediction) phase

4.5.2 Training phase

For training, the argument of the function $|\sin(.)|$ is considered over an interval $-1/\alpha_0$ to $+1/\alpha_0$ and divided into 23 equal divisions each of length (2/23) unit. Input values for the 24 input units are set by a sequence, namely, $[(-1)\gamma/\alpha_0, (-1 + 2/23)\gamma/\alpha_0, (-1 + 4/23)\gamma/\alpha_0, ..., 0, ..., (-4/23 +1)\gamma/\alpha_0, (-2/23 +1)\gamma/\alpha_0, (+1)\gamma/\alpha_0]$, where γ is picked from a set of 75 uniformly distributed pseudorandom numbers generated in the range $(-1/\alpha_0, +1/\alpha_0)$. Further, the input and the hidden layers are facilitated each with an additional bias unit clamped to -1, which is connected through a trainable weight to each unit in the hidden and output layer respectively. These bias units provide trainable offsets to the origins of the activation function for the units in the hidden and output layers.

The resulting output O_i of the network is compared with the teacher value (T_i) specified by the target (teacher) function. The error value resulting from the comparison of O_i and T_i as obtained by any one of the error-measure formulations under discussion, is used (via backpropagation mode) to adjust the weights through the gradient-descent method.

In the case of a conventional SE metric, the sign of the effective gradient is determined by a simple arithmetic difference between the target and output, so that the direction of the gradient descent is controlled by the feedback resulting from the comparison of the target *versus* output difference. The cross-entropy measures (ε_j) involving logarithmic functions are, however, strictly nonnegative. Therefore, they would not allow the effective gradient given by $\delta_{ij} = (\delta O_i / \partial X_i)\varepsilon_j$ (defined in Section 4.1) to change its sign in response to the target *versus* output differences resulting in a loss of the feedback which controls the weight modifications. To remedy this situation, the effective gradient calculated by using the cross-entropy error-measures is multiplied by ± 1, depending on the sign of the target-output difference. That is, the value of the effective gradient is specified with an appropriate sign, by: δ_{ij} = signum ($T_i − O_i$). The above procedure is carried out for the 75 different random values of γ generated, so as to complete one training epoch. The epochs are iterated until the network's output, as decided by the divergence measure used, reaches a desired value. In the present test studies, at least 500 iterations were done for each training cycle.

The value of the network parameter Q stemming from the nonlinear activity function $L_Q(.)$ was chosen so as to control the effectiveness of training and achieve stability in the network. Alternatively, if one chooses the hyperbolic tangent function, say, $\tanh(\Gamma X_i)$, the gain factor Γ can be changed to control the effectiveness of training). Further, the value of the learning coefficient η is altered to influence the convergence rates and the ultimate accuracy in the performance of the network. To assay the consequences of varying η and Q in deciding the network's performance, each of the distance measures under discussion was implemented in training the network (of Fig. 4.4) with the values of η ranging from 0.0001 to 0.2 and for Q values of 1/2, 2 and 10.

For each set of η and Q, the same random initializations of the interconnection weights were used for all the error-measures implemented for one training cycle (run). Then, the training runs were repeated, by changing the random initializations, 10 times for each case of error-measures and using a given set of η and Q. For every training run (consisting of 500 iterations) repeated, the consistency in convergence is checked. This ensures the reliable performance of the network vis-à-vis the error measures used and the training protocols adopted. Further, whenever the error measure magnitude exceeded 10^4 during the training, the network is declared to have diverged in the functioning.

❑   *Testing or prediction phase*

After each training phase is implemented with a given error-measure using a given set of {Q and η}, the learning convergence is observed. The reliability of convergence is confirmed via repeated, ten ensemble training cycles or runs (each with 500 iterations epochs). The performance of the network in predicting an output vis-á-vis the teacher function is then tested. The input units are addressed with values of x_i corresponding to 50 equally spaced arguments ($\theta_i = m\pi\alpha_0 x_i \gamma$, $\gamma = 1$ and $m = 0.5$) of the function $|\sin(m\alpha_0\pi x_i)|$ over the range $-1/\alpha_0$ to $+1/\alpha_0$. The predicted output and the teacher value at each of these equally spaced arguments (θ_is) are noted and compared. The deviations of the predicted function from the teacher function over the fifty points of θ_i are tabulated and the root mean-squared (rms) value of these deviations (E_d) was calculated for each error-measure implemented with a given set of {Q and η}.

Relevant to the results obtained in the testing phase as above, it is observed that for all the error-measures considered, increasing the value of Q, enhanced the tolerance of the network optimization to larger learning rates. It is also observed that for each error-measure, a minimum value of E_d in predicting the function in reference to the teacher function exists with a specific set of values of Q and η. Listed in Table 4.1 are values of the set {Q, η} for which the observed values of Es were minimum for each case of error-measures used in the testing phase. The corresponding minimum values of E_ds are also presented in Table 4.1. Further, those error-measures which do not facilitate convergence of the network during training are indicated in Table 4.1 as appropriate.

4.6 Results on Neural Network Training with Csiszár's Error-Measures

Using the values of Q and η (as in Table 4.1), the corresponding predicted functions and the teacher function considered (in reference to the various error-measures) are presented in Figs. (4.5-4.9). In each of these figures, the upper plots (a) and (b) refer to the training phase data and depict the error-measure (normalized with respect to the maximum value) *versus* the training epochs. A jagged curve (dashed line) in these upper plots represents the learning for a single training run; that is, run with one sample ensemble of training data *versus* iteration epochs. The learning curve data obtained by the ensemble average of the values gathered from 10 repeated runs is shown by a solid line in each of the upper plots. This average learning curve can be seen to follow a smooth convergence trend, confirming the consistency in achieving a reliable convergence. The bottom plots (I & II) of Figs. (4.5-4.9) refer to the predicted and teacher values of the function $|\sin(m\alpha_0\pi x_i)|$ over the range of $\alpha_0 x_i$ from -1 to $+1$ and with $m = 0.5$.

In the simulations performed, it was observed that not all the error-measures mentioned in the previous sections offer tangible solutions in predicting the network output close to the teacher function nor do they facilitate the network convergence.

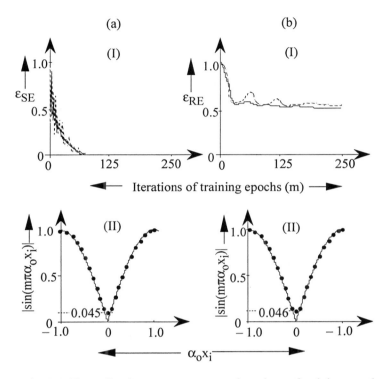

Fig. 4.5: Normalized output error versus iterations of training epochs
    (a) Computed results with square (SE) error measure
    (b) Computed results with relative (RE) error measure
        (*Note*: Data on Q and η as in Table 4.1)

—————   Learning curve of the ensemble average data from 10 training runs
------   Learning curve of a single training run
I & II: Predicted (. . .) and teacher (—) functions $|\sin(m\pi\alpha_o x_i)|$ versus $\alpha_o x_i$ with
        $m = 0.5$

In particular, the error-measures indicated below (also presented in Table 4.1) proved to be futile, and the network performance showed a divergence trend. (For these error-measures, no plots on learning curves or the predicted data on the function are furnished).

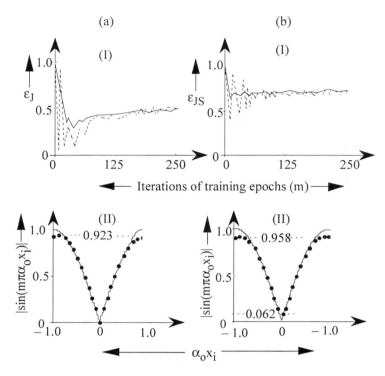

Fig. 4.6: Normalized output error versus iterations of training epochs
(a) Computed results with Jensen (J) error-measure
(b) Computed results with Jensen-Shannon (JS) error
measure ($\pi_1 = \pi_2 = 0.5$)
(*Note*: Data on Q and η as in Table 4.1)

——— Learning curve of the ensemble average data from 10 training runs
------ Learning curve of a single training run
I & II: Predicted (. . .) and teacher (——) functions: $|\sin(m\pi\alpha_o x_i)|$ versus $\alpha_o x_i$ with
$m = 0$

The plausible reasons for the failure of certain error-measures in being appropriate cost-functions for neural network applications are indicated against each case in the discussions of the following paragraphs. The performance of networks of neurons as information processing devices can only be gauged correctly by using appropriate information measures as performance qualities available in practice could be only partial. In such cases, especially in reference to the context of real neural activity measurements, empirical procedures to specify unbiased information measures using limited sample data are needed.

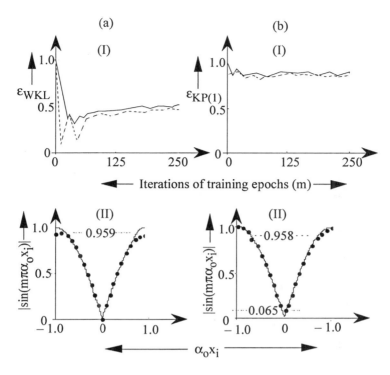

Fig. 4.7: Normalized output error versus iterations of training epochs
(a) Computed results with weighted and symmeterized Kullback-Leibler
(WKL) error-measure ($\mu = 0.5$)
(b) Computed results with Kapur type 1 (KP 1) error-
measure (a = 0.5)
(*Note*: Data on Q and η as in Table 4.1)
——— Learning curve of the ensemble   runs
------ Learning curve of a single training run
I & II: Predicted (. . .) and teacher (—) functions: $|\sin(m\pi\alpha_o x_i)|$ versus $\alpha_o x_i$ with
m = 0.5

However, data regularization is warranted to correct such information measures. In relation to such sample data situations, methods totality has been addressed by Abu-Mostafa [4.29-4.30]. The problems of learning an unknown function by putting together several pieces of partial information (hints) known about the function and incorporating such hint-based metrics in descent-learning strategies are considered in [4.29-4. 4.31]

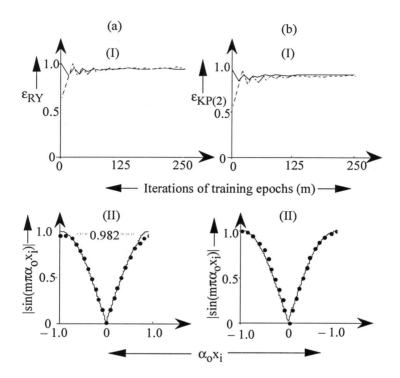

Fig. 4.8: Normalized output error *versus* iterations of training epochs
(a) Computed results with Rényi error-measure ($\alpha = 0.5$)
(b) Computed results with Kapur type 2 error-measure ($\alpha = 0.45$,
$\beta = 0.55$)
(*Note*: Data on Q and η as in Table 4.1)

―――― Learning curve of the ensemble   runs
------ Learning curve of a single training run
I & II: Predicted (. . .) and teacher (―) functions: $|\sin(m\pi\alpha_o x_i)|$ *versus* $\alpha_o x_i$ with
m = 0.5

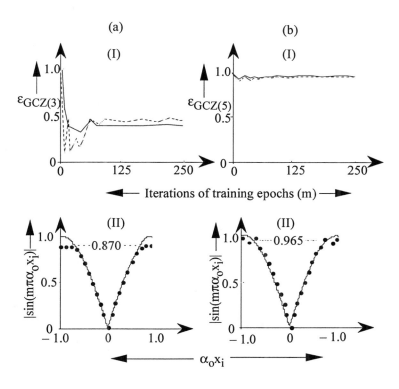

Fig. 4.9: Normalized output error *versus* iterations of training epochs
(a) Computed results with generalized Csiszár's type 3 (GCZ 3) error-measure
(a = 0.5)
(b) Computed results with generalized Csiszár's type 5 (GCZ 5) error-measure
(a = 1.0, b = 0.5., c = 0.5, d = 1.0)
(*Note*: Data on Q and η as in Table 4.1)

———   Learning curve of the ensemble   runs
------   Learning curve of a single training run
I & II: Predicted (. . .) and teacher (—) functions: $|\sin(m\pi\alpha_o x_i)|$ versus $\alpha_o x_i$ with
         m = 0.5

4.6.1 Unusable error-measures

Among the various error-measures discussed and listed in Table 4.1, those which do not allow the network to converge are found to be characteristically of two types. The first category refers to the asymmetric form of the cross-entropy functional with respect to p_i and q_i. Typically, it can be observed that the Kullback-Leibler measure (elucidated in terms of p_i and q_i and given by Eqn. (4.10) or (4.17)) is not symmetrized, and therefore,

181

proved to be an unusable error-measure. On the other hand, the other error-measures of the Kullback-Leibler family, namely, the Jensen/Jeffreys measure, the Jensen-Shannon measure and the weighted (symmetrized) Kullback-Leibler measure, enable network optimization.

The other measures of asymmetric type falling in the first category which do not allow the network to converge are generalized Csiszár's measures given by Eqns. (4.26) and (4.27), as well as that specified by Eqn. (4.28).

The second category of error-measures which are not feasible for applications in neural network optimization problems, due to their inability to facilitate the network convergence, are those based on the nonadditive entropy generalizations discussed in reference to Eqns. (4.20) and (4.21), sometimes referred to as *additivity discrimination* is concerned with random measurements that are independent under the probability distributions p and q. In reference to [4.26], the additivity property has a distinct meaning concerning the weighted mean of the entropies p and q. Thus, when used as cost-functions in ANN applications, the α-entropy measure developed by Harvda and Charvát (Eqn. (4.19)) and the α, β-entropy measure proposed by Sharma and Mittal (Eqn. (4.22)), allow the network to diverge. As a result, they do not facilitate target-seeking optimizations in the test ANNs.

Excluding these two categories of measures, the rest of the error-measures are found to be usable in the network optimizations tested. Relevant results pertinent to Figs. 4.5-4.9 are summarized as follows:

4.6.2 Usable error-metrics

(1) The results presented in Fig. 4.5a(I) correspond to the square error (SE) measure (Eqn. 4.7) and the convergence is seen to be attained at about the 80th iteration. The predicted values of the function closely follow the teacher function as can be observed in Fig. 4.5a(II). The maximum deviation is observed only at the point $\alpha_0 x_i = 0$ where the slope of the function reverses. This deviation is about 4.4% of the amplitude of the function. The rms value of the deviations (E_d) between the predicted and teacher values of the function computed at 50 equally spaced points over the entire argument (-1 to $+1$) is 0.0111 (Table 4.1).

(2) The RE metric (Eqn. (4.8)) also leads to a convergence within 80 iterations (Fig. 4.5b(I)) and the predicted function almost corresponds to the teacher function. Again, the maximum deviation in prediction (4.6%) is at $\alpha_0 x_i = 0$ and the value of E_d is 0.0134 (Table 4.1). The RE measure, though computed directly from O_i and T_i (unlike the Csiszár's family of errors which are deduced from the probabilities, namely $p_i(O_i)$ and $q_i(T_i)$ and not directly from O_i and T_i themselves), is implicitly a cross-entropy measure which is symmetric in its functional format by virtue of using the binary sets $\{\tilde{O}_i, \tilde{T}_i\}$ and $\{(1 - \tilde{O}_i), (1 - \tilde{T}_i)\}$. (See Eqn. (4.6)). Due to this

symmetry, the RE measure facilitates network convergence and the optimization sought as expected.

(3) The information-theoretic, Jensen error-measure (Eqn. (4.12)) of the Kullback-Leibler family also enables the network towards manageable convergence around the 50th iteration. However, the predicted values deviate from the corresponding values of the teacher function (where slope becomes zero) by 7.7%; and, $E_d = 0.0399$ (Table 4.1).

(4) The Jensen-Shannon error-metric (Eqn. (4.13b)) which is symmetrized in respect to p_i and q_i (like the J-measure) with $\pi_1 = \pi_2 = 0.5$ also offers the convergence around the 50th iteration of the training epoch. Here, the predicted values at the peak of the function differ from the teacher values only by 4.2%; and, at $\alpha_o x_i = 0$ by 6.2%. Further, $E_d = 0.0247$ (Table 4.1a).

(5) Though the Kullback-Leibler measure (Eqn. (4.10)) does not permit convergence (as mentioned earlier), its weighted and symmetrized version (given by Eqn. (4.14)) not only facilitates a fast convergence around the 50th iteration, but also offers a prediction of the function close to the teacher function. The deviation is about 4.1% at the peaks of the function (Fig. 4.7a); and $E_d = 0.0218$ (Table 4.1a).

(6) Among the Csiszár's family of the error-measures discussed, Kapur's type 1 measure (Eqn. (4.18)) leads to a very fast convergence (within the 40th iteration) with the predicted values (Fig. 4.7b(I)) close to the teacher values except at $\alpha_o x_i = 0$ where the deviation is 6.5%.) Further, $E_d = 0.0164$ (Table 4.1).

(7) Another Csiszár measure which offers an excellent network performance is the Rényi's divergence measure (Eqn. (4.23)) as could be evinced from the results depicted in Fig. 4.8a. The prediction of the function is as good as that obtained with ε_{SE} but the convergence is faster, around the 40th iteration. The maximum deviations (1.8%) observed in this case are at the peaks of the function and $E_d = 0.0116$ (Table 4.1b).

(8) The weighted version of Rényi's measure designated as Kapur's type 2 formulation (Eqn. (4.24)), also yields a superior network performance vis-á-vis in predicting the function (Fig. 4.8b) and, at the same time the convergence is realized within about 40 iterations. The rms value of the deviations in predicting the function, $E_d = 0.0204$ (Table 4.1) and the prediction performance is excellent, even at the peaks as well as at $\alpha_o x_i = 0$ where slope of the function changes significantly.

(9) The generalized Csiszár's measures (Eqns (4.26-4.28)) are based on the construction of a new set of probabilities namely: (p', q') where,

$$(p', q') = \left\{ \left[(ap_i + c)/(a + nc) \right]_{i=1,2,..,n}, \ \left[(aq_i + c)/(a + nc) \right]_{i=1,2,..,n} \right\}$$

with $a > 0$ and $c > 0$ and the choice of Φ is logarithmic. The results presented in Fig. 4.9a are in reference to $\varepsilon_{GCZ(3)}$ of Eqn. (4.28). The maximum deviations at the peaks of the function are 13% and $E_d = 0.0607$ (Table 4.1). Additional simulations performed show that the choice of the parameter namely, $a > -1$ in Eqn. (4.28), influences the closeness of the predicted network output versus the teacher values.

(10) Pertinent to other possible cases of Csiszár's measure, Eqn. (4.30b) offers a tangible distance measure designated as $\varepsilon_{GCZ(5)}$ for the network optimization problems under discussion. The results presented in Fig. 4.9a indicate the fast convergence around the 50th iteration, and low errors in the output predictions of the network which are close to the teacher values, except at the peaks of the function predicted, where the deviation is 3.5%. The value of $E_d = 0.1074$ (Table 4.1c).

In view of the algorithmic representations and discussions presented on the new set of error-measures adopted and from the results obtained from the relevant simulation studies, the following inferences are made.

4.6.3 Scope of the information-theoretic error-measures for neural network optimization

❑ The information-theoretic error-measures presented here offer an alternative set of cost-functions for ANN performance optimizations

❑ Considering the variety of formulations presented wlly spaced points e predicted auxiliary parameters such as α, β, a, b, c, etc., the scope of choosing an appropriate cost-function to classify a particular type of pattern/function by a neural network is enhanced effectively

❑ Faster convergence rate has been observed with a few of the measures adopted in comparison with the convergence rate realized with the conventional SE metric. (Examples: Kapur type 2 and Rènyi measures.)

❑ The other measures, namely, ε_{JS}, ε_{WKL}, $\varepsilon_{KP(1)}$, $\varepsilon_{GCZ(3)}$ and $\varepsilon_{GCZ(5)}$ also offer faster convergence relative to SE and RE measures, but render the predicted values to deviate slightly (by less than 8.7%) from the teacher values at arguments of the function where the slope of the function changes significantly

❑ ε_{RY} and $\varepsilon_{KP(2)}$ measures offer better accuracy with maximum deviations being 1.8% and $\approx$ 0%, respectively in comparison with SE and RE measures for which the maximum deviations observed are 4.5% and 4.7%, respectively.

In reference to the results presented, though the iterations were limited to 500, it can be expected that the differences observed in the predicted functions with respect to the teacher function could approach still lower values asymptotically, if the iterations were extended beyond 500. Further, some of the error-measures have auxiliary parameters (weighting factors and/or order parameters) each specified within certain bounds. The values of these parameters can be chosen within the specified bounds so as not to violate the symmetry and/or additive considerations of the entropy involved. Only specific values of these parameters were used in the present simulations performed and they are mentioned along with the results presented in Figs. 4.5 - 4.9 and in Table 4.1. It may be possible to obtain better (or worse!) results if the values of these parameters are changed. Relevant consequences are open questions that have to be addressed in future research. The momentum value chosen in the simulation can also be changed. This would control the learning and, hence the prediction, capability of the network, especially to achieve a better prediction for those inputs at which the pattern/function reverses its slope from positive to negative or from a finite value to zero, as conceived at the arguments of the $|\sin(.)|$ function, namely, $\alpha_o x_i = 0$ and at $\alpha_o x_i = \pm 1$ adopted in the present study.

❑   The error-measures discussed are appropriate for prediction problems in which the training data, namely, the inputs (x_i) and the corresponding teacher values (T_i) against which the network output O_i is compared are statistical in nature. In real world problems, for example, $\{x_i\}$ may represent a set of measured data under a noisy environment and $\{T_i\}$ may represent the corresponding set of measured values which are dependently or independently stochastical in reference to $\{x_i\}$. Therefore, it may be required to train the network with a set of stochastical variates such as $\{x_i\}$ and compare the output against another set of stochastical variables $\{T_iy$ by 4.2%; annetwork so trained, prediction of $\{O_i\}$ for a new set of $\{x_i\}$ (perhaps, in a different range of x_i as compared to the original training data) is feasible with the type of error-measures proposed in this work.

❑   Among the various error-measures proposed, and considering the optimum performance in terms of convergence rate and accuracy of prediction, it is contended at this stage that the choice of the error-measure is more dependent on the pattern / function to be classified.

❑   In the existing studies, a general cross-entropy based cost-function has been tried [4.32] in real-time experiments *via* log-likelihood measure deduced from the Kullback-Leibler measure [4.17] for binary targets in decoding artificial neural signals. The present study offers a broader scope of using information-theoretic cost-functions for such studies.

4.6.4. Note on the nonlinear activity function used

In the present study, the nonlinear activity function used was the Bernoulli function $L_Q(.)$. However, the conventional nonlinear sigmoids such as the hyperbolic tangent function with different gain factors can be used instead (without any loss of generality) when the information-theoretic error-measures discussed in this chapter are adopted. The reasons for using the Bernoulli function as a sigmoid in neural networks are based on certain stochastical justifiability considerations. As elaborated by Neelakanta et al. [4.14] in using conventional sigmoids, there is an arbitrariness in selecting the gain factor of the relevant function, for example, Γ in $\tanh(\Gamma x)$. When $L_Q(.)$ is adopted instead, the factor Q has the same role (of altering the slope of the sigmoid in the region between the squashing limits) as the gain factor Γ of conventional sigmoids, but gives a meaningful interpretation on the choice of Q. That is, Q is regarded as the order function of the stochastical state proliferation (from the input to the output section of the network) consistent with the statistical mechanics (spin-glass) considerations attributed to the systems constituted by interacting units, such as the neural network. Readers may refer to Chapter 2 and [4.5, 4.14, 4.33] for more details on the stochastical justifications in adopting $L_Q(.)$ *in lieu* of other sigmoidal functions. However, as stated earlier, one can use the conventional sigmoids as well while implementing the information-theoretic error-measures.

4.6.5. Gradient-descent algorithm

In the current work, only the gradient-descent algorithm was used in adjusting the weights of the interconnections which offers reasonably a fast convergence. However, it also renders the simulation rather computation-intensive. To achieve a faster convergence, the conjugate-gradient method has been proposed [4.34] for quadratic functions. It is an open question at this stage whether to study the feasibility aspects and *pros and cons* of adopting the conjugate-gradient technique while adopting the information-theoretic error-measures as the cost-functions.

4.6.6. Hidden layer considerations

Pertinent to the number of hidden layers used in the multilayer perceptron, one hidden layer is sufficient to achieve a desirable neural network performance. It has been, however, suggested that considering the network performance in learning certain complex patterns/functions, networks with more hidden layers may "outestimate" those with a lesser number of hidden layers [4.35]. Hence, in the present study, the simulation code was written to accommodate two hidden layers in preference to having a single hidden layer. But, a single hidden layer network will have a reduced architectural complexity and can be adopted, along with the information-theoretic error-measures under discussion, if the prediction performance is acceptable for the type for function/pattern being classified.

4.7 Symmetrized Information-Theoretic Error-Metrics

In training neural networks, it is demonstrated in the previous sections that a set of cross-entropy functions which can be identified in the information-theoretic plane are appropriate as error-measures. They can be used in a feedback loop towards training and optimization of prediction performance of pattern by a neural network. However, unless otherwise the chosen error-measure is in a symmetrized functional form, as well as that the entropy relations do not assume a nonadditive structure, the measures so considered could be futile in their performance requirements. Considered in the following sections, are ways of symmetrizing and/or balancing the error entropy measures so as to make them suitable as usable error-metrics in neural network applications. Typical formulations are developed and confirming results on the usability of such error-measures are presented *via* on-line simulation studies.

The essence of previous sections was to indicate the existence of a class of error-measures specified in the information-theoretic plane which can be regarded as suitable error-metric algorithms useful to supplement those adopted conventionally in neural network training strategies.

However, among the several available measures enumerated, analyzed, and classified under the realm of so-called Csiszár's measures, it was observed that the necessary and sufficient conditions for a member of Csiszár error-measure family compatible for neural network training are:

(i) The cross-entropy function should satisfy the essential and desirable conditions stipulated in the subsection 4.3.3.

(ii) The Φ-function used to construct the Csiszár measure should be twice-differentiable convex function for which $\Phi(1) = 0$.

(iii) The entropy function should be two-sided and constitute a balanced, symmeterized form.

(iv) The entropy relations should not be of nonadditive structure lest the relevant measures pose inherent asymmetry; and, therefore, do not satisfy the condition (iii).

The scope of the following sections is to consider the feasibility of symmetrizing (making two-sided and balanced) those error-measures which either explicitly pose one-sidedness or implicitly have a nonadditive structure.

Hence, those measures that failed to lead the test neural network (Fig.4.4) towards convergence are reconsidered and symmetrization techniques as appropriate are applied. Further, these symmetrized formulations developed are tested with the same test neural network (Fig. 4.4) and a meaningful approach towards symmetrization procedure is presented and discussed.

4.8 One-Sided Error-Measures and Implementation of Symmetrization

As an example of one-sided error-measure, the Kullback-Leibler measure given by Eqn. (4.17a) is reproduced here for discussion:

$$\varepsilon_{1(KL)} = \sum_i p_i \log(p_i/q_i) \tag{4.36}$$

where p_i and q_i represent the probability distributions of the network output $\{O_i\}$ and teacher values $\{T_i\}$ respectively. It was observed earlier that, when used in the test neural network, $\varepsilon_{1(KL)}$ proved not to be useful inasmuch as the network's performance could not be comprehended towards convergence. The divergence in the network's performance can be attributed due to the one-sided asymmetry in the algorithmic form of Eqn. (4.36).

It can be seen that Eqn. (4.36) represents a *directed divergence* of the probability p and q and not *vice versa*. That is, the measure specified by Eqn. (4.36) is one-directional or one-sided. As Blahut [4.15] calls it, Eqn. (4.36) is a simple measure of "discrimination". In this format, Eqn. (4.36) does not include the total pragramic aspects of the bilateral influence of p on q and q on p. Denoting Eqn. (4.36) as I(p/q), it refers to the mean information per observation from the network output process $\{O_i\}$ for discrimination in favor of $\{O_i\}$ against $\{T_i\}$. Similarly, one can specify the mean information per observation from the teacher process $\{T_i\}$ for discrimination in favor of $\{O_i\}$ against $\{T_i\}$; that is,

$$\varepsilon_{2(KL)} = -I(q/p) = \sum_i q_i \log(p_i/q_i) \tag{4.37}$$

which is same as Eqn. (4.17b). Hence, a divergence measure can be written as follows:

$$\begin{aligned} J(p/q) &= I(p/q) + I(q/p) \\ &= \sum (p_i - q_i) \log(p_i/q_i) \end{aligned} \tag{4.38}$$

which is same as the Jensen/Jeffreys error-measure ε_J indicated earlier (Eqn. 4.12). That is,

$$\varepsilon_J = \varepsilon_{1(J)} + \varepsilon_{2(J)} \equiv I(p/q) + I(q/p) \tag{4.39}$$

This divergence measure (Eqn. (4.39)) is invariant under the transformation of parameters. It measures the divergence between the hypotheses specified by the sets $\{O_i\}$ and $\{T_i\}$ and measures implicitly the difficulty of discriminating them (bilaterally). It is symmetric with respect to the probability measures p and q. It has all the properties of a distance (or metric) as defined on topology (subsection 4.3.3), except the triangle inequality property and is, therefore, not designated as a "distance" [4.16], but rather termed as a divergence measure [4.15].

The divergence inherits most of the properties of the discrimination. From a communication point of view, Blahut [4.15] indicates that "the symmetric form may appear more fundamental, but actually we have little need of it". However, as a neural network training error-measure, the symmetry becomes a necessary condition as is being deliberated here.

While the Kullback-Leibler measure (due to asymmetry) fails to let the test neural network converge towards the teacher values as discussed before, it should be noted that J-divergence, on the other hand, leads the network to the equilibrium status at the terminal stage of the iterative epochs of error feedback. Corresponding to I(p/q) and I(q/p), one can define the Φ-functions as follows:

$$I(p/q) = \sum q\Phi(p/q), \quad \Phi(p/q) = (p/q)\ell og(p/q) \tag{4.40}$$

$$I(q/p) = \sum p\Phi(q/p), \quad \Phi(p/q) = (q/p)\ell og(q/p) \tag{4.41}$$

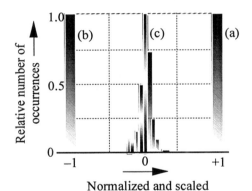

Fig. 4.10: Histograms of the number of occurrences of the error-values versus normalized and scaled error-values, obtained from the simulations of the network of Fig. 4.4

(a) One-sided Kullback-Leibler measure, ε_{1KL}
(b) One-sided Kullback-Leibler measures, ε_{2KL}
(c) Symmetrized measure $(\varepsilon_{1KL} + \varepsilon_{2KL}) = \varepsilon_J$

The symmetric and asymmetric aspects of I and J-measures can be visualized from the histograms furnished in Fig. 4.10 pertinent to the training results gathered from the test network (Fig. 4.4) using $\varepsilon_{1(KL)}$, $\varepsilon_{2(KL)}$ and $\varepsilon_J = \varepsilon_{1(KL)} + \varepsilon_{2(KL)}$.

(a)   (b)

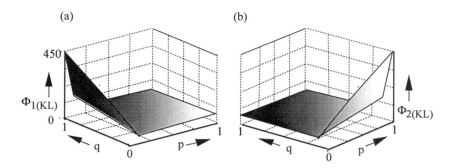

Fig.4.11: Φ-function of the one-sided error-measure versus the probabilities p and q.
(a) $\Phi_{1(KL)}$ and (b) $\Phi_{2(KL)}$

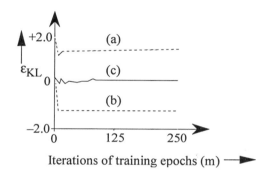

Iterations of training epochs (m) ➤

Fig. 4.12: Learning curves with a single training run: Dynamics of the error-measure
$(\varepsilon_{1KL} + \varepsilon_{2KL} + \varepsilon_j)$ vs. number of iterations of the training epochs (m).

(a) ε_{1KL} (one-sided/asymmetric)
(b) ε_{2KL} (one- sided/asymmetric)
(c) $(\varepsilon_{1KL} + \varepsilon_{2KL}) = \varepsilon_J$ (two-sided/ symmetric)
     (Note: ε values are scaled: $\varepsilon \cong 0 \Rightarrow$ Equilibrium/convergence status; and $\varepsilon = \pm$
     $1 \Rightarrow$ Divergent levels. Training parameters: Q = 0.5 and η = 0.001 in the
     simulations pertinent to the test network of Fig. 4.4)

The asymmetric measures $\varepsilon_{1(KL)}$ and $\varepsilon_{2(KL)}$ yield one-sided population
distributions clustered around the divergent limits of the error-values
(normalized and scaled) at $\pm$ 1 respectively; whereas, ε_J (normalized and
scaled) values are symmetrically disposed at the equilibrium (converging
limit) value of zero. Further, $\Phi_{\varepsilon 1(KL)}$ $\Phi_{\varepsilon 1(KL)}$ plotted against p and q (0 $\leq$
(p, q) $\leq$ 1) also illustrate the one-sided (asymmetic aspects) of the I-measure
as shown in Fig. 4.11.

190

It will be shown in Chapter 5 that $I(p/q)$ and $I(q/p)$ represent two competing error-measure species and the dynamics of the these competing processes are essential for the convergence anticipated. Otherwise, if only one species is considered, as illustrated in Fig. 4.12 *via* learning curves, the corresponding one-sided (asymmetric) values of the output error renders the network to have a divergent trend. That is, the network fails to be trained. Hence, the following can be inferred at this stage:

(1) The error-measure to be considered as useful for neural network training should be symmetrical or two-sided in respect to p and q.

(2) If the error-measure is asymmetric, it can be symmetrized by the following procedure:

Suppose $(p/q) = x$. Then an error-metric in Csiszár form can be written as

$$I(x) = \sum q\Phi(x) \tag{4.42}$$

Or, as:

$$I(1/x) = \sum p\Phi(1/x) \tag{4.43}$$

Considering the above two forms constituting the (asymmetric) components of the error-measure, the corresponding symmetric divergence measure can be written as

$$I_{sym}(x) = I(x) + I(1/x) \tag{4.44}$$

Using the above generalization, the various information-theoretic measures will be reconsidered for symmetrization purposes. Relevant results will be discussed later.

4.9 Efficacy of the Error-Measures in Neural Network Training

4.9.1 Square-error (SE) measure:

From Eqn. (4.7), $\varepsilon_{SE} = 1/2\sum(T_j - O_j)^2$. This measure is essentially constituted by the difference error-values $\pm (O_i - T_i)$. Hence, the contributions from $+ (O_i - T_i)$ and $- (O_i - T_i)$ jointly provide competing steps in the iterations process, leading the network towards convergence or the equilibrium status. The resultant histogram of the populations corresponding to the difference error $\pm (O_i - T_i)$ is depicted in Fig. 4.13. The learning curve and the performance of the network (Fig. 4.4) using the square-error in predicting a teacher function are shown in Fig. 4.5a(I) and Fig. 4.5a(II), respectively.

4.9.2 Relative entropy (RE) error-measure

Considering the ε_{RE} given by Eqn. (4.8), it can be observed that the bilateral symmetry between O_i to T_i or *vice versa* is rather implicit. It is simply a binary discrimination of two real numbers $\tilde{O}_i$ and $\tilde{T}_i$ (the tilde depicting each has been normalized and set as $0 \leq \tilde{O}_i \leq 1$ and $0 \leq \tilde{T}_i \leq 1$). It is a complex function in each var $\tilde{O}_i = \tilde{T}_i$. This divergence error-measure, therefore, yields a guaranteed convergence as discussed earlier. The relevant histogram on the number of occurrences of ε_{RE} over the training iterations and the converging characteristics is shown in Fig. 4.13. The learning curve of ε_{RE} is shown Fig. 4.5b(I). The prediction performance of the ε_{RE} is presented in Fig. 4. 5b(II).

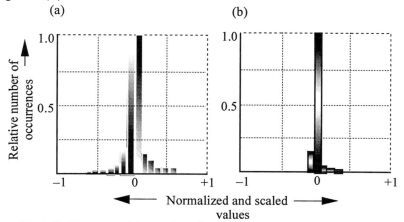

(a) (b)

Fig. 4.13: Histogram of the number of occurrences of the error-values *versus* normalized and scaled error-values

(a): Corresponding to the two-sided symmetric measure ε_{SE}, obtained from the simulations of the network of Fig. 4.4

(b): Corresponding to the two-sided symmetric measure ε_{RE}, obtained from the simulations of the network of Fig. 4.4

4.9.3 Kullback-Leibler family of error-measure

As discussed before, when considering $\varepsilon_{1(KL)} = I(p/q)$ or $\varepsilon_{2(KL)} = I(q/p)$, each being an asymmetric error-measure would not, per se, yield a convergence. On the contrary, by constructing the Jensen/Jeffreys error-measure, namely, $\varepsilon_J = \varepsilon_{1(KL)} + \varepsilon_{2(KL)}$, a symmetrization is realized. Therefore, competing species of error-values are presented to the network which enable the network to converge towards an equilibrium value. The computed results on the histograms of error-values and simulated results of learning curves presented in Figs. 4.10 and 4.12 (respectively) prove the efficacy of symmetrization pursued. The symmetric property of ε_J is also evident from the plots of $\varepsilon_{1(KL)}$, $\varepsilon_{2(KL)}$ and ε_J *versus* p and q as shown in Figs. 4.14 and 4.15a. The network's prediction performance can be ascertained from Fig. 4.6a(I).

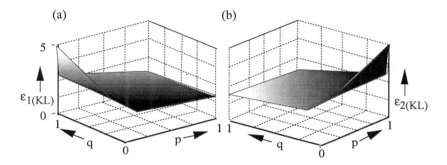

Fig. 4.14: The one-sided Kullback-Leibler functions which are the constituent
parts of ε_J versus the probabilities p and q.
(a) $\varepsilon_{1(KL)}$; (b) $\varepsilon_{2(KL)}$

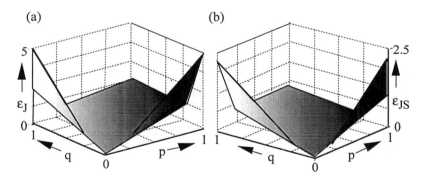

Fig. 4.15: (a) Symmetrized two-sided Jensen/Jeffreys ε_J
versus
the probabilities p and q
(b) Symmetrized error-measure $(\pi_1\varepsilon_{1KL} + \pi_2\varepsilon_{2KL}) = \varepsilon_{JS}$ versus the probabilities p and q

A modified version of the Jensen/Jeffreys measure is the Jensen-Shannon error-measure given by Eqn. (4.13) where π_1 and π_2 are fractions (such that $\pi_1 + \pi_2 = 1$) which allow selective weighting on the I(p/q) and I(q/p)

I(q/p) discrimination functions. Using Eqns. (4.40 and 4.41), the following formulation can be written:

$$\varepsilon_{JS} = [\alpha/(\alpha+\beta)]I(p/q)] + [\beta/(\alpha+\beta)]I(q/p)]$$
$$= \varepsilon_{1(JS)} + \varepsilon_{2(JS)} \qquad (4.45)$$

where $\alpha/(\alpha+\beta) \equiv \pi_1$ and $\beta/(\alpha+\beta) \equiv \pi_2$. The above equation (Eqn.4.45) is symmetrized only when $\pi_1 = \pi_2$, which under the condition $(\pi_1 + \pi_2) = 1$, refers to $\pi_1 = \pi_2 = 0.5$. Suppose the condition that $\pi_1 = \pi_2 = 1$ is imposed. Then, ε_{JS} is same as the Jensen/Jeffreys relation.

Per$\pi_1 \neq \pi_2$, and $(\pi_1 + \pi_2) = K_\pi$, a constant. The following symmetrization can be adopted:

$$\varepsilon_{JS}^{Sym}(\pi_1 \neq \pi_2) = \pi_1 I(p/q) + \pi_2 I(q/p) + \pi_2 I(p/q) + \pi_1 I(q/p)$$
$$= [\pi_1 + \pi_2][I(p/q) + I(q/p)]$$
$$= K_\pi \varepsilon_J \qquad (4.46)$$

which is, simply the J-measure scaled by a constant factor, K_π and is, therefore, automatically symmetric.

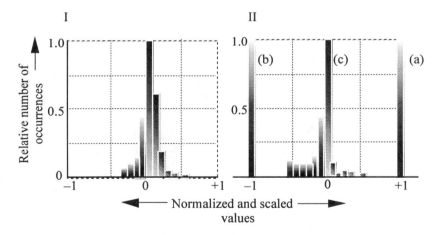

Fig. 4.16: Histogram of the number of occurrences of the error-values versus normalized and scaled error-values obtained from the simulations of the network of Fig. 4.4

I: Symmetrized measure $(\pi_1\varepsilon_{1KL} + \pi_2\varepsilon_{2KL}) = \varepsilon_{JS}$
II: (a) One-sided weighted a measure, ε_{1WKL}
      (b) One-sided weighted an measures, ε_{2WKL}
      (c) Symmetrized measure $(\varepsilon_{1WKL} + \varepsilon_{2WKL}) = \varepsilon_{WKL}$

The symmetric characteristics of ε_{JS} can be ascertained from the illustrations of a relevant histogram presented in Fig. 4.16(I) and the learning curves shown in Fig. 4.6b(I). The error function ε_{JS} versus the probabilistic p and q as plotted are Fig. 4.15b also signifies the symmetry involved. The prediction ability of the network using ε_{JS} is shown in Fig. 4.6b(II).

A weighted form of Kullback-Leibler error-measure is given by [4.16]:

$$\varepsilon_{1(WKL)} = \sum p_i \ell og[\mu p_i + (1 - \mu)q_i] \qquad (4.47)$$

or,

$$\varepsilon_{2(WKL)} = \sum q_i \ell og[\mu q_i + (1 - \mu)p_i] \qquad (4.48)$$

where $0 \leq \mu \leq 1$, and both are one-sided/asymmetrically directed error-measures. The symmetrized version of this measure is given by Eqn. (4.14). That is, $\varepsilon_{WKL} = \varepsilon_{1(WKL)} + \varepsilon_{2(WKL)}$. The computed histograms and simulated learning curves obtained by using this error-measure are presented in Figs. 4.16(II) and 4.17, respectively.

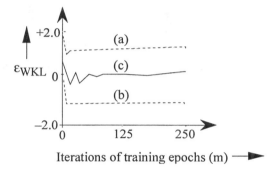

Iterations of training epochs (m) ➤

Fig. 4.17: Learning curves obtained with a single training run: Dynamics of the error-measure $(\varepsilon_{1(WKL)} + \varepsilon_{2(WKL)}) = \varepsilon_{WKL}$ versus number of iterations of the training epochs (m)
(Training parameters: Q = 0.5 and η = 0.001 in the simulations pertinent to the test network of Fig. 4.4)

(a) $\varepsilon_{1(WKL)}$ (One-sided/asymmetric)
(b) $\varepsilon_{2(WKL)}$ (One- sided/asymmetric)
(c) $(\varepsilon_{1(WKL)} + \varepsilon_{2(WKL)}) = \varepsilon_{WKL}$ (Two-sided/ symmetric)

(*Note*: ε values are scaled: $\varepsilon \rightarrow \varepsilon_0 = 0 \Rightarrow$ Equilibrium/convergence status; and $\varepsilon = \pm 1 \Rightarrow$ Divergent levels)

The prediction ability of the network utilizing this error-measure can be seen from the predicted function of Fig. 4.7a(II). Further, ε_{WKL} plotted as a function in Fig. 4.18 shows the symmetrization aspects of the function involved.

A modification of the Jensen/Jeffreys measure can be written as:

$$\varepsilon_{1MJ} = \Sigma\, q_i \phi_{1MJ}(p_i/q_i) \qquad (4.49)$$

where,

$$\Phi_{1MJ} = \left[(p_i/q_i) - \ell og(p_i/q_i) - 1\right] \qquad (4.50)$$

Or,

$$\varepsilon_{2MJ} = \Sigma\, p_i \Phi_{1MJ}(q_i/p_i) \qquad (4.51)$$

where

$$\Phi_{2MJ} = \left[(q_i/p_i) - \ell og(q_i/p_i) - 1\right] \qquad (4.52)$$

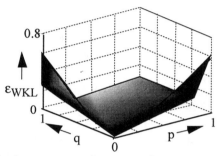

Fig.4.18: Symmetrized error-measure $(\varepsilon_{1WKL} + \varepsilon_{2WKL}) = \varepsilon_{WKL}$ *versus* the probabilities p and q

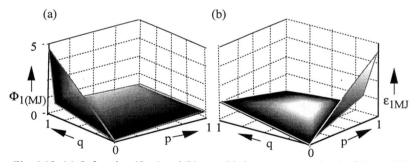

Fig. 4.19: (a) Φ-function (Φ_{1MJ}) and (b) one-sided error-measure (ε_{1MJ}) of the modified Jensen version *versus* the probabilities p and q

If the symmetrization procedure is applied, then

$$\varepsilon_{MJ} = \varepsilon_{1MJ} + \varepsilon_{2MJ}$$

$$= p_i \ell og(p_i/q_i) + q_i \ell og(q_i/p_i) \qquad (4.53)$$

which is the same as the Jensen/Jeffreys measure.

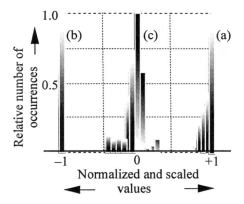

Fig. 4.20: Histograms of the number of occurrences of the error-values normalized and scaled *versus* error-values, obtained from the simulations of the network of Fig. 4.4

(a) One-sided modified Jensen measure, ε_{1MJ}
(b) One-sided modified Jensen measures, ε_{2MJ}
(c) Symmetrized measure $(\varepsilon_{1MJ} + \varepsilon_{2MJ}) = \varepsilon_{MJ}$

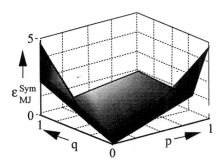

Fig. 4.21: Symmetrized error-measure $(\varepsilon_{1MJ} + \varepsilon_{2MJ}) = \varepsilon_{MJ}$ *versus* the probabilities p and q

It can be noted that $\varepsilon_{1(MJ)}$ or $\varepsilon_{2(MJ)}$ by itself is asymmetric. This can also be seen from the plots of, for example, ϕ_{1MJ} and ε_{1MJ} illustrated in Fig. 4.19.

When used in the test studies, neither $\varepsilon_{1(MJ)}$ nor $\varepsilon_{2(MJ)}$ has permitted the network to converge. However, as indicated above, relevant symmetrization renders this modified version as a J-measure (which is explicitly symmetric); and therefore, it is compatible for neural network applications. The set of figures (Figs. 4.20 to 4.22) illustrate the characteristics of this modified Jensen/Jeffreys measure in terms of the associated histogram, (Fig. 4.20), ε_{MJ} versus p and q characteristics (Fig. 4.21) and a learning curve (Fig. 4.22).

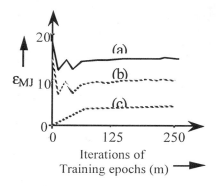

Fig. 4.22: Learning curve(s) obtained with a single training run: Dynamics of the error-measure $(\varepsilon_{1(MJ)} + \varepsilon_{2(MJ)}) = \varepsilon_{MJ}$ *versus* number of iterations of the training epochs (m). (Training parameters: Q = 0.5 and η = 0.001 in the simulations pertinent to the test network of Fig. 4.4)

(a) $\varepsilon_{1(MJ)}$ (One-sided/asymmetric)
(b) $(\varepsilon_{1(MJ)} + \varepsilon_{2(MJ)}) = \varepsilon_{MJ}$ (Two-sided/ symmetric)
(c) $\varepsilon_{2(MJS)}$ (One- sided/asymmetric)
     (*Note*: ε values are scaled: ε = ε₀ = 0 ⇒ Equilibrium/convergence status; and ε = ± 1 ⇒ Divergent levels)

After symmetrization, the performance of the test network of Fig. the test neural network (Fig.4. the $|\sin(.)|$ function. The result is shown in Fig. 4.23. The symmetrization facilitates a double-sided approach towards self-organization. It avoids the discourse of optimization veering away from the target. Thus, symmetrization fortifies the cost-functional attribute of the error-function in the robust search of the global minimum.

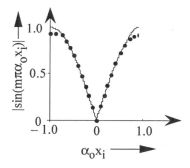

Fig. 4.23: Teacher function (——) and the predicted function (.....) obtained with the two-sided, symmetrized error-measure $(\varepsilon_{1(MJ)} + \varepsilon_{2(MJ)}) = \varepsilon_{MJ}$ using the perceptron network of Fig.4.4. (Network parameters used: Q = 0.5 and η = 0.001)

4.9.4 Generalized Jensen error-measure

Another generalized form of Jensen measure with an order parameter (α) can be written as follows:

$$\varepsilon_{1GJ} = \sum q_i \Phi_{1GJ}(p_i/q_i) \qquad (4.54a)$$

or, alternatively,

$$\varepsilon_{2GJ} = \sum p_i \Phi_{1GJ}(q_i/p_i) \qquad (4.54b)$$

with the corresponding ϕ-functions being (respectively):

$$\Phi_{1GJ} = (x^\alpha - x)/\alpha(\alpha - 1) \qquad (4.55a)$$

and

$$\Phi_{2GJ} = (1/x)^\alpha - (1/x)/\alpha(\alpha - 1) \qquad (4.55b)$$

where x = p/q and $\alpha \neq 0$ or 1.

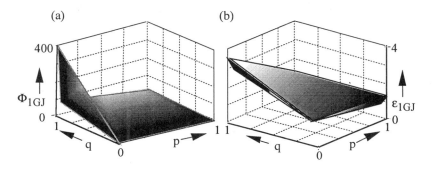

Fig. 4.24: (a) Φ-function (Φ_{1GJ}) and (b) one-sided error-measure (ε_{1GJ}) of the generalized Jensen version versus the probabilities p and q

While neither ε_{1GJ} nor ε_{2GJ} let the network converge, it was observed in the on-line training of 4.12 perceptron (Fig. 4.4) that a symmetrized version, namely $\varepsilon_{1GJ} + \varepsilon_{2GJ}$ constitutes a usable error-measure facilitating a convergence. Relevant results in respect to this measure are presented in Figs. 4.24 to 4.28. The concept of introducing an order parameter (say, α) refers to deducing an entropy function of degree α. In its generalized form Daróczy [4.36] developed a measure of uncertainty depicted by an entropy functional of an order which is subadditive and its characteristics justify its use in certain problems such as feature selection [4.18]

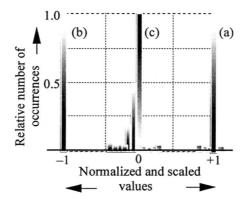

Fig. 4.25: Histograms of the number of occurrences of the error-values versus normalized and scaled error-values obtained from the simulations of the network of Fig. 4.4

(a) One-sided generalized Jensen measure, ε_{1GJ}
(b) One-sided generalized Jensen measures, ε_{2GJ}
(c) Symmetrized measure $(\varepsilon_{1GJ} + \varepsilon_{2GJ}) = \varepsilon_{GJ}$

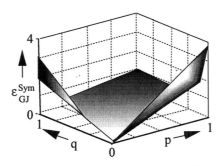

Fig. 4.26: Symmetrized error-measure $(\varepsilon_{1GS} + \varepsilon_{2GJ}) = \varepsilon_{GJ}$ *versus* the probabilities p and q

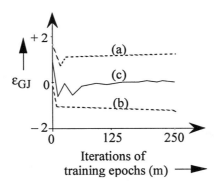

Fig. 4.27: Learning curve(s) obtained with a single training run: Dynamics of the error-measure $(\varepsilon_{1(GJ)} + \varepsilon_{2(GJ)}) = \varepsilon_{GJ}$ *versus* number of iterations of the training epochs (m)
(Training parameters: Q = 0.5 and η = 0.0001 in the simulations pertinent to the test network of Fig. 4.4)

(a) $\varepsilon_{1(GJS)}$ (One-sided/asymmetric)
(b) $\varepsilon_{2(GJ)}$ (One- sided/asymmetric)
(c) $(\varepsilon_{1(GJ)} + \varepsilon_{2(GJS)}) = \varepsilon_{GJ}$ (Two-sided/ symmetric)
     (*Note*: ε values are scaled: $\varepsilon = \varepsilon_0 = 0 \Rightarrow$ Equilibrium/convergence status; and ε
     $= \pm 1 \Rightarrow$ Divergent levels)

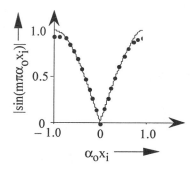

Fig. 4.28: Teacher function (——) and the predicted function (.....) obtained with the two-sided, symmetrized error-measure $(\varepsilon_{1(GJ)} + \varepsilon_{2(GJ)}) = \varepsilon_{GJ}$ using the perceptron network of Fig. 4.4. (Network parameters used: Q = 0.5 and η = 0.0001.)

4.9.5 Csiszár's family of error-metrics

As indicated in the earlier sections, the information-theoretic, cross-entropy based error-measure, in general, can be specified by Csiszár's form, namely $\varepsilon_1 = \sum q_i \Phi(p_i/q_i)$ or $\varepsilon_2 = \sum p_i \Phi(q_i/p_i)$.

The appropriate choice of ϕ decides various forms of error-measures as discussed in subsection 4.3.4. For example, $\Phi = (p/q)\ell og(p/q)$ leads to the Kullback-Leibler error-metric. As a special case of Csiszár's measures, the Kapur's type 1 error-measure given by Eqn. (4.18) presents a directed parameter a dependent algorithm; and, as a $\rightarrow$ 0, $\varepsilon_{KP(1)} \rightarrow \varepsilon_{1(KL)}$.

The symmetrization aspect of Kapur type 1 error-measure is achieved through the factor a > 0. For example, taking a = 0.5, the Φ-function of this error-measure given by Eqn. 4.56 with x = p/q plotted as a function of p and q in Fig. 4.29a. Though it appears to be one-sided, the function is of an almost constant value with respect to p and q, except at the limit values of p $\rightarrow$ 0, and q $\rightarrow$ 0. Here, the corresponding $\varepsilon_{KP(1)}$ *versus* p and q as plotted in Fig. 4.29b shows almost a two-sided symmetry. This is also confirmed from the histogram plot of Fig. 4.30 on the number of occurrences of the error value *versus* error values. The learning curve obtained with this error-measure as depicted in Fig. 4.7b(I) shows the convergence of the network by virtue of $\varepsilon_{KP(1)}$ being a two-sided symmetric measure; otherwise, the curve would have been divergent.

$$\Phi(x)\Big|_{\text{Kapur's type 1}} = (p/q)\ell og(p/q) - 1/a[1 + a(p/q)] + 1/a(1 + a)\ell og(1 + a)$$

$$(4.56)$$

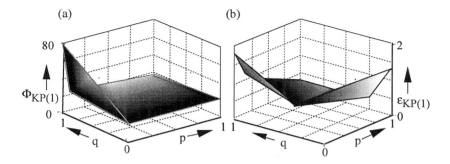

(a)   (b)

Fig.4.29: The Kapur's type 1 one-sided error-measure

(a) Φ-function *versus* the probabilities p and q
(b) $\varepsilon_{(KP1)}$ *versus* the probabilities p and q

Any asymmetry in $\varepsilon_{KP(1)}$ prevails only to a marginal extent, as decided by the value of a, and by the two constitutive parts of the expression of $\varepsilon_{KP(1)}$. In other words, $\varepsilon_{KP(1)}$ contains species of error-values which are almost balanced and competitive. Hence, they pull the network into the equilibrium status. The extent of balance or unbalanceis decided by the constant (a) chosen.

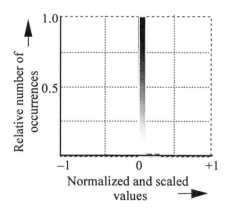

Fig. 4.30: Histogram of the number of occurrences of the error-values versus
normalized and scaled error-values, obtained from the simulations of the network of
Fig.4.4. Symmetric measure $\varepsilon_{KP(1)}$ with a = 0.5

In order to ensure a guaranteed/balanced condition so as to enable the error measure to train the network towards convergence, it is proposed here to adopt an appropriate value of a > 0. Results on function prediction presented in Fig. 4.7b(II) support the symmetric characteristics of $\varepsilon_{KP(1)}$ concept of symmetrization envisaged.

4.9.6 Symmetrized Havrda and Charvát error-measure

An asymmetric error-measure (due to Havrda and Charvát) is depicted by Eqn. 4.19. Corresponding $\varepsilon_{1(HC)}$ and $\varepsilon_{2(HC)}$ are inherently asymmetric as could be seen from Fig. 4.31.

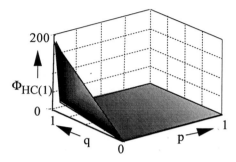

Fig. 4.31: Φ-function of the one-sided error-measure ($\varepsilon_{HC(1)}$) *versus* the probabilities p and q. (The corresponding Φ-function of ($\varepsilon_{HC(2)}$) *versus* p and q is the mirror reflection of the above pattern)

Hence, it is suggested that a symmetrized version of the Havrda-Charvát error divergence measure can be formulated as follows:

$$\varepsilon_{HC}^{Sym} = \varepsilon_{HC(1)} + \varepsilon_{HC(2)} \qquad (4.57)$$

where $\varepsilon_{HC(2)}$ corresponds to $\varepsilon_{HC(1)}$ to with p replaced by q and vice versa.

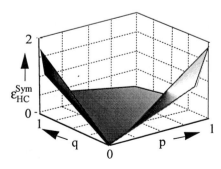

Fig. 4.32: Symmetrized error-measure ($\varepsilon_{HC(1)} + \varepsilon_{2HC(2)}$)= ε_{HC} versus the probabilities

p and q

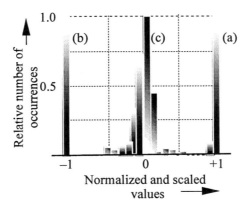

Fig. 4.33(a): Histograms of the number of occurrences of the error-values *versus* normalized and scaled error-values, obtained from the simulations of the network of Fig. 4.4

(a) One-sided Havrda-Charvát measure, $\varepsilon_{HC(1)}$
(b) One-sided Havrda-Charvát measure, $\varepsilon_{2HC(2)}$
(c) Symmetrized measure $(\varepsilon_{HC(1)} + \varepsilon_{2HC(2)}) = \varepsilon_{HC}$

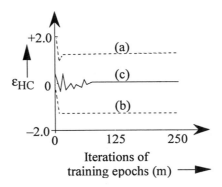

Fig. 4.33(b): Learning curves obtained with a single training run: Dynamics of the error-measure $(\varepsilon_{HC(1)} + \varepsilon_{HC(2)}) = \varepsilon_{HC}$ *versus* number of iterations of the training epochs (m). (Training parameters: Q = 0.5 and η = 0.001 in the simulations pertinent to the test network of Fig. 4.4)

(a) $\varepsilon_{HC(1)}$ (One-sided/asymmetric)
(b) $\varepsilon_{HC(2)}$ (One- sided/asymmetric)
(c) $(\varepsilon_{HC(1)} + \varepsilon_{HC(2)}) = \varepsilon_{HC}$ (Two-sided/symmetric)
Note: ε values are scaled: $\varepsilon = \varepsilon_0 = 0 \Rightarrow$ Equilibrium/convergence status; and $\varepsilon = \pm 1 \Rightarrow$ Divergent levels.)

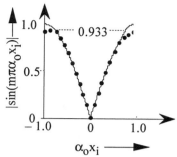

Fig. 4.34: Teacher function (—) and the predicted function (.....) obtained with the two-sided, symmetrized error-measure $(\varepsilon_{HC(1)} + \varepsilon_{HC(2)}) = \varepsilon_{HC}$ using the perceptron network of Fig. 4.4.

(Network parameters used: $Q = 0.5$, $\eta = 0.001$; $\alpha = 0.5$ and $\beta = 0.5$)

Results pertinent to the Havrda-Charvát error measure presented in Figs. 4.31 to 4.33 are self-explanatory. Again, $\alpha > 0$ (and not equal to 1) controls the extent of asymmetry; and, as $\alpha \to 0$, $\varepsilon_{HC(1)} \to \varepsilon_{1(KL)}$ and $\varepsilon_{HC(2)} \to \varepsilon_{2(KL)}$. The prediction characteristic of ε_{HC} as obtained *via* simulations from the network of Fig. 4.4 is shown in Fig. 4.34.

4.9.7 Symmetrized Sharma and Mittal error measure
Another two-parameteric (α and β) version of the directed divergence is due to Sharma and Mittal [4.26] and is given by Eqn.(4.22), with the conditions ($\alpha > \beta$, $\beta < 1$) or ($\alpha < \beta$, $\beta > 1$). In the limiting cases of $\alpha \to 1$ or $\beta \to 1$, this one-sided measure degenerates to Kullback-Leibler or Harvada-Charvát (one-sided) metrics as appropriate.

Inasmuch as ε_{SM} of Eqn. (4.22) is asymmetric, its symmetric version can be constructed as follows:

$$\varepsilon_{SM}^{Sym} = \varepsilon_{SM(1)} + \varepsilon_{SM(2)} \tag{4.58}$$

In the above expression (Eqn. (4.58)) $\varepsilon_{SM(1)}$ represents Eqn. (4.22) with $\Phi(x) = (x^{\alpha} - x^{\beta})/(\alpha - \beta)$ ($x = p/q$; $\alpha > 1$, $\beta \leq 1$; or, $\alpha < 1$, $\beta \geq 1$); and, $\varepsilon_{SM(2)}$ is obtained by substituting $x = q/p$. Further, both α and β control the asymmetry or skewness of the functions. However, regardless of the relative values of α and β, symmetrization renders the resulting error-metric to facilitate the neural network towards a convergence performance.

It should also be noted that despite Sharma-Mittal function (ε_{SM}) belonging to a nonadditive sum generalization, as discussed in an earlier section, symmetrization overcomes any associated incompatibility of these functions in being an error-metric (in symmetrized form, ε_{SM}^{Sym}) for neural network training in the information-theoretic plane. Figs. 4.35 to 4.37 are relevant results obtained for ε_{SM} and/or ε_{SM}^{Sym}. The predicted function using ε_{SM}^{Sym} in the network by Fig. 4.4 is shown in Fig. 4.38.

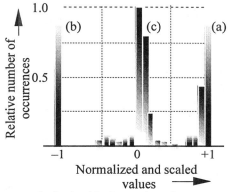

Fig. 4.35: Histograms obtained with the number of occurrences of the normalized and scaled error-values versus error-values with $\alpha = \beta = 0.5$, obtained from the simulations of the network of Fig. 4.4

(a)One-sided Sharma and Mittal measure, $\varepsilon_{SM(1)}$; (b) One-sided Sharma and Mittal measures, $\varepsilon_{SM(2)}$; (c) Symmetrized measure $(\varepsilon_{SM(1)} + \varepsilon_{SM(2)}) = \varepsilon_{SM}^{Sym}$

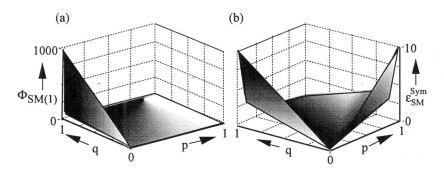

Fig. 4.36: Sharma and Mittal error-measure

(a) Φ-function of one-sided $\varepsilon_{SM(1)}$ *versus* the probabilities p and q
(The corresponding Φ-function of one-sided $\varepsilon_{SM(2)}$ *versus* p and q is the mirror-reflection of the above pattern).

(b) Symmetrized error-measure $(\varepsilon_{SM(1)} + \varepsilon_{SM(2)}) = \varepsilon_{SM}^{Sym}$ *versus* the probabilities p and q with $\alpha = \beta = 0.5$

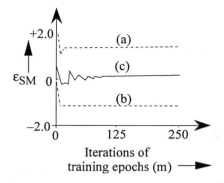

Fig. 4.37: Learning curves obtained with a single training run: Dynamics of the error-measure $(\varepsilon_{SM(1)} + \varepsilon_{SM(2)}) = \varepsilon_{SM}$ versus number of iterations of the training epochs (m) (Training parameters: $Q = 2.0$ and $\eta = 0.001$ in the simulations pertinent to the test network of Fig. 4.4)

(a) $\varepsilon_{SM(1)}$ (One-sided/asymmetric)

(b) $\varepsilon_{SM(2)}$ (One- sided/asymmetric)

(c) $(\varepsilon_{SM(1)} + \varepsilon_{SM(2)}) = \varepsilon_{SM}^{Sym}$ (Two-sided/ symmetric)

(*Note*: ε values are scaled: $\varepsilon = \varepsilon_0 = 0 \Rightarrow$ Equilibrium/convergence status; and $\varepsilon = \pm 1 \Rightarrow$ Divergent levels)

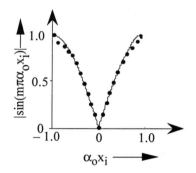

Fig.4.38: Teacher function (——) and the predicted function (.....) obtained with the two-sided, symmetrized error-measure $(\varepsilon_{SM(2)} + \varepsilon_{SM(2)}) = \varepsilon_{SM}^{Sym}$ using the perceptron network of Fig. 4.4 (Network parameters used: $Q = 2.0$ and $\eta = 0.001$)

4.9.8 Symmetrized Rényi measure

A measure due to Renyi which was developed as an information gain of order α was discussed in subsection 4.3.5. The expression given by Eqn.(4.23) is, symmetric depending on the order factor α. The corresponding Φ-functions of the constituent asymmetric parts of Eqn. (4.23) are given by

$$\Phi_{1(RY)} = \left[\exp(x^{\alpha}) - \exp(1)\right]/(\alpha - 1) \qquad (4.59a)$$

and

$$\Phi_{2(RY)} = \left[\exp(1/x^{\alpha}) - \exp(1)\right]/(\alpha - 1) \qquad (4.59b)$$

where $x = (p/q)$.

In order to make the relevant error-metric symmetrized, the following function is constructed as proposed earlier:

$$\varepsilon_{RY}^{Sym} = \varepsilon_{RY} = \varepsilon_{1(RY)} + \varepsilon_{2(RY)} \qquad (4.60)$$

(same as Eqn. (4.23))

with the corresponding Φ-function given by:

$$\Phi_{e(RY)} = \left[\exp(x^{\alpha} + 1/x^{\alpha}) - \exp(2)\right]/(\alpha - 1) \qquad (4.61)$$

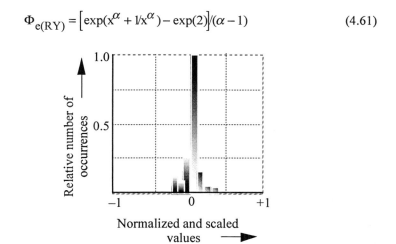

Fig. 4.39: Histogram of the number of occurrences of the error-values versus normalized and scaled error-values, obtained from the simulations of the network of Fig. 4.4. (Two-sided symmetric measure ε_{RY} with $\alpha = 0.5$)

The one-sided Rényi measures namely $\varepsilon_{1(RY)}$ and $\varepsilon_{1(RY)}$, also tend to the Kullback-Leibler measure as $\alpha \to 1$. The results corresponding to ε_{RY} are presented in Figs. 4.39 and 4.40. The learning curve and function prediction relevant to ε_{RY} are presented in Fig. 4.8a(I) and Fig. 4.8a(II), respectively.

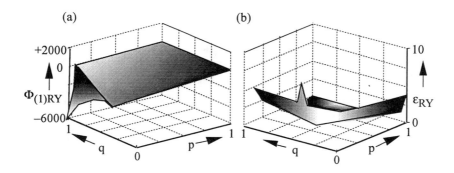

Fig. 4.40: The Réyni error-measure
(a) $\Phi_{1(RY)}$ corresponding to of the one-sided error-measure ($\varepsilon_{1(RY)}$) versus
   the probabilities p and q with $\alpha = 0.5$
   ($\Phi_{2(RY)}$ will be a mirror-reflection of $\Phi_{1(RY)}$)
(b) Réyni error-measure ε_{RY} versus the probabilities p and q for $\alpha = 0.5$

4.9.9 Symmetrized Kapur type 2 error-measure

Kapur [4.28] has developed another type of directed distance measure based on the exponential format of information/entropy functions. It is based on a weighted sum of two Réyni's measures. It has two parameters α and β which jointly determine the extent of asymmetry.

The expression given by Eqn. 4.24 refers to a Kapur type 2 error-measure which can be written modified in a more explicit form as:

$$\varepsilon_{1KP(2)} = \{[\beta/(\alpha - 1)\ell og[p^{\alpha}q^{(1 - \alpha)}] + \{[\alpha/(\beta - 1)\ell og[q^{\beta}p^{(1 - \beta)}]\}/(\alpha + \beta)$$

$$(4.62)$$

and

$$\varepsilon_{2KP(2)} = \{[\beta/(\alpha - 1)\ell og[p^{\alpha}q^{(1 - \alpha)}] + \{[\alpha/(\beta - 1)\ell og[q^{(1 - \beta)}p^{\beta}]\}/(\alpha + \beta)$$

$$(4.63)$$

where $(\alpha \neq \beta)$ and the corresponding Φ-functions are

$$\Phi_{1KP(2)} = [\exp(x^{\alpha}) - \exp(x^{\beta})]/(\alpha - \beta) \qquad (4.64)$$

and

$$\Phi_{2KP(2)} = [\exp(1/x^{\alpha}) - \exp(1/x^{\beta})]/(\alpha - \beta) \qquad (4.65)$$

where, as defined earlier $x = (p/q)$. Hence, a symmetrized measure can be constructed as follows:

$$\varepsilon_{KP(2)}^{Sym} = \varepsilon_{KP(2)} = \varepsilon_{1KP(2)} + \varepsilon_{2KP(2)} \qquad \text{(same as Eqn. (4.24))} \qquad (4.66)$$

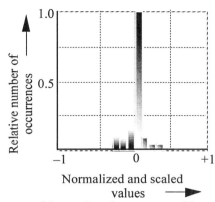

Fig. 4.41: Histogram of the number of occurrences of the error-values *versus* normalized and scaled error-values, obtained from the simulations of the network of Fig. 4.4.
(Two-sided symmetric measure $\varepsilon_{KP(2)}$ with $\alpha = 0.45$ and $\beta = 0.55$)

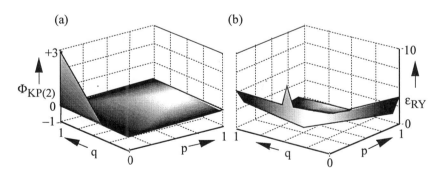

Fig. 4.42: The Kapur's type 2 error-measure

(a) Φ-function of the one-sided error-measure ($\varepsilon_{KP(2)}$) versus the probabilities p and q with $\alpha = 0.45$ and $\beta = 0.55$

• $\varepsilon_{KP(2)}$ versus the probabilities p and q with $\alpha = 0.45$ and $\beta = 0.55$

Kapur type 2 error-measure is similar to Réyni's measure except that it has two weighting factors, namely α and β instead of one as in the Réyni's case. With the stipulated condition, namely $\alpha \neq \beta$, it is attempted in the present

211

work to study the effect of balancing/unbalancing attributes imparted to the expression of Eqn.(4.66) by the chosen values of α and β.

For example, two cases are studied: $\alpha = 0.45$; $\beta = 0.55$. Presented in Figs.4.41 and 4.42 are the various results obtained. The learning curve and prediction performance are presented in Fig. 4.8b(I) and Fig. 4.8b(II), respectively. From these results the following is inferred: Even with an asymmetric choice of α and β, $\varepsilon_{KP(2)}$ enables the convergence of the test network.

4.10 Generalized Csiszár's Symmetrized Error-Measures

4.10.1 Error measure #1

In Section 4.3.5, a set of generalized representations of Csiszár's error-measures were defined and considered as feasible error-measures for neural network applications. They are defined in Eqns. (4.26) through (4.30). Presently, each of these measures is analyzed for the symmetry/asymmetry aspects and symmetrization is implemented in each case so as to elucidate their efficacy as usable error-metrics for neural networks in the information-theoretic plane.

Generalized Csiszár's measure #1 depicted by Eqn. (4.26) as $\varepsilon_{GCZ(1)}$ is in asymmetric form, inasmuch as, when $c \rightarrow 0$, it would reduce to the corresponding Kullback-Leibler expression. Hence, it was observed in Section 4.6 that $\varepsilon_{GCZ(1)}$ measure would fail to lead the network towards convergence.

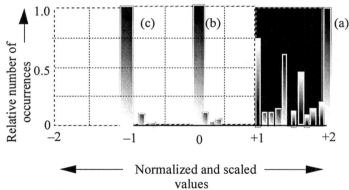

Fig. 4.43: Histograms of the number of occurrences of the error-values *versus* normalized and scaled error-values with c = 0.5, obtained from the simulation of the network of Fig. 4.4.

(a) One-sided general Csizár Type 1 measure, $\varepsilon_{GCZ(1)}$.

(b) One-sided general Csizár Type 1 measures, $\varepsilon^{*}_{GCZ(1)}$

(c) Symmetrized measure $\varepsilon_{GCZ(1)} + \varepsilon^{*}_{GCZ(1)} = \varepsilon^{Sym}_{GCZ(1)}$

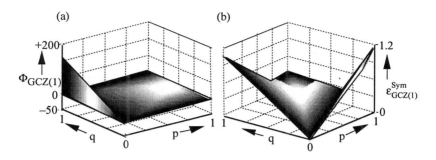

(a)   (b)

Fig. 4.44: The generalized Csiszár's error-measure 1

(a) Φ-function of the one-sided error-measure ($\varepsilon_{GCZ(1)}$) versus the probabilities p and q

(b) Symmetrized error-measure $\varepsilon^{Sym}_{GCZ(1)}$ versus the probabilities p and q with c = 0.5

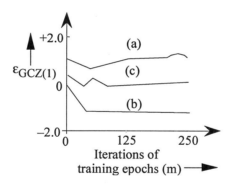

Fig. 4.45: Learning curves obtained with a single training run: Dynamics of the error-measure $\varepsilon^{Sym}_{GCZ(1)}$ versus number of iterations of the training epochs (m)

(Training parameters: Q = 10.0 and η = 0.001 in the simulations pertinent to the test network of Fig.4.4.)

(a) $\varepsilon_{GCZ(1)}$ (One-sided / asymmetric)

(b) $\varepsilon^{*}_{GCZ(1)}$ (One-sided/asymmetric)

(c) $\varepsilon^{Sym}_{GCZ(1)}$ (Two-sided/ symmetric)

(*Note*: ε values are scaled: $\varepsilon = \varepsilon_0 = 0 \Rightarrow$ Equilibrium/convergence status; and $\varepsilon = \pm 1 \Rightarrow$ Divergent levels; and c = 0.5)

Whenever asymmetry exists, if a symmetrization procedure is adopted, it would render the measures suitable as information-theoretic cost-functions for neural network applications. The proposed method, for example, is as follows:

$$\varepsilon_{GCZ(1)}^{Syss} = \varepsilon_{GCZ(1)} + \varepsilon_{GCZ(1)}^{*} \tag{4.67}$$

$$\varepsilon_{GCZ(2)}^{Syss} = \varepsilon_{GCZ(2)} + \varepsilon_{GCZ(2)}^{*} \tag{4.68}$$

where $\varepsilon_{GCZ(1)}^{*}$ and $\varepsilon_{GCZ(2)}^{*}$ are constructed by replacing p by q (or vice versa) in $\varepsilon_{GCZ(1)}$ and $\varepsilon_{GCZ(2)}$. The results corresponding to $\varepsilon_{GCZ(1)}$ measures (with and without symmetrization) are presented in Figs. 4.43 and 4.44. The learning curve and prediction performance of the network shown in Fig. 4.4 are illustrated in Fig. 4.45 and Fig. 4.46, respectively.

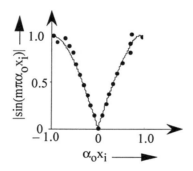

Fig. 4.46: Teacher function (——) and the predicted function (.....) obtained with the two-sided, symmetrized error-measure $\varepsilon_{GCZ(1)}^{Sym}$ with c = 0.5 using the perceptron network of Fig.4.4.
(Network parameters used: Q = 10.0 and η = 0.001)

4.10.2 Error measures # 2 and # 3
Relevant results are indicated in Fig. 4.47 and Fig. 4.48. The learning curve and the prediction performance of $\varepsilon_{GCZ(2)}^{Sym}$ error-measure with a = 0.5 and when used in the network of Fig. 4.4 are shown in Fig. 4.49 and Fig. 4.50, respectively.

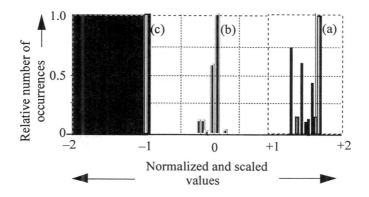

Fig. 4.47: Histograms of the number of occurrences of the error-values *versus* normalized and scaled error-values with a = 0.5, obtained from the simulations of the network of Fig. 4.4

(a) One-sided generalized Csiszár type 2 measure, $\varepsilon_{GCZ(2)}$.

(b) One-sided generalized Csiszár type 2 measure, $\varepsilon^*_{GCZ(2)}$

(c) Symmetrized measure $\varepsilon^{Sym}_{GCZ(2)}$

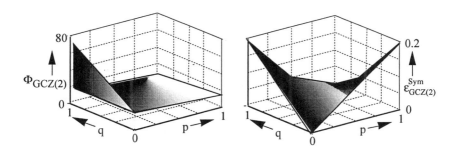

Fig.4.48: Generalized Csiszár error-measure # 2

(a) Φ-function of the one-sided error-measure ($\varepsilon_{GCZ(2)}$) versus the probabilities p and q with a = 0.5.(The corresponding ϕ-function of $\varepsilon^*_{GCZ(2)}$ is a mirror-reflection of ($\varepsilon_{GCZ(2)}$))

(b) Symmetrized error-measure $\varepsilon^{Sym}_{GCZ(2)}$ versus the probabilities p and q with a = 0.5

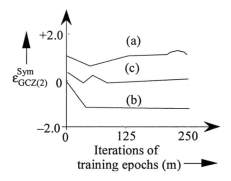

Fig. 4.49: Learning curves obtained with a single training run: Dynamics of the error-measure $\varepsilon_{GCZ(2)} + \varepsilon^*_{GCZ(2)} = \varepsilon^{Sym}_{GCZ(2)}$ versus number of iterations of the training epochs (m)

(Training parameters: Q = 2.0 and η = 0.001 in the simulations pertinent to the test network of Fig. 4.4)

(a) $\varepsilon_{1GCZ(2)}$ (One-sided/asymmetric)

(b) $\varepsilon^*_{GCZ(2)}$ (One- sided/asymmetric)

(c) $(\varepsilon_{1GCZ(2)} + \varepsilon_{2GCZ(2)}) = \varepsilon_{GCZ(2)}$ (Two-sided/symmetric)

(Note: ε alues are scaled: ε = ε_0 = 0 $\Rightarrow$ Equilibrium/convergence status; and ε = ± 1 $\Rightarrow$ Divergent levels and a = 0.5)

A third case of generalized Csiszár measure denoted as $\varepsilon_{GCZ(3)}$ in Eqn. (4.28) is symmetric due to its two constituent parts; and, the factor a > −1 decides the extent of asymmetry. Test results with a = 0.5 as presented in Figs. 4.51 and 4.52 confirm the suitability of this measure as an error-metric leading to convergence, by virtue of its built-in symmetry.

$$\phi\big|_{GCZ(3)} = (p/q)\ell og(p/q) - (1/a)\big[(1 + ap)/q\big]\ell og\big[(1 + ap)/(1 + aq)\big] \qquad (4.69)$$

This ϕ-function as shown in Fig. 5.53 shows only marginal asymmetry when p $\to$ 0 and q $\to$ 1.

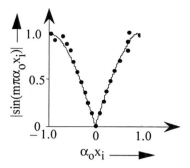

Fig. 4.50: Teacher function (——) and the predicted function (.....) obtained with the two-sided, symmetrized error-measure $\varepsilon_{GCZ(2)} + \varepsilon^*_{GCZ(2} = \varepsilon^{Sym}_{GCZ(2)}$ with a = 0.5 using the perceptron network of Fig. 4.4

(Network parameters used: Q = 2.0 and η = 0.001)

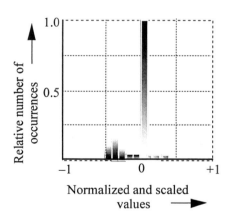

Fig. 4.51: Histogram of the number of occurrences of the error-values versus normalized and scaled error-values, obtained from the simulations of the network of Fig. 4.4.
(Two-sided symmetric measure $\varepsilon_{GCZ(3)}$ with a = 0.5)

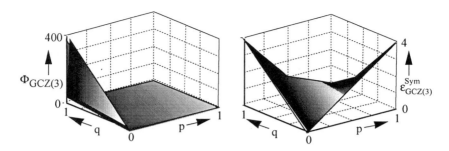

Fig.4.52: Csiszár's generalized error-measure # 3

(a) ϕ-function of the error-measure ($\varepsilon_{1GCZ(3)}$) with a = 0.5 *versus* the probabilities p and q

(b) Symmetrized error-measure $\varepsilon_{GCZ(3)}$ with a = 0.5 *versus* the probabilities p and q

4.10.3 Symmetrized generalized Csiszár error-measures # 4 and # 5

Two more versions of generalized Csiszár error-measures are given in the expression of Eqns. (4.29 and 4.30). Analysis of these measures in the earlier sections proved that $\varepsilon_{GCZ(5)}$ error-measure (Eqn. (4.29)) fails to be compatible for network training; whereas $\varepsilon_{GCZ(5)}$ proved to be successful. Again, the reasons for those observations are that $\varepsilon_{GCZ(4)}$ is inherently asymmetric whereas $\varepsilon_{GCZ(5)}$ has a symmetry, the extent of which is decided by the constants a, b, c, and d chosen. $\varepsilon_{GCZ(4)}$ can, however, be rendered symmetrical by the following algorithmic symmetrizations:

$$\varepsilon^{Sys}_{GCZ(4)} = \varepsilon_{GCZ(4)} + \varepsilon^{*}_{GCZ(4)} \tag{4.70}$$

$$\varepsilon^{Sys}_{GCZ(5)} = \varepsilon_{GCZ(5)} + \varepsilon^{*}_{GCZ(5)} \tag{4.71}$$

where $\varepsilon_{GCZ(4)} = \varepsilon^{*}_{GCZ(4)}$ with p replaced by q and *vice versa*.

Relevant to the above formulation of $\varepsilon^{Syss}_{GCZ(4)}$ the set of computed and/or simulated results are furnished in Figs. 4.53 to 4.56. The effect of symmetrization can be seen in the results presented.

The symmetrized version of $\varepsilon_{GCZ(5)}$ is given by Eqn. (4.71) and can be evaluated from the results shown in Figs. 4.57 and 4.58, as well as from the learning curve of Fig. 4.9b(I) and the prediction performance shown in Fig. 4.9b(II).

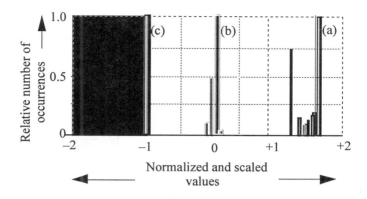

Fig. 4.53: Histograms of the number of occurrences of the error-values *versus* normalized and scaled error-values with a = b = 0.5, obtained from the simulations of the network of Fig. 4.4

(a) One-sided generalized Csiszár type 4 measure, $\varepsilon_{GCZ(4)}$.
(b) One-sided generalized Csiszár type 4 measures, $\varepsilon^*_{GCZ(4)}$.
(c) Symmetrized measure $\varepsilon^{Sym}_{GCZ(4)}$

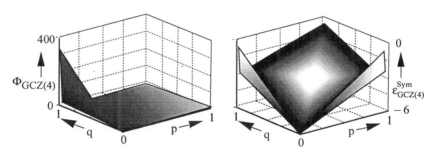

Fig. 4.54: Csiszár's generalized error-measure # 4

(a) Φ-function of the one-sided error-measure ($\varepsilon_{GCZ(4)}$) versus the probabilities p and q with a = b = 0.5
(Eqn. (4.29a) depicts $\varepsilon_{GCZ(4)}$ with x = p/q and a mirror-reflection, namely, $\varepsilon^*_{GCZ(4)}$ corresponding to $\varepsilon_{GCZ(4)}$ can be obtained by substituting x = q/p in Eqn. (4.29a))
(b) Symmetrized error-measure $\varepsilon^{Sym}_{GCZ(4)}$ with a = b = 0.5 versus the probabilities p and q

219

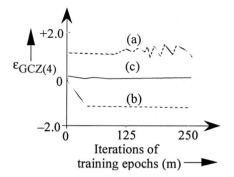

Fig.4.55: Learning curves obtained with a single training run: Dynamics of the error-measures versus number of iterations of the training epochs (m)
(Training parameters: $Q = 10.0$ and $\eta = 0.01$ in the simulation pertinent to the test network of Fig. 4.4)

(a) $\varepsilon_{GCZ(4)}$ (One-sided/asymmetric)

(b) $\varepsilon^{*}_{GCZ(4)}$ (One-sided/asymmetric)

(c) $\varepsilon^{Sym}_{GCZ(4)}$ (Two- sided/symmetric)

(*Note*: ε values are scaled: $\varepsilon = \varepsilon_0 = 0 \Rightarrow$ Equilibrium/convergence status; and $\varepsilon = \pm 1 \Rightarrow$ Divergent levels with a = b = 0.5)

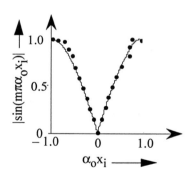

Fig. 4.56: Teacher function (—) and the predicted function (.....) obtained with the two-sided, symmetrized error-measure $\varepsilon^{Sym}_{GCZ(4)}$ with a = b = 0.5 using the perceptron network of Fig. 4.4
(Network parameters used: $Q = 10.0$ and $\eta = 0.0100$)

220

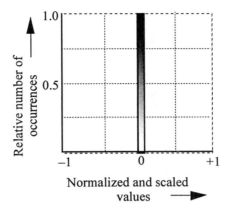

Fig. 4.57: Histogram of the number of occurrences of the error-values (with $a = 1.0$, $b = 0.5$, $c = 0.5$ and $d = 1.0$) versus normalized and scaled error-values, obtained from the simulations of the network of Fig. 4.4.
(Two-sided symmetric measure $\varepsilon_{GCZ(5)}$.)

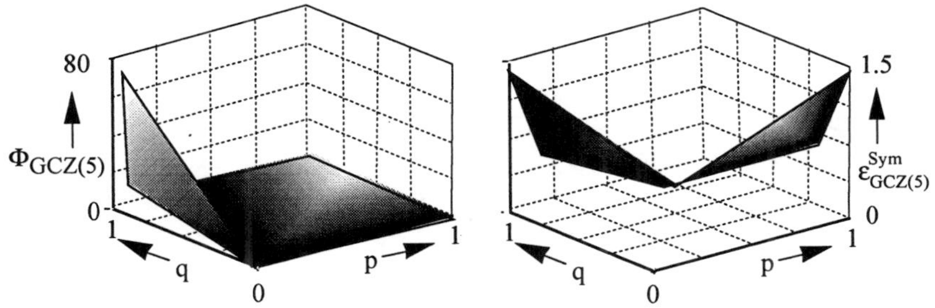

Fig. 4.58: Csiszár's generalized error-measure # 5

(a) Φ-function of the error-measure ($\varepsilon_{1GCZ(5)}$) with $a = 1.0$, $b = 0.5$, $c = 0.5$ and $d = 1.0$) *versus* the probabilities p and q

(b) Symmetrized error-measure $\varepsilon_{GCZ(5)}$ (with $a = 1.0$, $b = 0.5$, and $d = 1.0$) *versus* the probabilities p and q

(Note: The partial asymmetry in this plot is decided by the choice of a, b, c, and d)

4.11 Concluding Remarks

A set of error-measures defined in the information-theoretic plane compatible for neural network optimization problems *via* gradient descent learning using backpropagation algorithm is presented in this chapter. Designated as the Csiszár family of error-measures, these measures are based on cross-entropy (relative/mutual information) considerations vis-à-vis the statistics of the network's output and the teacher function. Hence, these measures refer to network optimization in the information-theoretic plane. Relevant on-line experimental simulations performed on a multilayered neural network (perceptron) indicate the following: The Csiszár's measures which are based essentially on constructing a cross-entropy functional (Φ) that satisfies certain constraints, can be formulated in a variety of forms. However, unless otherwise the chosen function (Φ) enables a symmetrized functional form of the cross-entropy, as well as that the associated entropy relations are of additive structure, the measures developed may not facilitate the network convergence. Omitting such functionals, however, a class of usable error-measures can still be comprehended. Typical formulations developed thereof, proved to be compatible error-measures for neural network optimization problems as confirmed by the test results obtained from the simulation studies performed on a multilayer perceptron.

The essence of this chapter is to indicate the feasibility of evolving a new class of error-measures exclusively in the information-theoretic plane. These error-measures can be regarded as candidate algorithms to represent a set of cost-functions useful in supplementing those adopted conventionally in neural network training strategies.

The information-theoretic error-metrics are also compatible for neural network training when the input pattern refers to a set of random values. For example, in [4.37] a neural network has trained successfully with random data collected from alluvial river geometry using Jensen-Shannon error-measure.

A code was written in C (Borland C^{++} for Windows, version 3.1) to simulate the network shown in Fig. 4.4 and implement training using the various algorithms on error-measures discussed in this chapter. Also, the same code facilitates the prediction of an input pattern.

The research efforts described in this chapter thus refer to formulating algorithms for the error-measure derivable from the Kullback-Leibler/Csiszár family of cross-entropy or mutual information functions. These error-measures are suitable for training neural networks in the information-theoretic plane. Further, a procedure can be adopted towards symmetrization of one-sided algorithms of Kullback-Leibler and/or Csiszár family of cross-entropy measures so as to enable them to be useful for neural network applications. The basis of symmetrization is as follows: The so-called discrimination function, derived as a cross-entropy measure, represents the expected value of the log-likelihood ratio concerning a probability distribution, namely q with respect to p. Conventionally, in communication/radar problems a metric is evolved as the constituting bounds on the probability of error in the hypothesis testing problems relevant to $q(T_i)$ vis-à-vis $p(O_i)$. However, in order to conceive a bilateral influence between the observations $\{O_i\}$ and $\{T_i\}$ (or vice versa) as required in neural network applications, the invariance of the parameters (in a law of chance) for all transformations of the parameters of O_i with respect to T_i or vice versa requires a symmetrization procedure. That is, if a cross-entropy-based discrimination function is defined, for example, *via* $I(p/q)$ or $I(q/p)$ (see Eqns. (4.40) and (4.41)), Jeffreys' suggestion [4.20] is to construct a J-measure which is a linear combination of $I(p/q)$ and $I(q/p)$. Such a measure is designated as stated earlier as the distance or divergence metric.

The question of applying the one-sided discrimination functions of Kullback-Leibler and/or Csiszár family as considered first in this chapter; has indicated that the one-sided measures, in general, are incompatible for neural network training. However, a few measures, as exceptions, with partial or marginal asymmetry did enable the network's performance to converge.

Further inferences that could be deduced from this study are:

1) Measures which are the totally one-sided (asymmetric) type of discrimination functions cannot be used as such in neural network training. Rather, they have to be symmetrized via the J-divergence (superposition of the two bilaterally asymmetric measures) procedure.

2) Those discrimination error-functions which do enable the network's performance to converge are seen to be totally or at least only marginally asymmetric, with the extent of asymmetry being decided by the associated constant coefficients. Therefore, these measures, which are largely symmetric, have been found to be useful error-metrics for the applications discussed, even without the symmetrizatiion procedure being implemented.

Table 4.1 presents the error measures and Table 4.2 lists the symmetrized formulations and their effectiveness as error measures for ANN applications.

Table 4.1: Error-measures adopted in the simulations performed $\{Q,\eta\}$ used for minimum E and results on the convergence performance

No.	Family of error measures and remarks	Notation used	$\{Q, \eta\}$ for minimum error	
			Q	η
I	Square-error Fig. 4.5 $(m = 0.5)$	ε_{SE}	0.5	0.001
II	Relative-entropy Fig. 4.5 $m = 0.5$	ε_{RE}	2.0	0.001
III				
1.	Kullback-Leibler	ε_{KL}	-	-
2.	Jensen Fig. 4.6 $m = 0.5$	ε_J	0.5	0.001
3.	Jensen-Shannon Fig.4.6 $m = 0.5;$ $\pi_1 = \pi_2 = 0.5$	ε_{JS}	10.0	0.010
4.	Symmetrical weighted Kullback-Leibler Fig. 4.7 $m = 0.5; \mu = 0.5$	ε_{WKL}	0.5	0.001
IV				
1.	Kapur type 1 Fig. 4.7 $m = 0.5; a = 0.5$	$\varepsilon_{KP(1)}$	0.5	0.001
2.	Havrda and Charvát	ε_{HC}	-	-

(Continued) ...

3.	Sharma and Mittal	ε_{SM}	-	-
4.	Rènyi	ε_{RY}	0.5	0.0001
	Fig. 4.8 m = 0.5; α = 0.5			
5.	Kapur type 2	$\varepsilon_{KP(2)}$	0.5	0.001
	Fig. 4.8 m = 0.5; α = 0.45 β = 0.5			
6.	Generalized Csiszár type 1	$\varepsilon_{CSZ(1)}$	-	-
7.	Generalized Csiszár type 2	$\varepsilon_{CSZ(2)}$	-	-
8.	Generalized Csiszár type 3	$\varepsilon_{CSZ(1)}$	0.5	0.001
	Fig. 4.9 m = 0.5 a = d =1.0 c = b = 0.5			
V				
1.	Generalized Csiszár type 4	$\varepsilon_{CSZ(4)}$	-	-
2.	Generalized Csiszár type 5	$\varepsilon_{CSZ(5)}$	2.0	0.010
	Fig. 4.9 m = 0.5; α = 0.5			

Table 4.2: Summary of error measures and symmetrization considerations

No.	Family of error measures	Type	Error measures
I	Square-error between output and teacher values (Parametric type)	Square-error ε_{SE}	ε_{SE}
II	Relative entropy between output and teacher values (Parametric type)	Relative entropy ε_{RE}	Symmetrized function? Yes ε_{RE}
III	Kullback-Leibler family (Cross-entropy types: Information-theoretic versions)	Kullback-Leibler $\varepsilon_{1(KL)}$ and $\varepsilon_{2(KL)}$	Symmetrized function? No. (For any value of Q and η) One-sided type
		Jensen ε_J	Symmetrized function? Yes. $\varepsilon_J = \varepsilon_{1(KL)} + \varepsilon_{2(KL)}$
		Jensen-Shannon ε_{JS}	Symmetrized function? Yes. $\pi_1\varepsilon_{1(KL)} + \pi_2\varepsilon_{2(KL)}$ $\pi_1 < 1.0; \pi_2 < 1.0$ and $(\pi_1 + \pi_2) = 1$
		Symmetrical and weighted Kullback-Leibler ε_{WKL}	Symmetrized function? Yes. ε_{WKL} with $0 \leq \mu \leq 1$
		Modified Jensen ε_{MJ}	Symmetrized function? No. (For any value of Q and η) Symmetrized version: $\varepsilon_{1MJ} + \varepsilon_{2MJ}$

(Continued)

		Generalized Jensen ε_{GJ}	Symmetrized function? No. (For any value of Q and η) One-sided type Symmetrized version $\varepsilon_{GJ} = \varepsilon_{1(GJ)} + \varepsilon_{2(GJ)}$ with $\alpha \neq 0, 1$
		Kapur type 1 $\varepsilon_{KP(I)}$ with a = 0.5	Symmetrized function? Yes.
IV	Csiszár family (Cross-entropy types: Information theoretic versions)	Havrda and Charvát ε_{HC}	Symmetrized function? No. (For any value of Q and η) Nonadditive entropy Symmetrized version: $\varepsilon_{HC} = \varepsilon_{1(HC)} + \varepsilon_{2(HC)}$ with $\alpha = 0.5$
		Sharma and Mittal ε_{SM}	Symmetrized function? No. (For any value of Q and η) Nonadditive entropy Symmetrized version: $\varepsilon_{SM} = \varepsilon_{1(SM)} + \varepsilon_{2(SM)}$ with $\alpha = 0.5$; $\beta = 1.5$
		Rènyi ε_{RY}	Symmetrized function ? Yes.
		Kapur type 2 $\varepsilon_{KP(II)}$	Symmetrized function ? Yes.
		Generalized Csiszár type1 $\varepsilon_{CSZ(1)}$	Symmetrized function? No. (For any value of Q and η) One-sided type Symmetrized version: $\varepsilon_{GCZ(1)}^{Sys} = \varepsilon_{GCZ(1)} + \varepsilon_{GCZ(1)}^{*}$ with c = 0.5
		Generalized Csiszár type2 $\varepsilon_{CSZ(2)}$	Symmetrized function? No. (For any value of Q and η) One-sided type Symmetrized version: $\varepsilon_{GCZ(2)}^{Sys} = \varepsilon_{GCZ(2)} + \varepsilon_{GCZ(2)}^{*}$ with a = 0.5

(Continued)

Generalized
Csiszár type3
$\varepsilon_{CSZ(3)}$ Symmetrized function? Yes.

V   Other versions of   Generalized
        generalized Csiszár's   Csiszár type 4   Symmetrized function? No.
        family of error $\varepsilon_{CSZ(4)}$ (For any value of Q and η)
        measures   One-sided type
                                                          Symmetrized version:
                                          $\varepsilon_{GCZ(4)}^{Sym} = \varepsilon_{GCZ(4)} + \varepsilon_{GCZ(4)}^{*}$
                                                 with a = 0.5   b = 0.5
                             Generalized
                             Csiszár type5
                               $\varepsilon_{CSZ(5)}$ Symmetrized function? Yes.
                                               a = 1.0; b = 0.5; c = 0.5;
                                                          d = 1.0

Bibliography

[4.1] Anderson, J. A., and Rosenfeld, E.: (Eds.) *Neurocomputing: Foundation of Research* (MIT Press, Cambridge, MA: 1988)

[4.2] Rosenblatt, F.: *Principle of Neurodynamics: Perceptron and the Theory of Brain Mechanisms* (Spartan Books, Washington, D. C.: 1961)

[4.3] Rumelhart, D. E. and McClelland, J. L. (Eds.): *Parallel Distributed Processing*: Vol. I (MIT Press, Cambridge, Ma: 1986).

[4.4] Shannon, C. E: A mathematical theory of communication. *Bell System Technical Journal*, 27, 1948, 329-423, 623-659.

[4.5] Neelakanta, P. S. and De Groff, D.: *Neural Networks Modeling: Statistical Mechanics and Cybernetic Perspectives* (CRC Press, Boca Raton, FL.: 1994).

[4.6] Solla, S. A., Levin, E. and Fleisher, M.: Accelerated learning in a layered neural network. *Complex Syst.*, 2, 1988, 625-640.

[4.7] Watrous, R. L.: A comparison between squared error and relative entropy metrics using several optimization algorithms. *Complex Syst.*, 6, 1992, 495-505.

[4.8] Baum, E. B. and Wilczek, F.: Supervised learning of probability distributions by neural networks. In D. A. Anderson (Ed.) *Neural Information Processing Systems*. (American Institute of Physics, New York: 1988)

[4.9] Ackley, D., Hinton, G. E. and Sejnowski, T. J: A learning algorithm for Boltzmann machines. *Cognit. Sci.,* 9, 1985, 147.

[4.10] Liou, C. Y. and Lin, S. L.: The other variant Boltzmann machine. *Proc. Joint Conf. Neural Networks* (June 18-22, 1990, Washington, D.C.), pp. I 449- I 454.

[4.11] Csiszár, I.: Information-type measures of difference of probability distributions and indirect observations. *Studia Sci. Math. Hungar*, 2, 1967, 299-318.

Information-Theoretic Aspects of Neural Networks

[4.12] Csiszár, I.: A class of measures of informativity of observation channels. *Perio. Math. Hungaria*, 2, 1972, 191-213.

[4.13] Csiszár, I. and Körner, J.: *Information Theory, Coding Theorems for Discrete Memoryless Systems*. (Akademiai Kaidó, Budapest: 1981).

[4.14] Neelakanta, P. S., Sudhakar, R. and De Groff, D.: Langevin machine: A neural network based on stochastically justifiable sigmoidal function. *Biol. Cybern.*, 65, 1991, 331-338.

[4.15] Blahut, R. E.: *Principle and Practice of Information Theory* (Addison-Wesley Publishing Co., Reading, MA: 1987).

[4.16] Kapur, J. N. and Kesavan, H. K.: *Entropy Optimization Principles with Applications* (Academic Press/Harcourt Brace Jovanovich Publishers, Boston, MA: 1992).

[4.17] Kullback, S.: *Information Theorey and Statistics* (John Wiley and Sons, New York, NY: 1959).

[4.18] Kullback, S. and Leibler, R. A.: On information and sufficiency. *Ann. Math. Stat.*, 22, 1951, 79-86.

[4.19] Österrecicher, F. and Vajda, I.: Statistical information and discrimination. *IEEE Trans. Inform. Theory*, 39, 1993,1036-1039.

[4.20] Jeffreys, H.: An invariant form for the prior probability in estimation problems. *Proc. Roy. Soc. Lon. Ser. A*, 186, 1946, 453-461.

[4.21] Lin, J.: Divergence measures based on Shannon entropy. *IEEE Trans. on Inform. Theory*, 17, 1991, 145-151.

[4.22] Aczél, J.: Measuring information beyond communication theory. *Inform. Mgmt. Process.*, 20, 1984, 383-395.

[4.23] Kapur, J. N.: Measures of uncertainty mathematical programming and physics. *J. of the Indian Soc. of Agri. Stat.*, 24, 1972, 47-66.

[4.24] Havrda, J. and Charvát, F.: Quantification method of classification processes. *Kybernetika*, 3, 1967, 30-35.

[4.25] Rényi, A.: On measures of entropy and information. In *Proceeding of the Fourth Berkeley Symposium of Mathematical Statistics and Probability*, Vol. 1, 1961, pp.547-561.

[4.26] Sharma, B. D. and Mittal, D. P.: New non-additive measures of entropy for discrete probability distributions. *J. Math. Sci.*, 10,1975, 28-40.

[4.27] Varma, R.S.: Generalizations of Renyí's entropy of order α. *J. Math. Sci.*, 1, 1966, 34-48.

[4.28] Kapur, J.N.: Generalized entropy of order α and β. *The Mathematics Seminar*, 4,1967, 78-94.

[4.29] Abu-Mostafa, Y. S.: Learning from hints in neural networks, *J. Complexity*, 6, 1990, 192-198.

[4.30] Abu-Mostafa, Y. S.: Financial market application of learning from hints. In: Refenes, A.N.(Ed.): *Neural Networks in Capital Markets* (John Wiley, England: 1994).

[4.31] Abu-Mostafa, Y.S.: A method for learning from hints. In: Hanson, J. et.al (Eds.): *Advances in Neural Information Processing Systems*, Vol.5 (Morgan Kaufmann 1993), pp. 73-80.

[4.32] Kjaer, T. W., Hertz, J. A. and Richmond, B. J. Decoding of cortical neuronal signals: Network models, information estimation and spatial tuning. *J. Comput. Neurosci.*, 1, 1994, 109-139.

[4.33] DeGroff, D., Neelakanta, P. S., Sudhakar, R. and Medina, F.: Liquid crystal model of neural networks. *Complex Syst.*, 7, 1993, 43-57.

[4.34] Bose, N. K. and Liang, P.: *Neural Network Fundamentals with Graph, Algorithms, and Applications*, (McGraw-Hill, Inc., New York, NY: 1996).

[4.35] Kosko, B.: *Neural Network and Fuzzy Systems: A Dynamical Systems Approach to Machine Intelligence.* (Prentice Hall, Englewood Cliffs, NJ: 1992).

[4.36] Daróczi, Z.: Generalized information functions. *Inform. Control*, 16, 1970, 36-51

[4.37] Hoffman, D. C.: *Artificial Neural Network Prediction of Alluvial River Geometry*. M. S. E. Thesis, Department of Ocean Engineering, Florida Atlantic University, Boca Raton, Florida, USA, August 1995.

[4.38] Neelakanta, P. S., Abusalah, S., De Groff, D., Sudhakar, R. and Park, J. C.: Csiszár's generalized error-measures for gradient-descent based optimizations in neural networks using the back propagation algorithm. *Connection Sci.*, 8,1996,79-114.

[4.39] Abusalah. S: Studies on Nonlinear Activity and Cross-Entropy Considerations in Neural Networks. Ph.D. Dissertation, Department of Electrical Engineering, Florida Atlantic University, Boca Raton, Florida, USA, April 1996.

Chapter 5

Dynamics of Neural Learning in the Information-Theoretic Plane

P.S. Neelakanta

Teleologically speaking...

(in **finalism**), *"the end itself acts as a cause* (driving the process to terminate) *at the predetermined end state;* ...

(in **vitalism**), *"some spritual agency directs the process to a predetermined end";* ... (and),

(in the **mechanism**), *"the process is determined by its properties at the beginning ...";* (it) *"involves the information channels conveying instructions to the effector telling it what to do and information back to the controller telling it how the effector has got on ...."*

-P.Calow
Biological Machines:
A Cybernetic Approach to Life

5.1 Introduction

Learning in reference to the real neural complex depicts progressive modifications occurring at the synaptic levels of the interconnected neurons. The presence of inherent intraneural disturbances or any extraneural noise in the input data and/or in the teacher values may, however, affect such synaptic modifications specified by the set of weighting vectors of the interconnections. The noise considerations, when translated to artificial neurons refer to inducing an offset in the convergence performance of the network in striving to reach the goal or the objective value via the supervised learning procedure implemented. The dynamic response of a learning network when the target itself changes with time can be studied in the information-theoretic plane. The relevant nonlinear (stochastical) dynamics of the learning process can be assessed by the Fokker-Planck equation, in terms of a conditional-entropy or mutual information based error-measure elucidated from the probabilities associated with the input and teacher (target) values. In this chapter, the logistic growth (evolutionary aspects) and certain attractor features of the learning process are described and discussed in reference to neural manifolds using the mathematical foundations of statistical dynamics. On-line simulation studies on a test multilayer perceptron are presented and the asymptotic behavior of accuracy, speed of learning *vis-à-vis* the convergence aspects of the test-error-measures are elucidated. When fuzzy attributes are

0-8493-3198-6/99/$0.00+$.50
© 1999 by CRC Press LLC

given to the neural variables, relevant learning dynamics have been specified by a *fuzzy Fokker-Planck equation* and discussed.

Pertinent to learning processes in the neural complex, it is well-known that synaptic modifications (specified by a vector array of adjustable weighting parameters $\mathbf{w}_1$) could be influenced by the inevitable presence of intraneural disturbances which would affect invariably the network's convergence towards equilibrium. Further, in the event of the input data and/or the teacher values themselves being stochastical, the corresponding extraneural influence may also augment the entropy of the system (real or artificial), facilitating the eventual veering of the network's output from the equilibrium value/stable state. Relevant neurodynamic considerations governing the variable $\mathbf{w}_1$ in artificial neural networks have been addressed by the author elsewhere [5.1] in terms of a stochastical differential equation of Langevin or Fokker-Planck type. Also, the dynamic states of the architectures, such as the Hopfield's network subjected to white noise (random) inputs, have been analyzed via Ito type stochastical differentiatial equations applied to the so-called "diffusion machine" [5.2].

Here, an alternative approach is presented to describe the learning dynamics of an artificial neural network in the presence of destabilizing factors caused by intra- or extraneural influences. The stochastical variable considered presently to model the relevant nonlinear neural dynamics refers to an error-measure parameter evaluated in the information-theoretic plane. Though a limited extent of neural dynamics considerations have been addressed in the information-theoretic plane pertinent to biological neurons [5.3,5.4], equitable study or considerations vis-á-vis artificial neural network are rather sparse. Amari et al. [5.5] have described a basic neural mainfold being embedded as a submanifold in the manifold of a general nonneural information-processing system. They have developed an "information geometry" method to apply the information-theoretic approach to learning dynamics and pattern classification problems. Further, the dynamics of an ensemble of learning processes in a changing environment which feeds the training inputs to the network has been described by Heskes and Kapper [5.6] via a continuous-time master equation.

In the present study, the approach presented is concerned with the logistic growth considerations pertinent to the network's learning process in the information-theoretic plane. Relevant to this proposed method, a cross-entropy (or mutual information) based distance measure (ε) is specified as a stochastical variable, the asymptotic behavior of which, with respect to time, is studied as a discourse of the learning process. It is given by the following relation.

$$\varepsilon = H_\varepsilon(p_\ell, q_\ell) = K \sum_{\ell=1}^{N} q_\ell \phi(p_\ell/q_\ell) \qquad (5.1a)$$

or,

$$\varepsilon = H_\varepsilon(q_\ell, p_\ell) = K \sum_{\ell=1}^{N} p_\ell \phi(q_\ell/p_\ell) \tag{5.1b}$$

where ϕ is a twice-differentiable convex function for which $\phi(1) = 0$ and K is a constant factor. This error-measure is adopted to train a neural network (depicted in Fig.5.1a) via a gradient-descent algorithm in the backpropagation mode [5.7]. The output of the network (Fig.5.1a), namely, O_ℓ at the ℓth, cell is assumed to have probabilities p_ℓ, and q_ℓ. This refers to a target set of probabilities with $1 = (1,2,3,...N)$ enumerating the number of the cells and thereby offering an index for the output units. The error function given by Eqn. (5.1) is same as the Csiszár error-measure [5.8] defined in the information-theoretic plane. When $\phi(y) = y\ell og(y)$ (with y = p/q_ℓ, or p/q_ℓ as appropriate), this measure is better known as the Kullback-Leibler measure [5.9] as indicated in Chapter 4.

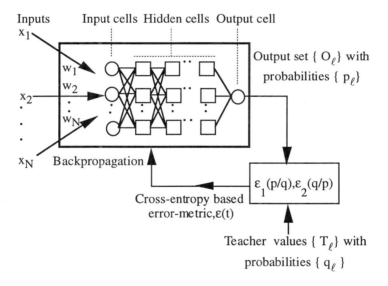

Fig. 5.1a: An artificial neural network trained via cross-entropy based error-metric in the information-theoretic plane using backpropagation mode

Bergström and Nevanlinna [5.10] have justified the entropy attribution to the activities of the neural complex in general, and also specifically to the real neurons. Their strategy is based on the considerations of the principle of conservation of total "neural energy", its distribution and the associated entropy. They have offered an operational definition of the *macrostate* of a neural system (in the same sense as in physical theromodynamic principles) and have associated it with the Shannon's concepts of information [5.11]. Disturbances in the real neural system caused by the enviroment have been perceived in [5.10] as forces enhancing the associated entropy (or uncertainity) and correspondingly reducing the information content which would otherwise enable physiological self-regulation.

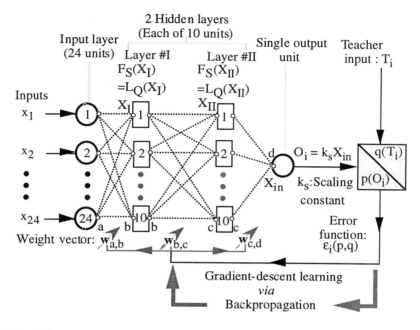

Fig. 5.1b: Test neural network: A multilayered perceptron simulated for learning strategies in the information-theoretic plane

These existing bases on the real-neural information-processing offer a direction to extend the entropy (or information theory) based concepts to optimization algorithms used in artificial neural networks. The error-measure indicated in Eqn.(5.1) is a time-dependent stochastical variable specified over the epochs of iterations performed towards convergence and mediated through feedback strategies (such as the backpropagation algorithm) in the network. Due to the presence of any intra- and/or extracellular disturbances, the associated information flow in the neural system would, however, degrade with time. Therefore, the proliferation of information across the network may even become obsolete or nonpragmatic due to the asynchronous (random) synaptic delays between the internal state variable being adjusted towards learning and the adjusting influence (information) imparted via the control-loop to the network by the error-measure. That is, an aging of neural information, or degenerative negentropy, may occur which would lead to a devalued (or a value weighted) knowledge with reduced utility (or pragmatics) being available to the converging efforts of the network striving towards the goal or objective function. The degradation so perceived in the neural information plane depends on the extent of asynchronous delays encountered when the control-loop (error) information arrives at the controlling section. That is, the asynchronously delayed error-measure fed back will have no pragmatic value inasmuch as the asynchronous characteristics will not reflect the true (natural) output state because the global state of the neural complex would have changed considerably by then [5.12].

5.2 Stochastical Neural Dynamics

The trajectory of the time-dependent neural process pertinent to the learning endeavor represents the evolution of ε with time in a stochastical nonlinear dynamic system. Hence, in reference to the variable $\varepsilon(t)$ describing the neural stochastical dynamics, a differential time-evolution relation can be prescribed for $\varepsilon(t)$ as follows:

$$d\varepsilon(t)/dt = F_1[\varepsilon(t)] \tag{5.2}$$

Or, it can be specified by a recursive discrete-time process as follows:

$$\varepsilon(t+1) = F_2[\varepsilon(t)] \tag{5.3}$$

where F_1 (or F_2) is a differentiable function. It can be noted that ε, in general, is an N-dimensional vector affiliated with the phase-space containing the time-evolution of the underlying neural process spanned by the N-dimensional state vector of a dynamic system. A portrait of the corresponding time-evolution of this system is, therefore, constituted by a set of trajectories in the N-dimensional phase-space. When the system reaches a state of permanent regime where the trajectories stay bounded, the corresponding invariant subset is termed as an "attractor", specifying a state of convergence towards a stochastical equilibrium.

Relevant to an error-measure ε adopted in network training so as to facilitate the neural complex to learn from the environmental inputs, the neural dynamics can be described by a *stochastical differential equation* of the general types given by Eqn.(5.2). Both the conventional types of error-measure such as the quadratic error-measure, as well as any error-measure which can be evaluated in the information-theoretic plane, (on cross-entropy basis as given by Eqn. (5.1)) can be considered to follow the paradigm of stochastical dynamics along the temporal passage of iterative epochs facilitated via feedback methods, until the error value is minimized. To assess the approach characteristics of the error parameter $\varepsilon(t)$ towards an equilibrium value (attractor) over a period of time (that is, over the iterations of learning epochs) in order to develop an explicit dynamic model, the following valid assumptions can be made:

- The parameters which decide the stochastical aspects of $\varepsilon(t)$ are confined within the basin of attraction

- The initial conditions of the stochastic process (ε_0, t_0) involved should be known *a priori* and specified appropriately

- The process is likely to be attracted to a stationary stochastic process whose probability density function (pdf) can be uniquely decided by the parameters of the original system variable, namely, ε

- In view of the above assumption, in the terminal attractor regime the pdf of $\varepsilon(t)$ does not vary with time as $t \rightarrow \infty$. That is, $\varepsilon_\infty(t) = \varepsilon(\ t \rightarrow \infty)$ is a stationary process

- As a first order approximation, the stochasticity of the dynamics of $\varepsilon(t)$ is influenced only at fixed times corresponding to each onset of iterative epochs facilitated by the feedback.

- The epochal iteration times are much larger than the periods of any fluctuations associated with $\varepsilon(t)$

- At the terminal stage, convergence of $\varepsilon(t)$ towards an equilibrium value of ε_∞ is assured only if the network's optimization efforts are constituted favorably by reinforcement error-information

- On the contrary, in the event of degenerating (or annhilating) error information being overwhelmingly present, the dynamics of $\varepsilon(t)$ will be divergent.

5.3 Stochastical Dynamics of the Error-Measure (ε)

The dynamics of $\varepsilon(t)$, in general, can be equated to a random walk process by virtue of the aforesaid assumptions and in view of the following considerations. Specifically, the endeavor of the network towards convergence when conceived in the information-theoretic plane refers to an *adaptation* process wherein the progressive acquisition of information leads to minimization of disorganization or eradication of uncertainty (entropy) of the network output *vis-à-vis* the teacher function. When the network has learned or adapted itself to the environmental inputs to the fullest extent, it does not need any more information since it retains no further uncertainties about the output against the teacher values. That is, a fully trained network may not perceive any further information because the output is maximally certain against the teacher value with which it is compared.

A heuristic time-dependent model of the goal-oriented, converging aspect of the neural complex versus time expressed in terms of $\varepsilon(t)$ as described above, can be depicted qualitatively in terms of the variance of the teacher function σ_T^2 and that of the network output σ_O^2. The evolution of error entropy $\varepsilon(t)$ can be specified by an envelope profile given by [5.13]:

$$\varepsilon(t) = (k/2)\ell og(1 + \sigma_O^2/m(t)\sigma_T^2) \qquad (5.4)$$

where $m(t)$ is the number of iterations over the time (t) which can be modeled as a simple case, by depicting $m = \alpha t$ where α is the number of iterations per unit time; and k is a constant as decided by the base of the logarithm. Hence,

$$\varepsilon = (k/2)\ell og(1 + \sigma_O^2/\alpha\sigma_T^2 t) \qquad (5.5)$$

In the initial time frame, at the commencement of network learning, the error information pertinent to the output (in reference to the teacher value) could be significantly different and, as such, the network has a large potential to receive information in tending towards the objective function. Therefore, the initial error information $\varepsilon(t_0 \rightarrow 0) = \varepsilon_0$ can be designated as the potential error information.

5.4 Random Walk Paradigm of $\varepsilon(t)$ Dynamics

As discussed before, the error-measure $\varepsilon(t)$, when specified in the information-theoretic plane, can be written in the form of Eqn. (5.1). More generally, it can be constructed by linearly combining two weighted parts of the Csiszár metric given by Eqns. (5.1a) and (5.1b) as elaborated in Chapter 4. That is,

$$\varepsilon = K_1 \sum_i q_i \phi(p_i/q_i) + K_2 \sum_i p_i \phi(q_i/p_i)$$

$$= \sum_i (\varepsilon_{1i} + \varepsilon_{2i}), \quad i = 1, 2, ..., m, ..., n \qquad (n \rightarrow \infty, t \rightarrow \infty)$$

$$(5.6)$$

where K_1 and K_2 are weighting factors. If $K_1 = K_2$, Eqn. (5.6) can be considered as symmetrized and balanced.

Designating each iterative epoch as of duration $\Delta\tau$, the total time involved in reaching the terminal state of dynamics (with $n \rightarrow \infty$) is taken as an integral multiple $n\Delta\tau = T_\infty$. Suppose, the potential energy associated with the system (which is being minimized) is taken as E. For each epoch of iteration, there is a corresponding energy configuration and the ensemble of which can be represented by a canonical Gibbs' distribution given by [5.12]:

$$\mathcal{P}_i(\varepsilon_{1i}) = C_1 \exp(-\Delta E_{1i}/E_R) \qquad (5.7a)$$

and/or by

$$\mathcal{P}_i(\varepsilon_{2i}) = C_2 \exp(-\Delta E_{2i}/E_R) \qquad (5.7b)$$

where E_R is a reference energy level and the normalization constants (partition functions) C_1 and C_2 are determined from the requirement, $\sum_i \mathcal{P}_i(\varepsilon_{1i}) = \sum_i \mathcal{P}_i(\varepsilon_{2i}) = 1$. Hence, $C_1 = C_2 = 1/M(T_\infty)$, where M is the total number of energy levels configured over the time T_∞. The corresponding configurational entropy associated with ε_{1i} or ε_{2i} is

$$S_T(\varepsilon) = -K\sum P_i \log P_i$$

$$= -K\sum [1/M(T_\infty)]\log[1/M(T_\infty)] = K\log[M(T_\infty)]$$

$$(5.8)$$

where, again K is a constant specified by the base of the logarithm. The number of ways (or realizations) the ensemble $M(T_\infty)$ can be divided into two groups of m_1 and m_2, corresponding to ε_1 and ε_2, respectively, without regard to order, is given by the binomial coefficient, namely:

$$\binom{M(T_\infty)}{m_1} = M(T_\infty)!/m_1! \, m_2! \qquad (5.9)$$

where $M(T_\infty) = (m_1 + m_2)$.

Inasmuch as the statistics concerning state-transitions associated with $\varepsilon(t)$ are governed by the Gibb's distribution (Eqn. (5.7)), the discourse of ε with time represents a time-homogeneous Markov chain. Further, the transitional epochal state of $\varepsilon(t)$ is decided by the configurational energy level ΔE_{1i} and/or ΔE_{2i} and it can be modeled by the concept of one dimensional random walk. Starting at $t = 0$ and taking steps of length $\Delta \tau$, let $\Delta \varepsilon^+$ and $\Delta \varepsilon^-$ be the reinforcing and annhilating information imparted respectively by the error fedback *via* the control-loop, with the probability of each being equal.

The random walk model enables the computation of the probability of achieving a specific information state at $t = m\Delta\tau$ after m iterative steps. That is, by considering $\Delta\varepsilon^+$ as the reinforcement information and $\Delta\varepsilon^-$ to represent the degenerating counterpart, the corresponding (proportionate) contributions occurring randomly with equal probabilities refer to the evolution process depicting the excursion of $\varepsilon(t)$ about the equilibrium value (ε_∞) *versus* the iteration of epochs performed. This is illustrated in (Fig. 5.2).

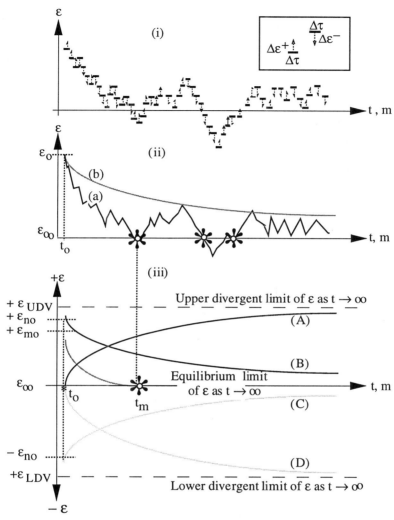

Fig. 5.2: Convergent and divergent modes of ε as a function
of time (t) or number of epochal iterations (m)

(i)   Random walk representation of $\Delta\varepsilon^+$ and $\Delta\varepsilon^-$ versus t or m
(ii)   Temporal trajectory of ε
        [(a) Actual trajectory crossing the equilibrium value of ε, (ε_∞) at specific attractor; (b) Envelope of the trajectory ε showing the asymptotic trend of ε at its terminal dynamics as t or m → ∞]
(iii)   Divergent and convergent profiles of ε versus time
        [(A) & (D): Diverging envelopes directed from positive and negative sides, respectively; (B) & (C): Converging envelopes directed from positive and negative sides, respectively]

In this random walk process, the current value of $\varepsilon(t)$ is decided by the potential level ΔE and therefore, the corresponding probabilities of the state of $\varepsilon(t)$ also depend on the current value of $\varepsilon(t)$. This energy-dictated random-walk process, as opposed to the *free diffusion process*, is a *force-field dependent diffusion process* and, therefore, corresponds to the *Ornstein-Uhlenbeck process* [5.14].

The transitional probability associated with the excursion of $\varepsilon(t)$ by $\Delta\varepsilon^{\pm}$ in the aforesaid one-dimensional random walk process commencing at an initial state depicted by $\varepsilon_0(t \rightarrow t_0) = (\varepsilon_0, t_0)$, is given by:

$$\mathscr{P}\left[(\varepsilon + \Delta\varepsilon^{\pm}, \ t + {}_m\Delta t)|(\varepsilon, \ t)\right]$$

$$= Transitional\ probability\ of\ \varepsilon(t)\ assuming\ the\ values:$$

$$\left.\begin{cases} \varepsilon + \Delta\varepsilon^{+} \\ or \\ \varepsilon + \Delta\varepsilon^{-} \end{cases}\right\} \ at\ m^{th}\ epoch\ or\ time$$

$$= 1/\left[1 + \exp[(\Delta E_{1m} - \Delta E_{2m})/E_R]\right] \tag{5.10}$$

For a given m, the possible values of ε (especially, for large values of t) would differ from each other by multiples of $2\Delta\varepsilon^{\pm}$, since changing ε (by $\Delta\varepsilon^{+}$ or $\Delta\varepsilon^{-}$) at any single step changes the final value of $\varepsilon(t)$ by that amount. Or, a probability $\mathscr{W}(\varepsilon; m)$ can be so defined such that $2\Delta\varepsilon^{\pm} \ \mathscr{W}(\varepsilon; m)$ refers to the probability of reaching ε after m in excursions. That is, $2\Delta\varepsilon^{\pm}\mathscr{W}(\varepsilon; m)$ is the probability reached in the interval $[(t = m\Delta\tau)] \leq t \leq [(t = m\Delta\tau + \Delta\tau)]$ after m steps. The relation between $\dot{\mathscr{w}}(\varepsilon,m)$ and $M(T_m)$ is, therefore: $2\Delta\varepsilon^{\pm}\mathscr{W}(\varepsilon) = M(T_m)(1/2)^m$, writing $\mathscr{W}(\varepsilon)$ for $\mathscr{W}(\varepsilon, m)$ and $T_m = (m\Delta\tau)$ for convenience.

It may be noted that any particular set $\{ \Delta\varepsilon_i^{+} \}$ or $\{ \Delta\varepsilon_i^{-}\}$, regarded now as defining a particular ensemble sequence of increments or decrements in ε in respect of each step in a random walk, has probability $(1/2)^m$ and there are $M(T_m)$ such sets which lead to the mth epoch at $t = T_m$. Inasmuch as $\mathscr{W}(\varepsilon)$ and $M(T_m)$ differ only by a coefficient (independent of T_m), the corresponding configurational entropy can be written as:

$$S_{T_m}(\varepsilon) = K\ell og\mathscr{W}(\varepsilon) \tag{5.11}$$

Further, for m $>>1$ and n $\rightarrow \infty$ and $(T_m/n\Delta\tau) = (\varepsilon/n\Delta\varepsilon^{\pm}) << 1$, the following approximation is valid:

$$\ell og[(m \pm \varepsilon/n\Delta\varepsilon^{\pm})/2] \approx (\varepsilon/n\Delta\varepsilon^{\pm}) - [(\varepsilon^2/2n^2(\Delta\varepsilon^{\pm})^2] + \ell og(m/2)$$

$$\tag{5.12}$$

Using the above approximation and applying *Stirling's formula*[1] to the expression $\ell og\,[M(T_m)]$, the following result is obtained:

$$W\,(\varepsilon\;;m) \cong [1/\{2\pi m(\Delta\varepsilon^{\pm})^2\}^{1/2}]\exp[-\varepsilon^2/2m(\Delta\varepsilon^{\pm})^2] \qquad (5.13)$$

That is, for $m \gg 1$ and $n \to \infty$ and $(\varepsilon/n\Delta\varepsilon^{\pm} \ll 1)$, the probability density function (pdf) describing the statistics of $\varepsilon(t)$ at m^{th} epoch is Gaussian with a mean $<\varepsilon> = 0$ and a variance $<\varepsilon^2> = m(\Delta\varepsilon^{\pm})^2$. In the considerations presented earlier, the equilibrium value is taken as ε_∞, and in reference to which, if ε is presumed to fluctuate, then $<\varepsilon> = 0$. Thus, the probability of the temporal statistics of ε, in a broad sense, refers to a superposition of m independent random variables and approaches a Gaussian distribution with zero mean about the equilibrium value and of a finite variance in the limiting stage of m approaching n. This is in concordance with the central limit theorem.

5.5 Evolution of $\varepsilon(t)$: Representation via the Fokker-Planck Equation

Pertinent to a given enviroment from which the network learns, if the time between learning steps is drawn from a Poisson distribution, the dynamics of an ensemble of learning processes have been described in [5.6] by a continuous-time master equation. Presently, the evolution of ε versus time, or number of iterative epochs, can be modeled as a forward equation (or the master equation) of a diffusion process (Fokker-Planck equation). This refers to the description of the transition probabilities of ε changing by $\Delta\varepsilon^+$ or $\Delta\varepsilon^-$ at each step with the initial conditions at the commencement of the iterative epochs being (ε_0, t_0). Such a description satisfies a stochastical differential equation given by [5.14]

$$d\varepsilon(t)/dt = \mu(\varepsilon, t) + \sigma(\varepsilon, t)\zeta(t) \qquad (5.14)$$

where $\zeta(t)$ is a random function such that $\int_0^t \zeta(\tau)d\tau$ imposes the attributes of a random walk to the variable $\varepsilon(t)$. In general, $\zeta(t)$ is a stationary, Gaussian, white noise suggesting that the dynamics specified by Eqn. (5.14) are driven by a stationary Gaussian process. The evolution of $\varepsilon(t)$ models a continuous Brownian motion. It has a pseudoderivative, namely, a time-derivative in a mean-squared sense. This pseudoderivative random process $\{d\varepsilon(t)/dt\}$ equals an ideal Gaussian white noise $\zeta(t)$ as given by Eqn. (5.14) where $\zeta(t)$ is zero mean and uncorrelated in time, but has a finite variance $<\zeta^2> < \infty$ for all t.

---

[1] *Stirling formula*: $\ell og(x!) = (x + 1/2)\ell og(x) + \ell og(\sqrt{2\pi} - x)$

Further, in the interval $(t_2 - t_1)$, the entity n $\Delta\tau\{\varepsilon, t_1 \le t \le t_2\}$ is a sample-continuous, second order Markovian process with $\mu(\varepsilon, t)$ and $\sigma(\varepsilon, t)$ being the Borel functions of $\varepsilon(t)$ specified within certain bounds.

Let the transition probability density function of $\varepsilon(t)$ be denoted by $2(\varepsilon, t \mid \varepsilon_0, t_0)$; and, $s(\varepsilon), -\infty < \varepsilon < +\infty$ is a Schwartz function of rapid descent. That is, s is infinitely differentiable, as well as, for any κ and λ, $|\varepsilon|^\kappa |f^{(\lambda)}(\varepsilon)| \to 0$ as $|\varepsilon| \to \infty$. Suppose an initial condition is imposed such that $2(\varepsilon, t \mid \varepsilon_0, t_0) = \delta(\varepsilon - \varepsilon_0)$, where ε_0 refers to the initial value of ε at the onset of the iterative process commencing at $t = t_0$ and,

$$\int_{-\infty}^{+\infty} s(\varepsilon) 2(\varepsilon, t \mid \varepsilon_0, t_0) d\varepsilon \to s(\varepsilon_0) \qquad t \to t_0, \quad \forall s \in S \qquad (5.15)$$

Subject to the above initial conditions, $2(\varepsilon, t \mid \varepsilon_0, t_0)$ satisfies the Fokker-Planck equation given by [5.14]

$$\partial 2 / \partial t = (1/2)\partial^2 [\sigma^2 2] / \partial \varepsilon^2 - \partial [\mu 2] / \partial \varepsilon \qquad (5.16)$$

where $2 \equiv 2(\varepsilon, t \mid \varepsilon_0, t_0)$, $\sigma \equiv \sigma(\varepsilon, t)$, $\mu \equiv \mu(\varepsilon, t)$ and $t_2 > t > t_0 > t_1$.

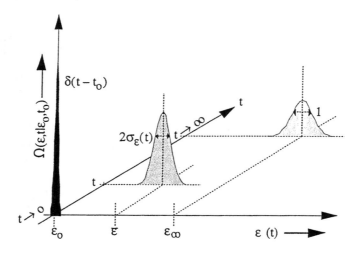

Fig. 5.3: Stochasticity of ε (expressed in terms of the transitional probability of $\Delta\varepsilon^\pm$) versus time, t

Suppose $\sigma^2(\varepsilon, t) = b(t)$ and $\mu(\varepsilon, t) = a(t)\varepsilon$. Specific to these prescriptions, the Fokker-Planck equation (Eqn. (5.16)) reduces to

$$\partial 2 / \partial t = (1/2)b(t)\partial^2 2 / \partial \varepsilon^2 - \partial a(t)\partial(\varepsilon 2)/\partial \varepsilon \qquad (5.17)$$

which has a solution given by [5.6]:

$$\mathcal{Z}(\varepsilon, t | \varepsilon_0, t_0) = (1/\sqrt{2\pi V^2}) \exp[-(\varepsilon - U\varepsilon_0)^2/2V^2] \tag{5.18}$$

The above expression depicts again the Gaussian nature of ε with mean and variance being U and V^2, respectively. They can be obtained explicitly by solving [5.14]: $d(V^2)/dt = \boldsymbol{\ell} + 2\boldsymbol{a}V^2$ and $dU/dt = \boldsymbol{a}U$ with the initial conditions $V^2(t_0) = 0$, and $U(t_0) = 1$.

The time-dependent wandering of ε under random force as given by Eqn. (5.14) can also be described by the following Langevin equation [5.15]:

$$d\varepsilon/dt + B_L\varepsilon = A_L(t) \tag{5.19a}$$

where $B_L = (2/\Delta\tau)$ and $A_L(t)$ is the random force function. The initial condition, namely, $\varepsilon(t \to t_0 \to 0) = \varepsilon_0$ specifies the solution of Eqn. (5.19a) as

$$\varepsilon(t) - \varepsilon_0 \exp(-2t/\Delta\tau) = \int_0^t [\exp[2(x - t)/\Delta\tau]]A(x)dx \tag{5.19b}$$

The corresponding solution for transition probability $\mathcal{Z}(\varepsilon - \varepsilon_0 e^{-2t/\Delta\tau})$ is given by

$$\mathcal{Z}(\varepsilon, t | \varepsilon_0, t_0) = 1/\{2\pi[1 - \exp(-4t/\Delta\tau)]\}^{1/2}$$
$$\times \exp\{-[\varepsilon - \varepsilon_0 \exp(-2t/\Delta\tau)]^2/2(1 - \exp(-4t/\Delta\tau)\}$$
$$= [1/(2\pi\sigma_2^2)^{1/2}] \exp[-(\varepsilon - \bar{\varepsilon}(t))^2/2\sigma_2^2(t)] \tag{5.20}$$

which approaches a delta-Dirac function as $t \to t_0 \to 0$ (Fig. 5.3); and, $\bar{\varepsilon}(t)$ depicts $\varepsilon_0 e^{-2t/\Delta\tau}$ and $\sigma_2^2 = [1 - \exp(-4t/\Delta\tau)]$.

Thus, the heuristic discussion on random walk model $\varepsilon(t)$ as presented earlier has led to the per unit time probabilities of the excursions of ε, namely, $\mathcal{W}(\varepsilon, m)$ along the temporal frame work of the iterative epochs. This modeled as a Gaussian process, is in conformity with the solution of stochastical differential equation representation of the transition probabilities of $\varepsilon(t)$ given by Eqn. (5.20). Again, as dictated by the initial conditions (ε_0, t_0) at the onset of the iterative epochs, it can be observed that the stochasticity of $\varepsilon(t)$ is decided by a Gaussian process along the passage of time in the attractor basin.

5.6 Logistic Growth Model of $\varepsilon(t)$

The temporal evolution of ε can also be modeled as a logistic growth with an oscillatory trend around its equilibrium (terminal) value (ε_∞), depending on the initial state of the evolutionary process. Suppose at the early stages of iterations of the error feedback imposed on the network, the value of ε is slightly less than the equilibrium value. This reduced value offers a niche to receive reinforcing information which augments the network's efforts towards optimization, leading to a further reduction in the cross-information based error at the output. However, after a few iterations, the stochasticity associated with the network-inputs may yield an output which could lead to negative information, with the result that an enhancement of the error measure causing an excursion of ε to drift away from the equilibrium value may be perceived. The causative actions changing ε towards or away from the equilibrium value could also, in general, be asynchronous in reference to the current state of $\varepsilon(t)$. They may be separated to an extent of a few iterative epochs depicting a delayed arrival in reference to t. Hence, the rate of change of ε caused by $\Delta\varepsilon^{\pm}$ is not a function of the present value $\varepsilon(t)$, but due to a past value $\varepsilon(t - t_d)$ where t_d is the delay time involved. The following growth model, in which the rate of change of $\varepsilon(t)$ is functionally dependent on a past value $\varepsilon(t-t_d)$ specifies the corresponding divergent and/or convergent aspect of $\varepsilon(t)$ from the equilibrium value.

$$[1/\varepsilon(t)]d\varepsilon(t)/dt = \Gamma(t) \tag{5.21}$$

where $\Gamma(t)$ depicts the growth rate function and it can be denoted by a simple linear form, as follows:

$$\Gamma(t) = [g - h\varepsilon(t - t_d)] \tag{5.22}$$

where g is the growth rate of ε without external influences and h represents the effect of augmentation in the values of ε.

Hence, the differential delayed equation depicting the logistic growth of $\varepsilon(t)$ is written as

$$d\varepsilon(t)/dt = \varepsilon(t)[g - h\varepsilon(t - t_d)] \tag{5.23}$$

For incremental changes in time (Δt), the above equation can be written as a second-order difference equation of the form:

$$\varepsilon(t + \Delta t) - \varepsilon(t) = (\Delta t)\varepsilon(t)[g - h\varepsilon(t - \Delta t)] \tag{5.24}$$

With the change of notations given by: $t = m\Delta t$, $\varepsilon(t) = \varepsilon(m\Delta t) = \varepsilon_m$, and $G = g\Delta t$ and $H = h\Delta t$, the discrete logistic equation at the mth epoch of iteration becomes

$$\varepsilon_{m+1} - \varepsilon_m = \varepsilon_m(G - H\varepsilon_{m-1}) \tag{5.25}$$

Any delay involved in the above process could be asynchronous, in respect to the current value of time, namely, t. Further, assuming the displacements of ε (namely, $\Delta\varepsilon^{\pm}$) are small and are confined to the proximity of the equilibrium value, a linear approximation of the discrete logistic equation is valid. That is, close to the equilibrium value of $\varepsilon \rightarrow \varepsilon_{\infty} = g/h$ (Fig. 5.4), and,

$$\varepsilon(t + \Delta t) - \varepsilon(t) = -G\,\varepsilon(t - \Delta t) \qquad (5.26)$$

Or, by denoting of $\varepsilon_m = (G/H) + \Delta\varepsilon_m^{\pm}(t)$ where $|\Delta\varepsilon_m^{\pm}(t)|$ being much smaller than the equilibrium value G/H, the following linearized (approximate) growth relation can be written

$$y_{m+1} - y_m \cong -Gy_{m-1} \qquad (5.27)$$

where $y_m = \Delta\varepsilon_m^{\pm}(t)$ is the displacement ε from the terminal value ε_{∞} at the mth epochal instant of iteration. Eqn. (5.27) can be recast in a more general form as

$$y_{m+1} + P_0 y_m + Q_0 y_{m-1} = 0 \qquad (5.28)$$

which is known as a constant coefficient difference equation. The solution to this equation is analogous to the solution of constant coefficient second order differential equations with the necessary conditions for a unique solution being the initial conditions of the first two values of y_0 and y_1 being available [5.16]. Such a solution can be, written as

$$y_m = C_1 r_1^{m} + C_2 r_2^{m} \qquad (5.29)$$

where the values of r are determined by the substitution in Eqn. (5.28). Hence, it follows that

$$r^{m+1} + P_0 r^{m} + Q_0 r^{m-1} = 0 \qquad (5.30)$$

Upon division by r^{m-1}, the following quadratic equation is obtained:

$$r^2 + P_0 r + Q_0 = 0 \qquad (5.31)$$

whose two roots are : $r_{1,2} = [-P_0 \pm (P_0^2 - 4Q_0)^{1/2}]/2$.

The arbitrary constants C_1 and C_2 in Eqn. (5.29) can be determined uniquely by the initial conditions leading to the following two independently solvable equations:

$$y_0 = C_1 + C_2 \qquad (5.32a)$$

$$y_1 = C_1 r_1 + C_2 r_2 \qquad (5.32b)$$

assuming that the two roots r_1 and r_2 are distinct; that is, $(P_0^2 - 4Q_0) \neq 0$.

Pertinent to Eqn. (5.27), the corresponding solution refers to that of a constant coefficient linear difference equation, namely, $r^2 - r + G = 0$ with the roots given by: $r_{1,2} = [1 \pm (1 - 4G)^{1/2}]/2$. The equilibrium state of ε at the terminal stage is stable if the solution of y_m does not grow as $t \to \infty$, that is, when $m \to \infty$ for any initial conditions, (ε_0, t_0). If $0 < G < 1/4$, then the roots are real, positive and less than 1. That is, $0 < r_{1,2} < 1$). Consequently, if $0 < G < 1/4$, $y_m = \Delta \varepsilon_m^{\pm}(t) \to 0$ as $t \to \infty$. Or, the excursion of ε vanishes and $\varepsilon(t)$ will approach the equilibrium value, namely, ε_∞ asymptotically. When $G < 1/4$, then the two corresponding roots are complex conjugates and hence the solution for $y_m(t)$ is given by:

$$y_m(t) = |r|^m [C_3 \cos(m\theta) + C_4 \sin(m\theta)] \qquad (5.33)$$

where $|r_1| = |r_2| = |r| = G^{1/2}$, $\theta = \arctan[(4G - 1)^{1/2}]$, and $m = (t/\Delta t)$.

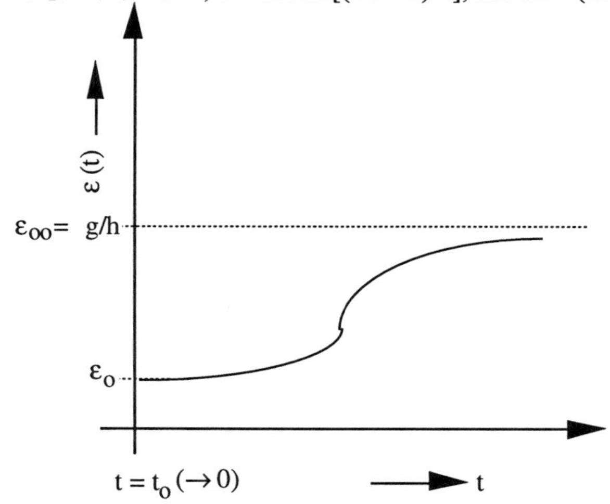

Fig. 5.4: Logistic growth of ε as a function of time (t) seeking the equilibrium value ε_∞ as $t \to \infty$

The above solution (Eqn. (5.33)) grows or decays as it oscillates, depending on $|r| = G^{1/2}$. When $1/4 < G < 1$, then the solution is a decaying oscillation. That is, $y_m \to 0$ and $\varepsilon(t)$ is convergent towards its equilibrium value. However, when $G > 1$, the equilibrium value may be reached quickly, but the instability would render an oscillatory growth around the attracted value leading to a divergent solution.

The divergent growth stems from the asynchronous occurrences of ε_{1i} and ε_{2i}, namely, the reinforcement information ($\Delta\varepsilon^+$) or the degenerating (annihilating) information ($\Delta\varepsilon^-$) fed back via the cross-information error-metric. Considering Eqn. (5.22), suppose g is a positive growth rate without any external constraints and h depicts a constraint stipulated by the network as a limiting factor. Let Δt_d refer to the time-delay due to the asynchronous attributes of ε_{1i} and/or ε_{2I} which can vary without any limit imposed. Then, $g\Delta t_d < 1/4$ would represent an extremely small delay permitting ε_{1i} or ε_{2i} to yield an entity amounting to $\Delta\varepsilon_m^+$ which constitutes a reinforcement information by canceling any divergent trend in the current value of ε, namely, $\varepsilon_m(t)$ at the mth epochal iteration. This is possible since $\varepsilon_m(t)$ and $\Delta\varepsilon^+$ occur synchronously due to negligible delay involved. This would guarantee consequently an eventual stabilization of $\varepsilon(t)$ at an equilibrium value, ε_∞. If the delay is specified by ($1/4 < g\Delta t_d < 1$), the function $\varepsilon(t)$ would oscillate with larger excursions, but would still seek ultimately the equilibrium value with the passage of time.

However, in the case of $g\Delta t_d > 1$, the oscillation would become divergent, destabilizing the optimization effort. That is, an entity $\Delta\varepsilon_m^\pm$ encountered by $\varepsilon_m(t)$ at the mth epochal instant of iteration would augment any divergent trend in the current value of $\varepsilon_m(t)$ predominantly. This can happen when $\Delta t_d \to \infty$ which represents either ε_{1i} or ε_{2i} is absent or unproportionately unbalanced and dissimilar (asymmetric) so that the destructive information component, namely, $\Delta\varepsilon^-$ dominates. Hence, with the reinforcement information contributed by ε_{1i} or ε_{2i} being absent, the chances of $\varepsilon(t)$ to diverge are increased. In other words, for an asymmetric (one-sided/unbalanced) error-metric represented *via* cross-information measures, either by ε_{1i} or ε_{2i} alone, the cumulative augmentation of $\Delta\varepsilon^+$ or $\Delta\varepsilon^-$ renders $\Delta\varepsilon_m^\pm$ to take over the dynamics of $\varepsilon_m(t)$. Therefore, the convergence of the network output towards the teacher value is not guaranteed.

The dynamics of $\varepsilon(t)$ with logistic functional growth characteristics as discussed above may cause the function $\varepsilon(t)$ to cross the equilibrium value (Fig. 5.2) at several instants of time. These crossings represent the bottom of the attractor basins attained repetitively during the iteration of epochs aimed at the convergence of a network's output towards the teacher values.

5.7 Convergence Considerations

5.7.1 Stochastical equilibrium

The crossings of $\varepsilon(t)$ trajectory at the equilibrium value (Fig. 5.6) represent conditions of instantaneous stochastical equilibrium states attained by the vector ε_j and depict a set of fixed-state point attractors (corresponding to steady-state conditions). Implicitly, at these points the stochastical equilibrium is specified by $\partial w/\partial t = 0$, where $\mathbf{0}$ is the null synaptic matrix (that is, the $\mathcal{R}^m$ null vector $\{0\}$), and, **w** is the coupling matrix adjusted through feedback via epochal iterations of the error function. In the sample space of the vector ε, $\partial\varepsilon/\partial t = 0$ denotes the stability or neural equilibrium with **0** representing the null vector of the changes in error activity. Globally, the neuronal stochastic stability is dictated by the steady-state conditions in the neuronal field, namely,

$$\partial\mathbf{w}/\partial t = \chi_t \tag{5.34}$$

and,

$$\partial\mathbf{x}/\partial t = \xi_t \tag{5.35}$$

where χ_t is a random vector from a Gaussian white random process $\{\chi_t\}$ which can be related to the random vector ζ_t used in Eqn. (5.14) to model the stochasticity of $\varepsilon(t)$. The neural state vector **x** has an associated independent Gaussian white noise process denoted by ξ_t. Eqns. (5.34) and (5.35) represent the stochastic equilibrium conditions vis-à-vis the neuronal state vector **x** and the synaptic state matrix **w**. Both **x** and **w** hover in a Brownian motion about (fixed/deterministic) equilibrium or terminal attractor value as m (or t) $\rightarrow \infty$. They reach the state of stochastic equilibrium only when the random vectors χ and ξ alter them temporally.

As mentioned earlier, the dynamics of neural parameters being pursued here correspond to the fluctuations of the error-metric $\varepsilon(t)$ which is computed presently in the information-theoretic plane. It offers competitive feedback information which either reinforces or destroys the current informatic status at time t (or at the mth iterative epoch) of the adjustments imparted to the network *via* the weight matrix **w**. Therefore, the relation given by Eqn. (5.34) can be written modified as

$$\partial(\Delta\varepsilon)/\partial t = \chi_t \tag{5.36}$$

where $\Delta\varepsilon$ is the fluctuating vector component of the error-metric vector set $\{\varepsilon_i\}$. That is, in elucidating the stochastic equilibrium of the neural networks, Eqn. (5.36) can be adopted *in lieu* of Eqn. (5.34).

In essence, the error-metric fed back, and the corresponding corrective algorithms pursued in the information-theoretic plane, can be regarded as those pertinent to a competitive learning strategy in the information-theoretic plane. It is also a differential learning pursuit. That is, only when a change occurs in $\varepsilon(t)$, namely, $\Delta\varepsilon^{\pm}$, learning takes place as per the competitive information, provided by $\Delta\varepsilon^+$ or $\Delta\varepsilon^-$. The learning process associates itself with an indicator function to flag when the learning is augmentative/reinforcing as a result of $\Delta\varepsilon^+$ or, whether the learning is degenerated due to the addition of annihilating information $\Delta\varepsilon^-$ facilitated via feedback.

For a guaranteed convergence, the weight adjustment would require that the error-metric (a distance measure) specified in terms of the cross-entropy (mutual information) parameter of Eqn. (5.1), computed in terms of the pdf of the output (p_i) and that of the teacher values (q_i) should be a balanced (equally weighted) and symmetrized (two-sided) function. The characteristics of such functions are described in the following definitions and theorems and are verified by simulation studies as discussed in Chapter 4.

5.7.2 Definitions and theorems

Definition # 5.1: Let $p = \{ p_1, p_2, p_3,..., p_k \}$ and $q = \{ q_1, q_2, q_3,..., q_N \}$ denote two complete set of probabilities $(\sum_i p_i = \sum_i q_i = 1, i = 1, 2,..., N)$ representing the *a priori* probability distribution of the discrete random output $\{O_i\}$ of the neural network and that of the teacher source $\{T_i\}$, respectively, and (p_i, q_i) correspond to the ith iterative epoch in the network training schedule. Or, when the network output and the teacher values are specified as continuous variables, p and q refer to the respective probability density functions such that $\int p\,dp = \int q\,dq = 1$.

Definition # 5.2: The relative entropy or cross/mutual information I(q|p) of O with respect to T is defined by the expression:

$$I(q|p) = \sum_i q_i \ell og(q_i/p_i) = \varepsilon_{1KL} \tag{5.37}$$

Likewise, I(p|q) refers to the cross-entropy of T with respect to O. Hence,

$$I(p|q) = \sum_i p_i \ell og(p_i/q_i) = \varepsilon_{2KL} \tag{5.38}$$

As indicated before and elaborated in Chapter 4, the transformations expressed by Eqns. (5.37) and (5.38) are known as Kullback-Leibler measures [5.17] and represent the amount of information contributed by T about O and the amount of information contained in O about T, respectively. That is, they refer to the mathematical expectation of the transinformation about (or, directed divergence of) each outcome of T *versus* O and O *versus* T respectively. Hence, it follows that

$$I(q|p) = < (prior\ uncertainity)_p - (posterior\ uncertainity)_q >$$
$$= \sum_i q_i \{[-\ell og(p_i)] - [-\ell og(q_i)]\}$$

$$\tag{5.39}$$

Likewise,

$$I(p|q) = < (prior\ uncertainity)_q - (posterior\ uncertainity)_p >$$
$$= \sum_i p_i \{[-\ell og(q_i)] - [-\ell og(p_i)]\}$$

$$(5.40)$$

Definition # 5.3: The distance between two probability distributions refers to a divergence measure between them and is given by the Kullback-Leibler-Jensen metric [5.18] defined as follows:

$$J(p|q) = I(p|q) + I(q|p) = \varepsilon_{KL-J} \qquad (5.41)$$

The J-measure refers to the divergence or the discrimination between the hypotheses $\mathcal{H}_O$ and $\mathcal{H}_T$ (constituted by O and T respectively) or between p and q and implicitly represents a measure of difficulty in discriminating between them. That is,

$$\varepsilon_{KL-J} = J(p|q) = I(q|p) - [-I(p|q)]$$
$$= \sum_i p_i \ell og(p_i/q_i) - \sum_i q_i \ell og(p_i/q_i) = \varepsilon_{1KL} + \varepsilon_{2KL}$$

$$= \sum_i (p_i - q_i)\ell og(p_i/q_i) \qquad (5.42)$$

Definition # 5.4: f-divergence — If $\Phi(x)$ is a convex function for $x > 0$, with $\Phi(1) = 0$, then the f-divergence (f depicting the function Φ) of a distribution p or q is defined in a two-sided form of Csiszár error-measure [6.8] with weighting factors K_1 and K_2 as:

$$\varepsilon_{CZ} = I^f(p|q) = K_1 \sum_i q_i \phi(p_i/q_i) + K_2 \sum_i p_i \phi(q_i/p_i) \qquad (5.44)$$

where $\phi(x) = x log(x)$, $\varepsilon_{CZ} \rightarrow \varepsilon_{KL-J}$; and, ε_{CZ} is a more generalized version than ε_{KL-J}.

Theorem # 5.1: If the f-divergence as defined by Eqn. (5.44) is to be considered as a feasible error-metric (ε_{CZ}) in training a multilayered neural network, then, the necessary condition is that $I^f = \varepsilon_{CZ}$ should be two-sided, and bounded. That is,

$$(I^f = \varepsilon_{CZ}) = K_1 \sum_i q_i \phi(p_i/q_i) + K_2 \sum_i p_i \phi(q_i/p_i)$$
$$= (\varepsilon_{1CZ} + \varepsilon_{2CZ}) \qquad (5.45)$$

In Eqn. (5.45), when the weighting factors are equal (when $K_1 = K_2$), it is the condition for balanced symmetrization.

Proof: This theorem can be proved by the geometrical notions of Pythagorus as follows:

The relative entropy based error-metric $\varepsilon(p|q, q|p)$ behaves intuitively like the square of the Euclidean distance norm, though $\varepsilon(p|q)$ itself represents no geometrical measure. For a convex set ε in $\mathcal{R}^m$, let $\mathcal{A}$ be a point outside the set and $\mathcal{B}$ be the point in the set closest to $\mathcal{A}$, and $\mathcal{C}$ be any other point in the set. Then, the angle between the lines $\mathcal{BA}$ and $\mathcal{BC}$ must be obtuse, which implies via the Pythagorus theorem that $\ell_{\mathcal{AC}}^{2} \geq \ell_{\mathcal{AB}}^{2} + \ell_{\mathcal{BC}}^{2}$ where ℓ represents the linear distance. Hence, the convergence of ε towards an infimum in the 1 norm refers to the minimum distance between the two distributions. That is, *infimum* of $\ell_{\mathcal{AC}}^{2} = (\ell_{\mathcal{AB}}^{2} + \ell_{\mathcal{BC}}^{2})$; or $\varepsilon_{CZ}(p|q, q|p) = (\varepsilon_{1CZ} + \varepsilon_{2CZ})$ where $\varepsilon_{1CZ} = K_1 \sum q_i \phi(p_i/q_i)$ and $\varepsilon_{1CZ} = K_1 \sum p_i \phi(q_i/p_i)$.

Theorem # 5.2: The sufficient condition that $I^f = \varepsilon_{CZ}$ represents an error-metric for training a neural network is that both constituent parts of ε, namely, the syntactic values of ε_1 and ε_2 should be nonzero. This nonzero condtion implies that the corresponding semantics imparted to the network via feedback through corrective algorithm, add meaningful information to the weight adjustments in the multilayered network during each iterative epoch. This should lead to an eventual convergence of the output error towards an equilibrium value ε_∞.

Proof: The constituent parts of ε_{CZ}, namely, $\varepsilon_{1CZ} \in (0, \Delta\varepsilon^\pm]$ and $\varepsilon_{2CZ} \in (0, \Delta\varepsilon^\pm]$ carry messages of relative importance and are applied to the system dynamics via feedback which allows the state variable $\varepsilon_{CZ}(t)$ to converge towards an attractor (ε_∞) at a given k^{th} instant of iteration.

Let ε_{1CZ} supply a "message of relative importance" given by: $[\mathcal{M}_{1k}/\mathcal{M}]$ at the kth instant; and $M = \sum (M_{1k} + M_{2k})$. At the same instant, the corresponding message of relative importance imparted by ε_{2CZ} is given by $[\mathcal{M}_{2k}/\mathcal{M}] = [1 - (\mathcal{M}_{1k}/\mathcal{M})]$, since the semantic aspects of ε_{1CZ} and ε_{2CZ} complement with each other.

Considering the total messages delivered over k iterations, it is given by

$$M_{Total}/M = \sum_k [M_{1k}/M] + \sum_k [1 - M_{1k}/M] \qquad (5.46)$$

The above relation (Eqn. (5.46)) specified in terms of a controlling (cybernetic) information parameter of the network say, C_ε, which results from the error fedback by the control-loop, can be written in reference to the kth iteration as:

$$C_{ek} = (I_{e^+}{}^k + I_{e^-}{}^k) \Rightarrow \{\Delta\varepsilon^\pm\} \qquad (5.47)$$

Here, I_{e^+} can be regarded as the reinforcing information which directs the output error towards equilibrium value, ε_∞, and I_{e^-} refers to annihilation information which leads the system dynamics to diverge from the equilibrium. Dominance of $I_{\varepsilon-}$ implies an information deficiency and, conversely, the overwhelming influence of $I_{\varepsilon+}$ means information augmentation perceived by the system dynamics in the pursuit of equilibrium or an attractor value. As Haken describes [5.19], a parameter such as $C_{\varepsilon k}$ either "sensitizes" or "desensitizes" the convergence process, depending on the dominance of the messages delivered by $I_{\varepsilon+}$ or $I_{\varepsilon-}$, respectively, at any given k^{th} instant in the network optimization strategy.

Considering $\Delta\varepsilon^+$ and $\Delta\varepsilon^-$, (deliberating $I_{\varepsilon+}$ or $I_{\varepsilon-}$) as dichotomous events, their repeated occurrences constitute Bernoulli trials with binomial distribution. Suppose ε is single-sided. That is, ε_{CZ} is assumed to be constituted by ε_{1CZ} or ε_{2CZ} alone. The corresponding number occurrence of $\Delta\varepsilon^+$ and $\Delta\varepsilon^-$ will be unbalanced significantly over $n \to \infty$ iterations (trials). This unbalanced condition would let either $\Delta\varepsilon^+$ or $\Delta\varepsilon^-$ dominate as $n \to \infty$, offering one-sided information to the control dynamics. Hence, the system will diverge positively or negatively from the equilibrium (ε_∞) depending on the dominance of $\Delta\varepsilon^+$ and $\Delta\varepsilon^-$, respectively (Fig.5.2).

On the other hand, if both ε_{1CZ} and ε_{2CZ} are present (two-sided/symmetrized representation of ε_{CZ}), the Bernoulli events of $\Delta\varepsilon^\pm$ are shared by ε_{1CZ} and ε_{2CZ} over n iterations. This would lead to a balanced state of $\Delta\varepsilon^+$ and $\Delta\varepsilon^-$ sharing almost $n/2$ iterations each. Due to the stochasticity of the system, however, $\Delta\varepsilon^+$ may dominate marginally in number of occurrences ($= n/2 + \Delta$, $\Delta << n/2$) so that $\Delta\varepsilon^-$ occurs $(n/2 - \Delta)$ times. This unbalanced state of sharing the iterations with a marginal dominance in number by $\Delta\varepsilon^+$ will augment the necessary information cumulatively as would be required for convergence (from negative side). Likewise, if $\Delta\varepsilon^-$ occurs $(n/2 + \Delta)$ times, then the convergence will be directed from the positive side (Fig. 5.2).

Existence of both ε_1 and ε_2 as constituent parts of ε is, therefore, a necessary condition needed for the network to converge. Hence, the competitive aspects of $I_{\varepsilon+}$ and $I_{\varepsilon-}$ facilitated by the dichotomous occurrences of $\Delta\varepsilon^\pm$ will decide ultimately the convergence towards attractors and it can be realized by proper choice of network parameters, such as the learning coefficient, and by adopting symmetrically weighted ε_1 and ε_2.

5.7.3 On-line simulation and results

Shown in Fig. 5.1b is a test neural network (multilayer perceptron) with 24 input units, 2 hidden layers (each with 10 units) and a single output. This is same as the network considered in the simulation studies of Chapters 4 and 5. It was trained to recognize a teacher function $|\sin(x)|$ using the error-metrics given by Eqns. (5.37), (5.38) and (5.41). Presented in Figs. 5.5-5.7 are the results pertinent to: (i) Histograms of error-value distributions Fig. 5.5) and (ii) trajectories of the errors as functions of the number of epochal iterations performed (Fig.5.6). (iii) The actual and simulated test function $|\sin(x)|$ *versus* the argument x is presented in Fig. 5.7. The simulated function (in Fig. 5.7) corresponds to the symmetric error-metric given by Eqn. (5.41) which converged to the equilibrium status as depicted in Fig. 5.6.

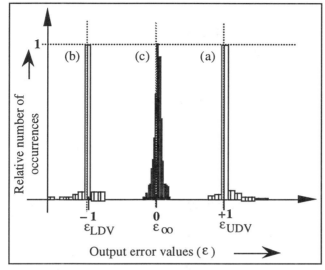

Fig. 5.5: Histograms of output error value distributions corresponding to:
    (a) One-sided Kullback-Leibler error-measure, ε_{1KL} : Eqn. (5.37)
    (b) One-sided Kullback-Leibler error-meausre, ε_{2KL}: Eqn. (5.38)
    (c) Symmetrized Kullback-Leibler-Jensen error-measure, ε_{KL-J} : Eqn. (5.41)

Note:
    Relative number of occurrences in each case refers to the value normalized with respect to the maximum value; and, as t → ∞, the lower divergent value, ε_{LDV} of ε_{2KL}, the equilibrium limit of ($\varepsilon_{1KL} = \varepsilon_{1KL} + \varepsilon_{2KL}$), namely, ε_{∞} and the upper divergent value, ε_{UDV} of ε_1 are set at −1, 0, +1, respectively

The asymmetric error-metrics (given by Eqns. (5.37) and (5.38)) failed in leading the network's performance towards convergence. Their trajectories, as can be seen from Fig. 5.6, veered off from the equilibrium value with the discourse of iterations. Thus, this simulation study confirms the need for a symmetrized error-measure for neural network learning applications in the information-theoretic domain. In the simulation studies performed, the learning coefficient was taken as η = 0.01 and the nonlinear sigmoidal function was the Bernoulli function $L_Q(x)$ with Q = 10 [5.1]. They were chosen so as to realize the minimum root-mean squared value of the deviations of the predicted function from the teacher function at 50 equally spaced arguments.

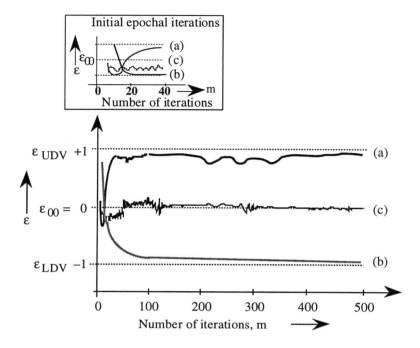

Fig. 5.6: Computed trajectories of ε versus number of iterations, m
with the test network of Fig. 5.1b subjected to simulation studies
(a) For ε_{1KL}, Eqn. (5.37) ; (b) For ε_{2KL}, Eqn. (5.38); (c) For $\varepsilon_{KL\text{-}J}$, Eqn. (5.41)

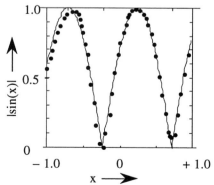

Fig. 5.7: Actual (———) and simulated (....) test functions

5.8. Further Considerations on the Dynamics of $\varepsilon(t)$

5.8.1 Competing augmentative and annihilative information species
As discussed in the previous section, the control dynamics of a neural network are dictated by the competition of reinforcing information I_{e+} and by the annihilating counterpart I_{e-}. The eventual convergence (or divergence) of network performance is decided by the dominance of $\Delta\varepsilon^+$ or $\Delta\varepsilon^-$ facilitated by ε_1 and ε_2 constituents of the error-metric. Depicting the population of $\Delta\varepsilon^+$ species by n_1 and that of $\Delta\varepsilon^-$ species by n_2, the growth of any one species, which eventually decides the convergence or divergence of the network performance, can be presumed to depend on the population of both species. That is, the dynamics of n_1 and n_2 can be represented in terms of arbitrary functions $\mathscr{U}$ and $\mathscr{Z}$ as follows:

$$dn_1/dt = \mathscr{U}(n_1, n_2) \tag{5.48a}$$

$$dn_2/dt = \mathscr{Z}(n_1, n_2) \tag{5.48b}$$

Correspondingly, both populations may affect each other negatively, so that the interaction between the species is competitive. That is, the growth rate of each species will be retarded due to the presence of the other. From Eqn. (5.48), it follows, by eliminating the explicit dependence on the time factor, t, that:

$$dn_2/dn_1 = \mathscr{U}(n_1, n_2)/\mathscr{Z}(n_1, n_2) \tag{5.49}$$

which represents the phase-plane representation of $\Delta\varepsilon^+$ and $\Delta\varepsilon^-$.

The two competing species, namely, $\Delta\varepsilon^+$ and $\Delta\varepsilon^-$, are virtually identical in their information values but differ only in dictating the convergence process to occur in opposite directions. As presumed earlier, attributing a marginal unbalance to the competition so that $\Delta\varepsilon^+$ is dominant (by letting $n_1 > n_2$), the following explicit equations can be specified in lieu of Eqn. (5.48):

$$dn_1/dt = n_1(a_1 - b_1n_1 - c_1n_2) \tag{5.50a}$$

$$dn_2/dt = n_2(a_1 - b_1n_2 - c_2n_1) \tag{5.50b}$$

and the corresponding phase-plane equation is given by:

$$dn_2/dn_1 = n_2(a_1 - b_1n_2 - c_2n_1)/n_1(a_1 - b_1n_1 - c_2n_2) \tag{5.51}$$

Inasmuch as $n_1 > n_2$, the interaction between $\Delta\varepsilon^+$ and $\Delta\varepsilon^-$ is strongly competitive in the sense that the interaction terms, $-c_1n_1n_2$ and $-c_2n_1n_2$ are

257

greater (as $n_1 \rightarrow n_2 \rightarrow n/2$) than the self-interaction terms $(-b_1n_1{}^2)$ and $(-b_1n_2{}^2)$. Thus, $c_1 > b_1$ and $c_2 > b_1$. Further, since $c_1 > c_2$, the resulting conditions lead to the following inequality: $c_1 > c_2 > b_1$, in which case an equilibrium state can be reached. Sketching the phase-plane diagram as shown in Fig.5.8, it is seen that the isoclines are straight lines with positive n_1 and n_2 intercepts and the equilibrium states are each marked with an asterisk on the diagram. Designating the equilibrium population as n_{1E} and n_{2E},

$$n_{1E} = (a_1c_1 - a_1b_1)/(c_1c_2 - b_1{}^2) \tag{5.52a}$$

$$n_{2E} = (a_1c_2 - a_1b_1)/(c_1c_2 - b_1{}^2) \tag{5.52b}$$

Analysis pertinent to the stability of this problem [5.20], leads to the *principle of competitive exclusion*, meaning only one species can survive ultimately. The solution curves for this problem can be sketched as shown in Fig. 6.8, by classifying the equilibrium points on the basis of the following considerations:

1. Coexistent equilibrium population is a saddle point, being always unstable.
2. A species that eliminates its competition is a stable node.

Thus, the unbalanced parts of $I_{\epsilon+}$ and $I_{\epsilon-}$ contributed by $\Delta\epsilon^{\pm}$ constituents (of the coexisting ϵ_1 and ϵ_2 terms) in the error-metric feedback towards network training, can facilitate a stable control/dynamics with an eventual equilibrium of the system or seeking the convergence towards the objective function.

The optimization in neural networks, therefore, implies implicitly the convergence of the learning process mediated by a cost-function such as $\epsilon(t)$ to an attractor. The locations of these attractors and their basins in the phase-space are dictated by the weight modifications; that is, by the iterative adjustments of w_{ij} as a result of supervised learning envisaged. The corresponding nonlinear dynamics follow a random-walk paradigm based information flow. Convergence towards the attractor also refers to the trend in the network's performance approaching stored vectors/memory configurations temporally.

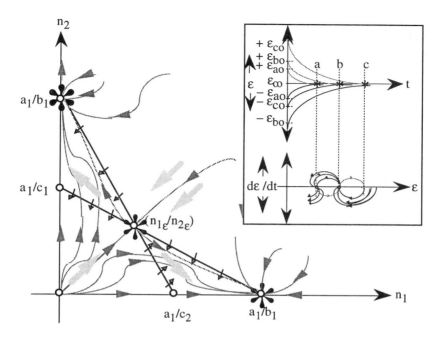

Fig. 5.8: Phase-plane diagram in reference to equilibrium dynamics of interaction between two nearly identical, competing species of error-metrics

5.8.2. Terminal attractor dynamics of error-metrics

The one-dimensional random-walk characteristics, $\varepsilon(t)$, as discussed earlier, can also be extended to study the terminal epochs of the dynamics which can be evolved by means of a procedure [5.21-5.24] as follows. Suppose the dynamics of $\varepsilon(t)$ are represented by a set of nonlinear ordinary differential equations. Further, suppose the dynamical system as it converges towards the objective sought, acquires a coherent structure with no specific inputs from the outside other than from the error information feedback; and, the terminal stochastics $\varepsilon(t)$ are presumed to represent a stationary process with a specified probability distribution function (pdf).

The network's state of disorganization as it proceeds towards extinction is a gradual, slow process rendering the envelope of the envelope $\varepsilon(t)$ as a continuous monotonic function of time. The corresponding logarithmic law (of Eqn. (5.4) akin to Weber-Fechner law) is in conformity with this monotonic decrease with time. The excursion of ε *versus* t curve as $t \to \infty$ and $\varepsilon \to \varepsilon_\infty$ (equilibrium value) depicts the asymptotic terminal dynamics of the learning process. In general, the probabilistic structure of the process involved towards network optimization as set by the random initial conditions will infuse evolutionary attributes to the process at the commencement of the iterations of the feedback epochs. However, as $t \to \infty$, the convergence towards the equilibrium value/attractor, that is, the network output approaching the teacher value, is facilitated by the competition of $I_{\varepsilon+}$ and $I_{\varepsilon-}$, and it would correspond to a stationary stochastic process. Also, in this limiting case, this stationary state within the basin of attraction may become independent of the initial conditions which had set forth the nonlinear dynamics of ε versus time.

Suppose the time derivative of $\varepsilon(t)$ is written as $\partial\varepsilon_i/\partial t = u_i^{\upsilon} (\varepsilon_1, \varepsilon_2, ..., \varepsilon_n)$, $i = 1, 2, ..., n$. Here, all the derivatives of the functions $\{u_i\}$ namely, $|\partial u_i/\partial\varepsilon_j|$, are bounded. That is, $|\partial u_i/\partial\varepsilon_j| < \infty$ in accordance to the so-called Lipschitz condition. This restriction would enable a unique solution for $\varepsilon(t)$ under a given set of initial conditions. Further, $\upsilon < 1$ defines the implicit order of the system. Specific to the assumed form of ε as above, it follows that

$$\left|\partial(\partial\varepsilon_i/\partial t)/\partial\varepsilon_j\right| = \upsilon u^{\upsilon-1}(\varepsilon_1, \varepsilon_2, ..., \varepsilon_n)\left|\partial u_i/\partial\varepsilon_i\right|$$

$$\to \infty \qquad \text{when } \partial\varepsilon_i/\partial t \to 0 \qquad\qquad (5.53)$$

which implies that the *Lipschitz condition* will be violated at the so-called equilibrium points where $\partial\varepsilon_i/\partial t = 0$. These equilibrium points represent the attractors if the real parts of the eigenvalues of the matrix, namely $\left\|\partial u_i/\partial\varepsilon_j\right\|$, are negative [5.21]. Otherwise, for positive values of some eigenvalues, the equilibrium values correspond to repellers from which the time-dependent process $\varepsilon(t)$ will try to escape or diverge. The terminal attractor refers to the point of the convergence wherein the energy concentration becomes unbounded in the neighborhood of the attractor. The other characteristics of the dynamics of $\varepsilon(t)$ are

- Dynamics of $\varepsilon(t)$ are inherently irreversible. That is, being a coherent process, ε versus t cannot be comprehended via time-reversal
- The dynamics of $\varepsilon(t)$ are also stochastical, governed by a set of stochastical dynamics equation(s). Its random attributes stem from the random initial conditions. However, the differential operator of the process would remain deterministic.

The optimization endeavor associated with the neural network, can be described by associated dynamics of the probability density function of $\varepsilon(t)$ as well. That is, as discussed earlier, the dynamics of $\varepsilon(t)$ can be modeled by one-dimensional restricted random walk model of the stochastical variable ε which is terminally attracted (as $t \to \infty$) to a stationary process with a Gaussian statistics of mean value, say, μ_∞ and variance σ_∞^2. The relevant pdf written explicitly is $f_\varepsilon(\varepsilon) = Z[(\varepsilon - \mu_\infty)/\sigma_\infty]$ where Z is the standard normal density function of $\varepsilon(t)$.

The one-dimensional random walk pertinent to ε can be written in a general form as [5.21]:

$$\dot{\varepsilon} = \partial\varepsilon/\partial t = \Gamma_0\{\Psi[\Omega\Psi(\varepsilon)]\}^{\upsilon}\Psi(\omega t) \tag{5.54}$$

where Γ_0, Ω, and ω are constants and ψ is a periodic function. The equilibrium points correspond to $\varepsilon_m{}^o$ ($m = ..., -1, -2, 0, +1, +2, ...$) which are eigen-roots of $Y(\varepsilon) = \psi[\Omega\psi(\varepsilon)] \equiv 0$. Over a range of the interval of ε containing $\varepsilon_m{}^o$ where the set of all points ε satisfy the inequality $\left|\varepsilon - \varepsilon_m^0\right| <$ e $\rightarrow 0$, $\zeta(\varepsilon) = Y^{-1}(\varepsilon)$ and it is continuous everywhere. The superscript o depicts the equilibrium states throughout. Also, using a new variable $\mathcal{f} = \psi(\varepsilon)$, it follows that $\mathcal{f}_m{}^o = mK/\Omega$ (where K is a constant) and the corresponding distance between the equilibrium points are $\mathcal{f}_m{}^o - \mathcal{f}_{m-1}{}^o = K/\Omega$ which is independent of the number of steps, m.

Further as discussed in the earlier sections, over the iterative epochs n $\rightarrow$ ∞, the competition between the reinforcing and annihilating information added to the system, via feedback of the two-sided error-metric, $\varepsilon = \varepsilon_1 + \varepsilon_2$, would enable $\varepsilon(t)$ to cross the equilibrium value repeatedly with the passage of time (Fig.5.2). This repetitive attraction and repulsion of $\varepsilon(t)$ from the equilibrium value is emulated approximately by choosing $\psi(\varepsilon)$ in Eqn. (5.54) as a periodic function. This also sets $\varepsilon(t)$ as a random-walk stochastical process wherein changes could occur only at fixed times [5.21].

Relevant to the above random walk model, the probability density function $f_\mathcal{f}(\mathcal{f},t)$ is governed by the following difference equation:

$$f_\mathcal{f}(\mathcal{f}, t + \pi/\omega) = (1/2)f_\mathcal{f}(\mathcal{f} - K/\Omega, t) + 1/2 f_\mathcal{f}(\mathcal{f} + K/\Omega, t) \tag{5.55}$$

Denoting $h = K/\Omega$, $\tau = \pi/\omega$, $\phi_t{}^{\pm}$ $f_\mathcal{f}(\mathcal{f},t) = f_\mathcal{f}(\mathcal{f}, t \pm \tau)$ and $\phi_\mathcal{f}^{\pm} = f_h(\mathcal{f} \pm h, t)$, Eqn. (5.55) can be written in operator form as [5.21]:

$$\left[\phi_t{}^+ - (\phi_h{}^+ - \phi_h{}^-)/2\right]f_\mathcal{f}(\mathcal{f}, t) = 0 \tag{5.56}$$

The $\phi_t{}^{\pm}$ and $\phi_h{}^{\pm}$ are known as time and space operators. They are related to the conventional partial differential operators as follows: $\phi_t = e^{\tau(\partial/\partial t)}$ and $\phi_h = e^{\tau(\partial/\partial\mathcal{f})}$.

In Eqn. (5.55), if $\omega \rightarrow \infty$, it refers to the total time period and total length of the random-walk being very large in comparison to the time shift (π/ω) or the spatial shift $h = K/\Omega$, respectively. That is, as $\omega \rightarrow \infty$, τ, h $\rightarrow$ 0. Under this condition ($\omega \rightarrow \infty$), Eqn. (5.55) reduces to the Fokker-Planck equation of the type:

$$\partial f_\mathcal{f}(\mathcal{f}, t)/\partial t = (1/2)D^2\partial^2 f_\mathcal{f}(\mathcal{f}, t)/\partial\mathcal{f}^2 \tag{5.57}$$

where $D^2 = K^2/\tau\Omega^2$, and the above equation (Eqn. (5.57)) can be specified validly within certain bounds, say $|\phi| \le |\Psi(\varepsilon)| \le 1$.

The boundary and initial conditions pertinent to the abovementioned Fokker-Planck equation are $\partial f_{\phi}(\phi, t)/\partial y|_{\phi=1,-1} = 0$, $f_{\phi}(0, \phi) = f_{\phi}^0(\phi) \ge 0$ and $\int_{-1}^{+1} f_{\phi}^0(\phi)d\phi = 1$. By virtue of these boundary and initial conditions $f_b(t, f) \to F_0$ (a constant) prescribed within the bounds $|\phi| \le 1$ as $t \to \infty$. This restriction translated to the original variable ε leads to $f_\varepsilon(\varepsilon) = F_0|\phi|$ and an inequality $(-\varepsilon_{Low} < \varepsilon < +\varepsilon_{High})$ specifies the corresponding limits of F_0. Further, within the bounds of $|\phi| \le 1$, ε would always approach a stochastic process.

Hence, the stochastical equation of the dynamics of $\varepsilon(t)$ can be generalized by the requirement that its solution may have a stochastic attractor with the prescribed pdf $f_\varepsilon(\varepsilon)$ subject to the constraints $f_\varepsilon(\varepsilon) = 0$ for $(-\infty < \varepsilon < \varepsilon_{Low})$ and $(\varepsilon_{High} < \varepsilon < +\infty)$ with $\int_{\varepsilon_{Low}}^{\varepsilon_{High}} f_\varepsilon(\varepsilon)d\varepsilon = 1$.

In view of the above considerations, instead of assuming a periodic function, an arbitrary function, namely, $\chi(\varepsilon)$ can also be introduced in lieu of Ψ in Eqn. (5.54). Hence,

$$\partial \varepsilon/\partial t = \Gamma_0\{\Psi[\Omega\chi(\varepsilon)]\}^\upsilon \Psi(\omega t) \tag{5.58}$$

However, $\chi(\varepsilon)$ must be constrained such that $\phi(-\varepsilon_{Low}) = -1$ and $\phi(-\varepsilon_{Low}) = +1$ and $\chi(\varepsilon) = 2\int_{\varepsilon_{Low}}^{\varepsilon_{High}} f_\xi(\xi)d\xi = 1$. Further, $f_\varepsilon(\varepsilon) = F_0|(\partial \phi/\partial \varepsilon)| = F_0(d\chi/d\varepsilon)$.

Suppose the dynamic system is attracted to a zero mean, Gaussian process with σ_ε as the standard derivation. Then,

$$f_\varepsilon(\varepsilon) = Z(\varepsilon/\sigma_\varepsilon) = \left(1/\sqrt{2}\sigma_\varepsilon\right)\exp(-\varepsilon^2/2\sigma_\varepsilon^2) \tag{5.59}$$

where Z is the standard normal density function as mentioned earlier. In order that this process is restricted in the sense that $f_\varepsilon(\varepsilon) = 0$ for $(-\infty < \varepsilon < \varepsilon_{Low})$ and $(\varepsilon_{High} < \varepsilon < +\infty)$, one can consider a truncated normal process, namely:

$$\overline{Z}(\varepsilon/\sqrt{2}\sigma_\varepsilon) = \begin{cases} Z(\varepsilon/\sqrt{2}\sigma_\varepsilon) & \text{if } \varepsilon_{Low}/\sqrt{2}\sigma_\varepsilon < \varepsilon/\sqrt{2}\sigma_\varepsilon < \varepsilon_{High}/\sqrt{2}\sigma_\varepsilon \\ 0 & \text{otherwise} \end{cases}$$

$$\tag{5.60}$$

Hence, defining the pdf of ε as a truncated Gaussian process, f_r:

$$f_r = [f_\varepsilon(\varepsilon)]_T = \mathrm{erf}_T(\varepsilon/\sqrt{2}\sigma_\varepsilon) = \left(2/\sqrt{\pi}\right) \int_0^{\varepsilon/\sqrt{2}\sigma_\varepsilon} \overline{Z}(u) du$$

(5.61)

where the subscript T indicates explicitly, the truncated aspect of the gaussian process. The Fokker-Planck equation for $f_r(r,t)$ can be written as

$$\partial f_r/\partial t = (D^2/2)\partial^2 f_r/\partial r^2 \qquad -N < r < +N$$

(5.62)

where $(- N, + N)$ are the boundary values of r corresponding to $\varepsilon \in [\varepsilon_{Low}, \varepsilon_{High}]$.

Eqn. (5.62) represents a dynamical system whose solution is attracted to a stochastic process with a pdf $\overline{Z}$. For a sufficiently large N, it will approximate a Gaussian process with a finite mean and a finite variance. The solution of Eqn. (5.62) subject to boundary conditions, namely, $(\partial f_r / \partial r)_{r=+N} = (\partial f_r / \partial r)_{r=-N} = 0$, is feasible when the function Ψ is expressed explicitly as a periodic function (such as a sine function) and $\upsilon = 1/(2n+1)$ with $n \to \infty$ where n is an integer. Discretization of υ in terms of the integer values of n specifies explicitly, the discrete occurrences of the critical points and quantifies the associated periodicity.

Suppose σ_I^2 represents the variance known *a priori* at the onset of the statistical forces. With the progression of time, this variance evolves as a function of time given by $\int \varepsilon^2 f_\varepsilon(\varepsilon,t)d\varepsilon$ and limited within the bounds of the variable $\varepsilon \in [\varepsilon_{Low}, \varepsilon_{High}]$. At the terminal stage $(t \to \infty)$, $\sigma_\varepsilon^2(t)$ will approach a constant value σ_∞^2 as decided by its initial value. That is, the dynamical system will approach a stochastic attractor with a pdf $\overline{Z}(\varepsilon/\sigma_\infty)$ as $t \to \infty$, provided that the initial conditions remain within the basin of attraction, namely $|r_\ell| < N_\ell$. In this respect, the convergence effort mediated by $\varepsilon(t)$ can be regarded as a well-organized, competitive task between reinforcing and annihilating information entities, leading asymptotically to terminal dynamics "driven by a global rhythm", manifesting as the occurrence of (more or less) periodic attractors. The associated probabilistic structure corresponds to a random-walk paradigm. The rhythmic occurrence of attractors correspond to the sequential characterization attributed by Rumelhart [5.25] to acceptance, rejection or replacement of information in the neurodynamic system.

5.9 Dynamics of Fuzzy Uncertainty

5.9.1. Fuzzy uncertainty and related error-metrics
The previous sections addressed the various considerations of neural dynamics pertinent to the convergence of the network's performance towards

the objective function in the information-theoretic plane specified in terms of crisp parameters.

In this section, the feasibility of developing a neural dynamics model depicting the network's march towards the convergence elucidated in terms of fuzzy uncertainty is presented. Probabilistic uncertainty associated with crisp variables lead to a set of error-metrics in the informatic-theoretic plane and the related convergence dynamics of the network refers to the minimization of one of these error-metrics. Whereas, considerations of fuzzy uncertainties pertinent to sets with vaguely defined boundaries also facilitate a dynamics model to depict the convergence dynamics in a fuzzy neural network. Again, dynamics of fuzzy uncertainty can be studied in the information-theoretic plane.

Uncertainty of a fuzzy set in the context of probabilistic framework has been quantified by Zadeh [5.26] who defined a measure (f_H^{Za}) associated with a fuzzy set $A \in \{x_i\}$ characterized by the membership function set $\{\mu(x_i)\}$ and probabilities $\{p_i\}$. It is given by

$$f_H^{Za}(A) = -\sum_{i=1}^{n} \mu(x_i)(p_i \ell o g p_i) \qquad (5.63)$$

which being similar to Shannon's entropy function is the entropy associated with the set A and weighted by the fuzzying index (membership function), $\mu(x_i)$. Alternative measures of fuzzy entropy are due to Deluca and Termini [5.27] given by

$$f_H^{DT}(A) = -\sum_{i=1}^{n} \mu(x_i)\ell o g[1 - \mu(x_i)] + [1 - \mu(x_i)]\ell o g[\mu(x_i)]$$

$$(5.64)$$

and due to Pal and Pal [5.28] given by

$$f_H^{PP}(A) = -\sum_{i=1}^{n} \mu(x_i)\exp[1 - \mu(x_i)] + [1 - \mu(x_i)]\exp[\mu(x_i)]$$

$$(5.65)$$

In training fuzzy neural networks via a backpropagation learning algorithm, the error function employed conventionally is similar to the quadratic error function adopted in crisp neural networks. Suppose Y_K^M and T_K^M represent the output fuzzy set and the target fuzzy set respectively for the training sample M. Then, the quadratic fuzzy error-metric is given by

$$f_{e_{SQ}} = (1/2)\sum (T_K^M - Y_K^M)^2 \qquad (5.66)$$

Instead of the fuzzy quadratic error-metric, the following (fuzzy) relative entropy (mutual information) error-metric is proposed here on the basis of fuzzy entropy concepts due to Zadeh [5.26]

$$f_{e_{RE}} = \sum \mu_Y(T_i)q_i\phi(p_i/q_i) + \sum \mu_T(T_i)p_i\phi(q_i/p_i) \tag{5.67}$$

where $Y = \{Y_i\}$, $T = \{T_i\}$ and μ_Y and μ_T are membership functions of Y and T respectively. Further, $\{p_i\}$ and $\{q_i\}$ are sets of probabilities of $\{Y_i\}$ and $\{T_i\}$ respectively. The function ϕ is the same as the Csiszár function defined in Eqn. (5.44) and discussed in Chapter 4; and, Eqn.(5.67) is similar to the symmetrized version of Csiszár error-metric elucidated for crisp neural architecture.

5.9.2 Dynamics of the mutual information error-metric

The dynamics of the error-metric $f_{\varepsilon_{RE}}$, namely, $f_{\varepsilon_{RE}}(t)$ versus t, when adopted in the training iterations of a fuzzy neural network, would follow the same considerations of a competitive (but fuzzified) trend of seeking the equilibrium (attractor) as in the crisp case discussed in the previous sections.

In the fuzzy domain, the competing species, namely $\Delta\varepsilon^+$ and $\Delta\varepsilon^-$, are fuzzified. Their discourse along time should correspond to the same random-walk paradigm as in the case of the crisp values of $\Delta\varepsilon^\pm$. The corresponding differential equation depicting the convergence of the fuzzy $\varepsilon(t)$ towards the equilibrium value ε_∞ is fuzzy in the sense that the corresponding solution represents the probability of the fuzzy dichotomous events $\Delta\varepsilon^\pm$. This version of fuzzy differential equation which yields a probability solution belongs to a distinct family a differential equations defined by Zadeh [5.26]. This is different from the interval-value calculus based differential equation discussed in Chapter 3.

Zadeh [5.26] defines the probability of a fuzzy event A as the expectation of the associated membership function μ_A.

$$P(A) = \int \mu_A(x)f_x(x)dx$$
$$= E(\mu_A) \tag{5.68}$$

where $f_x(x)$ is the probability distribution of x. When the fuzzy set A is time-dependent, the grade of membership in the set A can be regarded as time-dependent such that $\mu_A = \mu_A(x,t) \in [0,1] \ \forall x,t$. Then,

$$P_A(t) = \int \mu_A(x,t)f_x(x)dx \tag{5.69}$$

and $\int_{-\infty}^{+\infty} f_x(x)dx = 1$.

As discussed in the earlier sections, the random walk paradigm can be represented by a diffusion equation. When it is adopted to describe the

265

convergence process, namely, the continuous-time simulated annealing process, towards a global minimum via iterative endeavors associated with a stochastical neural network, the method has been termed as a diffusion machine [5.28].

In a fuzzy neural network, the output, which is not defuzzified, when compared with a target fuzzy set, will yield an error-metric such as those described in Chapter 3 pertinent to the information-theoretic plane. An error-metric of this kind constitutes a fuzzy set $\{E_f(\varepsilon)\}$ for which membership functions $^f\mu(\varepsilon)$ can be assigned. Inasmuch as ε is a function of time along the discourse of iterative feedback adopted in training the network, the associated $^f\mu(\varepsilon)$ can be also be regarded as a time-dependent variable. Writing $\varepsilon(t)$ as ε_t and $^f\mu[\varepsilon(t)]$ as $\mu_t(\varepsilon)$ for simplicity, the temporal trajectory of $\mu_t(\varepsilon)$ with stochastical attributes can be described by a random walk paradigm characterized by a stochastic differential equation which has the built-in (or inherent) notions of fuzziness. The solution of such an equation, namely $^f\mu(\varepsilon)$ when adopted, Eqn. (5.68) should yield the probability of the fuzzy set of events $\varepsilon(t)$ in the error domain E^f namely, $P_{E^f}(\varepsilon,t)$. Kandel and Byatt [5.29] have considered relevant fuzzy differential equations as a class by itself yielding solutions which represent probabilities of fuzzy events in the perspectives of Zadeh [5.26]. It is again stressed here that these fuzzy differential equations are not the same as those based on interval-value calculus discussed in Chapter 3.

Now consider $^f\mu(\varepsilon,t)$, at the initial stages of iterations, that is at $t \rightarrow 0$. $^f\mu(\varepsilon,t) \rightarrow$ maximum value ($\rightarrow$ 1), depicting the extent of association of error-metric values with $\varepsilon \gg \varepsilon_\infty$ is the largest. However, as t increases with the progression of iterative epochs, the error-metric values change and converge towards ε_∞. Or, as $t \rightarrow \infty, ^f\mu(\varepsilon,t) \rightarrow$ minimum value (= 0). A corresponding decay process of $^f\mu(\varepsilon,t)$ with respect to time can be represented by

$$^f\mu(\varepsilon,t) = {}^f\mu(\varepsilon_0,0)\exp[-k_\mu(\varepsilon)t] \tag{5.70}$$

The above formalism of $\mu(\varepsilon, t)$ should be consistent with the condition that $\int_{-\infty}^{+\infty} {}^f\mu(\varepsilon,t)d\varepsilon = 1$.

It was indicated earlier (Eqn. (5.18)) that the pdf describing the statistic of $\varepsilon(t)$ at the nth epoch of iteration is a mean zero Gaussian process (in reference to the equilibrium value ε_∞ taken as 0) and a variance $<\varepsilon^2> = m(\Delta\varepsilon^\pm)^2$ where $\Delta\varepsilon^\pm$ is the dichotomous reinforcing/annihilating information imparted by the error fedback *via* the control-loop in the information-theoretic plane. Denoting $<\varepsilon^2>$ as σ_ε^2, the pdf of $\varepsilon(t)$ can be written at a mean reference of ε_∞ as

$$f_\varepsilon(\varepsilon) = (1/2\pi\sigma_\varepsilon^2)\exp[-(\varepsilon - \varepsilon_\infty)^2/2\sigma_\varepsilon^2 \tag{5.71}$$

Consistent with the *Zadeh mean-condition*, namely, $\int {}^f\mu(\varepsilon,t)f(\varepsilon)d\varepsilon = 1$, combining Eqns. (5.68) and (5.67), the following relation can be written explicitly as

$$P_{E^f}(t) = \left[\frac{{}^f\mu(\varepsilon_0,0)}{2\pi\sigma_\varepsilon^2}\right]\int\exp[-k_\mu(\varepsilon)t]\exp[-(\varepsilon - \varepsilon_\infty)^2/2\sigma_\varepsilon^2]d\varepsilon$$

$$= \left[\frac{{}^f\mu(\varepsilon_0,0)}{2\pi\sigma_\varepsilon^2}\right]\int\exp[-k_\mu(\varepsilon)t - (\varepsilon - \varepsilon_\infty)^2/2\sigma_\varepsilon^2]d\varepsilon \tag{5.72}$$

Further, the conditons $P_{E^f}(t) \to 0$ as $t \to \infty$ and $P_{E^f}(t) \to 1$ as $t \to 0$ set ${}^f\mu(\varepsilon_0,0) \equiv 1$, which is in concordance with the assumption made earlier.

Thus, Eqn. (5.72) offers an explicit discourse of the probability $P_{E^f}(t)$ of the fuzzy error-metric set $E_f = \{\varepsilon\}_f$ with time. This probability of the fuzzy set E_f is dependent implicitly on the grade membership function ${}^f\mu(\varepsilon)$ (which is also time-dependent) via the Zadeh average relation defined earlier. Denoting $P_{E^f}(t) = \int_\Omega f_{E^f}(t)dt$ in the domain Ω, the associated stochasticity (Fokker-Planck) relation of the following type can be written:

$$\partial f_{E^f}(t)/\partial t = (1/2)\partial^2[\sigma_\mu^2 f_{E^f}]/\partial^{2f}\mu - \partial[m_\mu p f_{E^f}]/\partial^f\mu \tag{5.73}$$

which can be designated as the *fuzzy Fokker-Plank equation* in the perspective of Zadeh defining a probability *vis-a-vis* fuzzy uncertainty. In Eqn. (5.73), the parameters σ_μ^2 and m_μ consitute the variance and mean of ${}^f\mu$ as specified by the governing stochasticity of ${}^f\mu(t)$ given by

$$d^f\mu(t)/dt = m_\mu({}^f\mu,t) + \sigma_\mu({}^f\mu,t)\psi_\mu(t) \tag{5.74}$$

where ψ_μ is a random function such that $\int_0^t \psi_\mu(t)dt$ impose the random walk attributes to ${}^f\mu$ similar to the conditions presented in respect of a crisp random variable in Eqn. (5.14).

5.10 Concluding Remarks

The essence of this chapter is two-fold: 1) It portrays the dynamics of the learning process in neural networks; and, 2) the relevant portrayals are displayed in the information-theoretic plane. Within the broad scope of the aforesaid considerations, the major inferences and/or conclusions which can be gathered from the analyses presented are as follows:

- The stochastical dynamics associated with the neural learning process can be in the information-theoretic plane, as it can be done in the parametic space plane
- The error-metric specified in the information-theoretic plane should be a symmetrical, balanced version, as discussed in Chapter 4, in order to facilitate guaranteed converging dynamics of the network
- The relevant dynamics can be specified in terms of a class of error-metrics of the network, which can be elucidated in the information-theoretic plane for the purpose of network learning optimization using the said error-metrics (ε) as feedback entities. Hence, the relevant dynamics refer to ε *versus* time (t) over which the epochs of iterations of error feedback are performed to achieve the convergence
- The associated stochasticity models the dynamics of $\varepsilon(t)$ in terms of probability function *versus* time as governed by the Fokker-Planck diffusion equation
- The dynamics of $\varepsilon(t)$ can be specified by a logistic growth model depicting equilibrium conditions
- The convergence or divergence aspects of ε with the passage of time, or along the iterative epochs of error-feedback, depends on the competitive role played by augmentative and annihilating information imparted to the system by the error information fed back
- Corresponding values of $\Delta\varepsilon^+$ and $\Delta\varepsilon^-$ (deviatory measures of ε from the equilibrium value ε_∞) constitute dichotomous events repeated along the discourse of iterations performed. This Bernoulli process, has binomial distribution on a discrete basis. As n (number of events of $\Delta\varepsilon^\pm$) $\rightarrow$ 0, this distribution becomes a Gaussian process. Excessive unbalance between $\Delta\varepsilon^+$ and $\Delta\varepsilon^-$ would lead to divergence in the network's performance. A near-balanced state would, however, enable convergence
- In the terminal region, that is, as t $\rightarrow$ ∞, the convergence endeavor could set $\varepsilon(t)$ as a stationary process [5.21]. Apart from this terminal attractor status, during the discourse of ε versus t, the error-metric value may also cross the equilibrium value ε_∞ at several instants of time, each representing an attractor in the basin of convergence
- The network dynamics are also extended for fuzzy variables and the relevant stochasticity versus time is shown to follow a fuzzy Fokker-Planck equation with the pdf of the fuzzy error-metric as the variable. This pdf is related to the grade membership function associated with the fuzzy attributes.

Bibliography

[5.1] Neelakanta, P. S., Sudhakar, R. and De Groff, D.: Langevin machine: A neural network based stochastically justifiable sigmoidal function. *Biol. Cybern.*, 65, 1991, 331-338.

[5.2] Kesidis, W. : Analog optimization with Wong's stochastic neural network. *IEEE Trans. Neural Networks*, 6, 1995, 258-260.

[5.3] Usami, H., Masaki, S. and Sato, R.: The stochastic properties of basic neuron populations as information processing system. *Biol. Cybern.*, 29, 1978, 167-179.

[5.4] Bergstrîm, R. M.: Neural macrostates. *Synthese*, 17, 1967, 425-443.

[5.5] Amari, S., Kurata, K. and Nagoaka, H.: Information geometry of Boltzmann machines. *IEEE Trans. Neural Networks*, 3, 1992, 260-271.

[5.6] Heskes, T. M. and Kappen, B.: Learning processes in neural networks. *Phy. Review A*, 4, 1991, 2718-2726.

[5.7] Park, J. C., Neelakanta, P.S., Abusalah, S., De Groff, D. and Sudhakar, R.: Information-theoretic based error-metrics for gradient-descent learning in neural networks. *Complex Syst.*, 9, 1995, 287-304.

[5.8] Csiszár, I.: A class of measures of informativity of observation channels. *Perio. Math. Hungaria*, 2, 1972, 191-213.

[5.9] Kullback, S. and Leibler, R. A.: On information and sufficiency, *Ann. Math. Stat.*, 22, 1951, 79-86.

[5.10] Bergstrîm, R. M. and Nevanlinna, O.: An entropy model of primitive neural systems. *Intern. J Neuroscience*, 4, 1972, 171-173.

[5.11] Shannon, C. E. : A mathematical theory of communication. *Bell Syst Tech. J.*, 27, 1948, 623-659.

[5.12] Neelakanta, P. S. and De Groff, D.: *Neural Network Modeling: Statistical Mechanics and Cybernetic Perspectives* (CRC Press, Boca Raton, FL: 1994), Chapter 8.

[5.13] Norwich, K. H.: *Information, Sensation and Perception* (Academic Press, Inc., San Diego, CA: 1993), Chapter 9.

[5.14] Wong, E: *Stochastic Process in Information and Dynamic Systems* (R. E. Krieger Publishing Co., Huntington, NY: 1979), Chapter 4.

[5.15] Langevin, P.: Sur la Théorie du movement brownien, *C.R. Hebd Séances Acad. Sci.*, 146, 1908, 503-533.

[5.16] Haberman, R.: *Mathematical Models* (Prentice-Hall, Inc., Englewood Cliffs, NJ: 1977), Section 40.

[5.17] Kullback, S.: *Information Theory and Statistics* (John Wiley and Sons, New York, NY: 1959)

[5.18] Lin, J.: Divergence measures based on Shannon entropy. *IEEE Trans. Inform. Theory*, 37, 1991,145-151.

[5.19] Haken, H.: *Information and Self-Organization* (Springer-Verlag, Berlin: 1988).

[5.20] Haberman, R.: *Mathematical Models* (Prentice-Hall, Inc., Englewood Cliffs, NJ: 1977), Section 54.

[5.21] Zak, M.: Terminal attractors in neural networks. *Neural Networks*, 2, 1989, 259-274.

[5.22] Zak, M.: Terminal attractors for associative memory in neural networks. *Phy. Letts.*, *A*, 133, 1988, 18-22.

[5.23] Zak, M.: Spontaneously activated system in neurodynamics. *Complex Syst.*, 3, 1989, 471-492.

[5.24] Zak, M.: Physical models of cognition, *Intern. J. Theor. Phys.*, 33, 1994, 1113-1161.

[5.25] Rumelhart, D. and McClelland, J. L. (Eds): *Parallel Distributed Processing* (The MIT Press, Cambridge, MA.: 1987), Vol.1, pp.12.

[5.26] Zadeh, L. A.: Fuzzy sets, *Inform. Control*, 8, 1965, 338-354.

[5.27] Deluca, A. and Termini, S.: A definition of nonprobabilistic entropy in the setting of fuzzy sets theory. *Inform. Control*, 21, 1972, 301-312.

[5.28] Pal, N. R. and Pal, S. K.: Higher order fuzzy entropy and hybrid entropy of a set. *Inform. Sci.*, 61, 1992, 211-231.

Chapter 6

Informatic Perspectives of Complexity Measure in Neural Networks

P. S. Neelakanta, J. C. Park and D. De Groff

> " (Complexity) is the totality of interactions
> within a (large system) "
>
> M. H. van Emden

6.1 Introduction

A neural complex, real or artificial, is an embodiment of a massively connected set of neurons and represents a cellular automaton "trained to learn" and predict *via* endeavors managed by a set of protocols involving collection, conversion, transmission, storage and retrieval of information. The training or learning effort is to recognize and counter-balance the effects of the cellular disturbances (noise) present in the neural system which may tend to disorganize the system's convergence towards an objective function mediated through learning protocols. The extent of disorganization caused by such disturbances can be specified by a disorderliness parameter set by the maximum entropy considerations as discussed in the earlier chapters. Such an entropy functional also depicts implicitly the degree of complexity of the system in spatiotemporal domains. Therefore, the disorderliness in the neural complex can be specified by a *complexity metric* concurrent to the entropy paradigm adopted conventionally. Using this metric-parameter as an error-measure (or cost-function), a control strategy, such as the backpropagation based gradient-descent method, can be developed to train a multilayered perceptron. The present chapter offers relevant theoretical considerations, algorithmic representations and simulation results.

6.2 Neural Complexity

The cybernetic issues relevant to the neural complex refer to the task of learning envisaged through adaptive feedback in order to achieve an "orderliness" by overcoming the influence of randomness resulting from the intra- and/or extraneural disturbances. Such disturbances, would otherwise promote a deviatory response from the objective, or the target response that the network is striving to achieve. The extent of disorganization so promoted can be generalized in terms of a performance deficiency or an error-parameter. Also, the disorganization perceived can be correlated to the randomness and the network complexity. This complexity restricts the formation of the "assembly" of neurocellular units in the real neural complex in which the synaptic modifications are introduced by input-induced neurotransmitters as dictated by the regulating feedback involved. Sustaining such modifications, endured as "patterns" in the spatiotemporal domain of the neural complex is however, often hampered by the inherent neurocellular disturbances. In other words, the neural complexity is dictated implicitly by the effects of neurocellular disturbances. The "size" of the neural assembly depicting the extent of neurocellular complexity which participates in the training schedule is dependent functionally on the organization deficiency encountered due to the counteracting efforts (or the disturbances) present in the web of biological neurons.

Similar attributes can also be specified in respect to artificial neural networks constructed to "learn", recognize and predict the patterns. In general, as indicated in Chapter 4, the disorderliness in the neural complex, real or artificial, can be estimated in terms of the deviation of a selected variable, say y_j , with reference to a specified standard or objective, y_T. Geometrically, if y_T refers to the target vector pinpointing the center of a region of orderliness, around this center a quasi-ordered region can be considered wherein the orderliness is maintained within a specified statistical bound (Fig.6.1). The disorderliness associated with the variable y_j can be assessed in terms of $\Delta(y_j)$, the distance from the center of orderliness to the boundary of the quasi-ordered region. Orderliness here refers to absence of randomness in the subspace and quasi-ordered region refers to the target subspace wherein the randomness becomes vanishingly small on approaching the center of the subspace. Therefore, the disorderliness at Y_j realization can be written as [6.1]

$$Y_j = (y_j - y_T) - \Delta(y_j) \qquad (6.1)$$

where $(y_j - y_T)$ represents the error-vector. Normally, Y_j is rendered dimensionless by normalizing it in terms of a reference (disorderliness) value.

The goal-associated positional entropy vis-à-vis the disorderliness Y_j can be designated by H_{y_j} at the jth elementary subspace in an entropy space Ω_H . In Fig.6.1, the Ω_s-space refers to the global extent of neural complexity. ΔW_s can be regarded as the neural subassembly of Ω_s-space which has been triggered to endure the synaptic modifications, but is counteracted by the disturbances present in the system. Hence, ΔW_s in Ω_s-space in Fig. 6.1 can be considered as a subset of the neural complex striving to achieve a reduced entropy or, "simplicity", by gaining (learning) pertinent information so as to yield an output close to the target value, despite the presence of counteracting disturbances.

As mentioned earlier, the randomness or disorderliness of the neural system refers implicitly to the system *complexity* C_s, as observed from the system exterior, and depicts the "variety" associated with the system subsets

272

and/or microsubsets constituted by the neural assembly defined earlier. Mathematically, C_S can be represented by

$$C_S = \phi(N,\nu) \qquad\qquad C_S \in \Omega_{C_S} \Leftrightarrow \Omega_S \qquad\qquad (6.2)$$

where N is the number of cellular assemblies and ν is their (associated) *variety*, namely the synaptic modifications vis-à-vis excitatory/inhibitory states. The function ϕ measures implicitly the uncertainty or entropy due to the complexity and, therefore, could be in general, logarithmic in the perspectives of Shannon's law [6.2]. Further, Ω_{C_S} is the complexity domain equivalence of Ω_S and Ω_H in Fig.6.1.

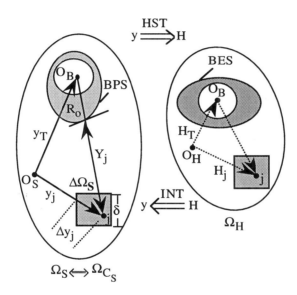

Fig.6.1: Parameter spread-space (Ω_S) depicting implicitly a complexity-metric space (Ω_{C_S}) and entropy-space (Ω_H) of the neural complex

[HST: Shannon transform; INT: Inverse transform; CDT: Complex domain transform; ICDT: Inverse complex domain transform O_B: Locales of the objective function in the specific domains; BPS: Boundary of the quasi-order of parameter spread space; BES: Boundary of the quasi-order region of entropy space (O_S and O_H are arbitrary reference origins in $\mathbf{W}_S$ and $\mathbf{W}_H$ planes, respectively)]

The uncertainty indicator of the complexity should be linked to the disorderliness function by a criterion that, in the event of the system tending to be well-ordered or least complex, through minimization of disorderliness, with the condition $Y_j \to 0$, it should be accommodated by an appropriate functional relation between Y_j and C_s.

Further, by considering the geometrical space representing the disorderliness, as illustrated in Fig. 6.1, the variate y defines not only a measure of uncertainty (deterministic or probabilistic), but it also refers explicitly to the a priori probability of a sample space within the region.

The collective influence of disorderliness and hence the complexity, therefore, determines the extent of entropy (or uncertainty) associated with the organizing efforts in the neural system in achieving a specific control goal. That is, the net complexity and disorderliness function bears information on the degree of disorganization defined earlier.

If organizational deficiency is attributed to every synaptic coupling, across N cellular units of the assemblies involved, its overall value of deficiency can be stipulated by:

$$O_D = \sum_{j=1}^{N} p_j f_o(\Theta_{Y_j})$$

(6.3)

where p_j refers to the probability of encountering the jth cellular subassembly, in the neural spatial complex, wherein the disorganization is observed, f_o is a function to be determined, and Θ_Y is a *disorderliness parameter* defined by the relation:

$$\Theta_{Y_j} = Y_j W_{Y_j} + C_Y$$

(6.4)

where Ω_{Y_j} is a weighting function measuring the deviation of Y_j of the jth realization from those of other realizations of the state variable. C_Y is a conditional coefficient which sets $f_o(\Theta_{Y_j}) = 0$ when $Y_j = 0$. In writing the above relation, it is presumed that the system is *ergodic* with the entire ensemble of the parameter space having a common functional relation f_o.

Thus, the disorganization is an ensemble-generalized characteristic of the disorder in the spatiotemporal state of the neural system, weighted primarily by the probability of encountering a cell with a random behavior and, secondarily, by its *relevance* . The relevance of the disorder to the jth situation is decided both by the functional relation d_o common to the entire ensemble, and by the additional weight Ω_{Y_j} accounting for the deviation of Y_j with the corresponding disorderliness of realizations other than j.

The functional relation f_o pertinent to the organizing (control) endeavors of a neural complex refers to the sensitivity of a higher order level of goal towards the degree of failure in attaining the subgoals considered. Such a failure or deviatory response arises due to the *entropy* of the system. Therefore, as stated earlier, O_D can be written more explicitly as an entropy function in terms of *Shannon's law* as:

$$O_D = - \Lambda \sum_{j=1}^{N} p_j \, log(\Theta_{Y_j}) \qquad (6.5)$$

where Λ is a constant of proportionality as decided by the base of the logarithm chosen.

Considering both temporal and spatial disorganizations associated with the interconnected neurons, a superposition leads to the following general expression to represent the cumulation of the effects:

$$O_D = \left[\bigcup_{\alpha}^{\ell} \tau_\alpha \right] \left[\bigcup_{\beta}^{\mu} r_\beta \right] \left[\bigcup_{\gamma}^{N} W_\gamma d(Y_\gamma) \right] \qquad (6.6)$$

where $\bigcup$ represents the union operation in the spatiotemporal domains over ℓ instants of time, μ realizations at N cellular assemblies, d is a factor which denotes explicitly the disorderliness perceived at the jth unit; and, τ, r, W represent the corresponding weights, respectively.

From the foregoing discussions, it is evident that the disorganization of a neural complex is the consequence of:

- Spatial factors conceived in reference the random locations in the neuronal assembly
- Temporal characteristics as decided by the random occurrence of state transitions across the interconnected cells in the subassemblies considered
- Stochastical attributes of the neural disturbances
- Combinatorial aspects arising from the number and variety of the participating subsets of assemblies wherein the neural activity is perceived.

Referring to the spatiotemporal disorderliness described earlier, let the a priori probability p_j denote the probability of occurrence of synaptic modifications (in time and space) of the jth neural assembly. When $p_j \to 1$, it amounts to a total orderliness with $|Y_j| \to 0$. Likewise, $p_j \to 0$ sets a total disorderliness with $|\mathbf{Y}_j| \to \infty$. The above conditions are met simultaneously by the following coupled relations:

$$|\mathbf{Y}_j| = (1/p_j) - 1 \qquad (6.7a)$$

$$\Theta_{Y_j} = 1/p_j \qquad (6.7b)$$

Therefore, the extent of disorganization can be written as:

$$O_D = \sum_{j=1}^{\mu} p_j \ell og(1/p_j) = - \sum_{j=1}^{\mu} p_j \ell og(p_j) \qquad (6.8)$$

which is again in the standard form of the Shannon's statistical measure of entropy.

The structural or combinatorial aspect of disorganization pertains to the number of alternatives, such as the paths of proliferating state-transitions or *traces,* in the interconnected network of cellular assemblies. Usually, only μ out of such μ_j alternatives are warranted to confirm a total orderliness. Therefore, the disorderliness is written as

$$|\mathbf{Y}_j| = (\mu_j/\mu) - 1 \qquad (6.9)$$

so that $|\mathbf{Y}_j| \to \infty$ for $\mu_j \to \infty$ with m being a constant and $|\mathbf{Y}_j| \to 0$ for $\mu_j \to \mu$. The corresponding measure of disorganization written in conformity with Shannon's measure of entropy simplifies to

$$O_D = \ell og(|\mathbf{Y}_j|) \qquad (6.10)$$

In a learning endeavor, the neurocybernetic system attempts to regulate, via feedback techniques, the state of the system corresponding to, say, the state of the jth realization being nondeviatory with respect to a target state. Designing the generalized state of the system vector by y_j in the jth realization from the target vector y_T, the corresponding vector diversion is represented as indicated earlier by $(y_j - y_T)$, and $|y_j - y_T| = Q_j$ represents the magnitude of the error vector. The bounded two-dimensional spread-space, Ω_S, includes all such likely diversions as depicted in Fig. 6.1. It can be decomposed into elementary subspaces, $\Delta\Omega_S$, which can be set equal to δ^2, where δ represents the one-dimensional quantizing extent of the subspace.

Suppose the a priori probability (p_j) of finding the vector $\mathbf{y}_j$ at the jth elementary subspace is known, along with the distance of this j^{th} realization from the goal (target), namely, Q_j. The corresponding diversion ensemble of the entire spread-space $\{\Delta\Omega_S\}$ can be written as

$$\{\Delta\Omega_S\} = \begin{bmatrix} Q_1 & Q_2 & \cdots & Q_k \\ p_1 & p_2 & \cdots & p_k \\ W_1 & W_2 & \cdots & W_k \end{bmatrix} \tag{6.11}$$

where $\sum_{i=1}^{\kappa} p_i = 1$, and the number of total realizations constituted by equal subspaces over the domain Ω_S is $\kappa = \Omega_S/\Delta\Omega_S$. Assuming equal weights, namely, $W_1 = W_2 = ... = W_\kappa = 1$, a goal-associated positional entropy H_y can be defined as follows:

$$H_y = \sum_{i=1}^{\kappa} p_i \, \ell og[W_i\{Q_i - |\Delta(\mathbf{y}_j)| + 1\}] \tag{6.12}$$

As indicated in Fig. 6.1, the function H defines a new entropy space (Ω_H) as distinct from the parameter space Ω_S. That is, each value of H_y in the entropy space could be mapped onto the parameter spread-space and denoted on an one-to-one basis by H_y' (Q_i). This value of H' represents the expected mean diversion from the goal in the spread-space Ω_S. Further, the spaces Ω_S and Ω_H are *affinely similar*. That is, $\Omega_S \to \Omega_H$ permits both translational and rotational transformations. The goal-related entropy of Eqn.(6.12) satisfies the conditions specified in the following lemmas [6.1]:

Lemma 1: $H_y = 0$, if all $|\mathbf{y}_j - \mathbf{y}_T| = Q_i \leq |\Delta(\mathbf{y}_j)|$ or if $p = 0$ for all
$|\mathbf{y}_j - \mathbf{y}_T| - |\Delta(\mathbf{y}_j)| > 0$ $\qquad\qquad$ (6.13)

Lemma 2: $H_y \to 0$, for the ensemble $p > 0$, if $|\mathbf{y}_j - \mathbf{y}_T| - |\Delta(\mathbf{y}_j)| \to 0$,
and $H_y \to \infty$, if $|\mathbf{y}_j - \mathbf{y}_T| - |\Delta(\mathbf{y}_j)| \to \infty$ $\qquad$ (6.14)

Lemma 3:   At $p = 1/\kappa$,
$$H_y = -(1/\kappa)\sum_{i=1}^{\kappa} \ell og\{1/[|y_i - y_T| - |\Delta(y_i)|]\} + \varepsilon_\kappa \tag{6.15}$$

where,

$$\varepsilon_\kappa = (1/\kappa) \sum_{i=1}^{\kappa} \log\{[| \, y_i - y_T \, | - | \, \Delta(y_i) \, |]/[| \, y_i \, - \, y_T \, | - | \, \Delta(y_i) \, | \, + \, 1\}$$

and $\varepsilon_k \rightarrow 0$, if $|y_j - y_T| - |\Delta(y_j)| \gg 1$.

Lemma 4:   Sum of two entropies satisfying the conditions of independence and summation in the spread-space of the state-vector leads to:

$$H_{y_{(1,2)}} = \log\{H_{y_1}[| \, \mathbf{y}_1 - \mathbf{y}_T |] + H_{y_2}[| \, \mathbf{y}_2 - \mathbf{y}_T |] + \varepsilon_{(1,2)}$$

$$(6.16)$$

where, $\varepsilon_{(1,2)} = \log\{[H_{y_1}(Q_1) + H_{y_2}(Q_2)]/[H_{y_1}(Q_1) + H_{y_2}(Q_2) + 1]\} \rightarrow 0$
for $H_y(Q_j) \gg 1$ with $Q_j = | \, \mathbf{y}_j - \mathbf{y}_T |$, $(j = 1, 2)$

Lemmas 1 and 2 represent the intuitive concept of neural disorganization in the state of being controlled towards the goal. If an ideal control is perceived, it strikes the well-ordered target domain in all realizations specified by $H_y = 0$. Diversions from this ideality of the ensemble with an increasing or decreasing trend enable H_y to increase or decrease, respectively.

Lemma 3 stipulates that, in the event of equiprobable diversions, the relation between the spread-space of the state vector and the entropy space is logarithmic with an error $\varepsilon_\kappa \rightarrow 0$ for $| \, \mathbf{y}_j - \mathbf{y}_T | \gg 1$.

The entropy associated with a target-seeking endeavour is not additive. That is, the goal-associated entropies cannot be added or subtracted directly in the entropy space. However, these superposition operations can be performed in the spread-space of the state vector. After the necessary extent of such operations are executed in the spread-space, the consequent effects can be translated to the entropy space.

Lemma 4 specifies the rule of additivity in the spread space, pertaining to the independent goal-associated position entropies with an accuracy set by $\varepsilon_{(1,2)} \rightarrow 0$ with $H_y(Q_{1,2}) \gg 1$.

6.3 Complexity Measure

The "complexity" vis-à-vis neural networks generally refers to the strain on the computational power of the network as dictated by the architectural considerations and spatiotemporal aspects of neural state proliferation across the network. It includes the associated nonlinearity and the transfer functional attributes of the network. In the perspectives of MacGregor [6.3], relevant characteristics of the neural complex could be specified in terms of the so-called "beds" and "realizations". Specifically, the stochastical theory of multiple "traces" or subassemblies of neural interconnections in biological systems has been addressed comprehensively by MacGregor [6.3], who associates the concepts of *beds* and *realizations* in describing the memory borne by the traces. The bed or a set of neurons fire specific sequences which represent the items of information denoting physiological variations (or realizations) at the anatomical sites. Thus, the population of excitatory and inhibitory neurons taken as an ordered sequence of sets constitutes the bed of a trace. A realization is an observable physiological manifestation of an underlying bed. It consists of an ordered sequence of subsets of active neurons, some of which are members of the corresponding sets of the bed and which fire over a given time interval in the temporal correspondence that exists among the cells of the bed.

The architectural aspect of neural complexity in reference to artificial neural networks has also been specified in terms of the "size" of the network representing the number of cellular elements, and the "depth" of the network measuring the longest path from any input gate(s) to the output gate(s). This occurs when the gates are arrayed in layers so that all the gates in the same layer compute concurrently, as in the perceptron version of artificial neural networks [6.4].

In a naive manner, Haken [6.5] describes complex systems as those "composed of many parts, or elements, or components which may be of the same or different kinds. The components or parts may be connected in a more or less complicated fashion." This description matches appropriately in characterizing biological neural systems as well as artificial neural networks, considering the associated learning process at the macroscopic level.

Curtis [6.6] describes complexity as "an attribute of the interaction between two systems that describes the resources one system will expend in interacting with the other system." When translated to the neural complex, the above definition of complexity aptly describes the intricacies of a neural assembly in which one neuron ("system") expends its state in deciding the state of another neuron ("system") in a complex manner through spatiotemporal interactions.

Quantitatively, to compare different complex systems or to evaluate any given complex system, a quantitative entity, say, "a measure of the algebraic degree of complexity" or "algorithmic complexity" is necessary. For example, in the classical Turing computational machine, the minimum length of a program and of initial data is presented as the degree of complexity of the machine [6.1]. But, the feasibility of evaluating this degree of complexity (vis-à-vis, Turing machine) is rather questionable in view of *Goedel's theorem* which indicates the nonfeasibility of solving for a minimum program and a minimum number of initial data *via* a universal fashion. This is true in almost all complex systems. However, attempts have been made to find at least approximate methods of quantifying the complexity using information theory [6.7], statistical mechanics [6.8], equilibrium strategies applied to Markov processes [6.9] etc. Presently, a convenient approach, as suggested by Ferdinand [6.10] and based on maximum entropy considerations, is applied to the neural complex to evaluate a quantifiable parameter that estimates the associated complexity.

Another way of formalizing complexity is due to Kolmogorov [6.11], who evolved the concept of algorithmic complexity on the basis of generalized entropy considerations. His concept is important for its ability to distinguish between random and regular sequences. He has enunciated several axioms concerning the complexity aspects of an output sequence resulting from an input sequence and processed by an algorithm implemented on a real or theoretical automaton. The concepts of Kolmogorov complexity can be viewed as a basis for information theory without recourse to probability concepts and also as a theoretical foundation for probability itself. As stated by Li and Vitànyi [6.12], Kolmogorov complexity can be interpreted as the amount of information contained in an object about itself. Typical applications of Kolmogorov's results refer to grammar complexity of computer and natural languages measured *via* syntactical measures through the general principles of Shannon's perspectives of information (negative entropy). Kolmogorov's complexity has also been applied to study the one-dimensional strings of biological origin, similar to strings originating from computer and/or natural languages [6.13].

Notwithstanding the aforesaid avenues of quantifying the system complexity, pragmatic algorithms representing the neural complexity are rather sparse. Hence, indicated here is a systematic formalism of evaluating a complexity parameter using the maximum entropy considerations, due to Ferdinand [6.10], adopted conveniently to correlate the output error stemming from the neural learning strategies with the "complexity" of the neural system. That is, a complexity parameter is attributed and synonymously prescribed as a cost-function to the learning algorithm pertinent to an artificial network. Relevant considerations are as follows.

The disorderliness associated with the neural complex specified in terms of a goal-associated j^{th} neural subassembly or positional entropy H_{y_j} of

Eqn.(6.12) corresponds implicitly to the *maximum entropy functional*. With $N \geq 0$ ($N \in M$) being the participating subassemblies in the organization effort against the disorderliness posed and M being the maximum number of disordered subsets, p(N) denotes a probability of existence of such N disordered subsets; and $p(N) \geq 0$. Also, $\Sigma Np(N) = 1$, ($N = 0,1,..., M$). Corresponding mean value of the disordered system expected can be

specified by a constraint $k > 0$ where $k = \sum_{N} N\, p(N)$, and the associated

entropy functional (in Shannon's sense) is given by:

$$H(p) = - \sum_{N=0}^{M} p(N)\ell n \, p(N) \quad \text{nats} \tag{6.17}$$

where $\ell n(.)$ denotes the natural logarithm.

Using the Lagrange multipliers a and b, one can define a Lagrangian (L) so as to maximize the entropy function subject to the constraints indicated before.

$$L = - \sum_{N}^{M} p(N)\ell n[p(N)] - a \sum_{N}^{M} p(N) - b \sum_{N}^{M} Np(N) \quad (N = 0, 1, ..., M) \tag{6.18}$$

Differentiating Eqn.(6.18) with respect to p(N) and setting the resulting derivatives identically equal to zero, the following set of equations are realized:

$$1 + a + \ell n \, p(N) + bN = 0 \quad \text{for all N} \tag{6.19}$$

The Lagrange multipliers can be determined from the constraints $\sum_{N}^{M} p(N) = 1$ and $\sum_{N}^{M} Np(N) = k$. The first constraint yields,

$$\exp[-(a+1)] \sum_{N}^{M} \exp(-bN)) = 1 \tag{6.20}$$

and the second constraint leads to

$$k = \exp[-(a+1)] \sum_{N}^{M} N\exp(-bN) - \sum_{N}^{M} N\exp(-bN))/ \sum_{N}^{M} \exp(-bN) \tag{6.21}$$

Let the following new parameter be defined for convenience [6.10]:

$$s = \exp[-b] \tag{6.22}$$

Hence, it follows that

$$s\frac{d}{ds}\{\ell n[S(s,M)]\} = k > 0 \tag{6.23}$$

where $S(s,M) = \sum_{N}^{M} s^{N}$ and,

$$\ell n[S(e^{-b}, M)] > 0 \qquad\qquad (6.24)$$

Therefore, the possible solution of Eqn. (6.23) is unique and positive. Also,

$$\frac{\partial^2 H(p)}{\partial p(N)\partial p(u)} = 0 \qquad \text{if } N \neq u$$

$$= \frac{-1}{p(N)} \qquad \text{if } N = u \qquad\qquad (6.25)$$

stipulates that the Hessian matrix of second derivatives of $H(p)$ is negative definite and $H(p)$ is, therefore, maximized. Also, because of the concavity property, the matrix represents the global maximum value. By performing the summation of Eqn. (6.20), one can obtain:

$$\begin{aligned}
\exp(a+1) &= (1 - s^{M+1}) / (1 - s) & s < 1 \\
&= (s^{M+1} - 1) / (s - 1) & s > 1
\end{aligned}$$

$$(6.26)$$

Also, from the set of equations of (6.19) it follows that

$$\begin{aligned}
p(N) = \exp[-(1+a)] \exp[-bN] &= s^N (1-s)/(1-s^{M+1}) & s < 1 \\
&= s^N (s-1)/(s^{M+1} - 1) & s \geq 1
\end{aligned}$$

$$(6.27)$$

The above relation (Eqn. (6.27)) indicates that $p(N)$ maximizes the entropy functional and describes the probability of having N disorderly subsets in the neural complex of total size, M. That is, $p(N)$ depicts the probability distribution which maximizes the disorderly entropy H_y.

The introduction of a constraint in the Lagrange multiplier operation *via* k is equivalent to the energy constraint in the energy functional optimization problems. The correspondence between k and the new variable s in the above derivations is similar to the introduction of the activity variable in the statistical mechanics.

Calculation of the mean number of k is as follows: k is the expected value of N, namely,

$$k = s\frac{d}{ds}[\ell n\{S(s, M)\}] \qquad\qquad (6.28a)$$

Explicitly,

$$k = \frac{s}{(1-s)} - \frac{(M+1)s^{M+1}}{\left(1s^{M+1}\right)} \qquad s < 1 \qquad (6.28b)$$

$$= M - \left[\frac{1/s}{(11/s)} \frac{(M+1)(1/s)^{M+1}}{1(1/s)^{M+1}} \right] \qquad s \geq 1$$

(6.28c)

Denoting k by the ensemble average E[s,M], the entropy H(s,M) of the disorder neural subsets can be specified as follows:

$$H(s,M) = \ell n \, S(s,M) - [s,M] \, \ell n \, (s) \qquad (6.29)$$

With relevant substitutions it follows that

$$H(s,M) = \ell n \left[\frac{(1-s^{M+1})}{(1-s)} \right] - \left[\frac{s}{(1-s)} - \frac{(M+1)\,s^{M+1}}{1-s^{M+1}} \right] \ell n \, (s)$$
$$\text{for } s < 1$$

$$= \ell n \left[\frac{(1-\alpha^{M+1})}{(1-\alpha)} \right] - \left[\frac{\alpha}{(1-\alpha)} - \frac{(M+1)\,\alpha^{M+1}}{1-\alpha^{M+1}} \right] \ell n \, (s)$$
$$\text{for } s < 1$$

(6.30)

where $\alpha = 1/s$.

The function E[s,M] is strictly monotonic in respect to s. For s > 0, it is also a strictly monotonic function of M. The entropy of the disordered neural subsets is a positive function increasing monotonically with respect to M for all values of $s \geq 0$. It also increases monotonically with s for s < 0, but decreases monotonically with s for s > 0.

The parameter s defined by Eqn.(6.22) can be regarded as a measure of complexity of the neural system. For a given size of a neural assembly, it depicts the extent of disorderliness expected. As such, when s = 0, the neural assembly constitutes a "simple" subsystem of the total complex with an expected value of E[0, M] equal to zero. The other extreme situation refers to $s \to \infty$, in which case, the system is totally complex with E[∞, M] = M.

When the number of disordered subsets, namely, $M \to \infty$, the associated complexity refers to the entire universe of the neural complex and the corresponding expected value E[s,∞] for a unit size of the assembly can be deduced using the following relations for $s \geq 1$:

$$S = \left[\frac{(1-\alpha^{M+1})}{(1-\alpha)} \right] \left[\frac{1}{\alpha M} \right] \qquad (6.31)$$

$$s \frac{d}{ds} [\ell n(S)] = -\frac{\alpha}{1-\alpha} \left(\frac{1}{\alpha} \right) + \frac{(M+1)\alpha^{M+1}}{1-\alpha^{M+1}} \left(\frac{1}{\alpha^{M+1}} \right) \qquad (6.32)$$

In the limit $M \to \infty$, the functions associated become nonanalytic at $s = 1$. Explicitly,

$$[E(s,M)]_{M\to\infty} = \begin{cases} s/(s-1) & s < 1 \\ M/2 & s = 1 \\ \alpha - \alpha/(1-\alpha) & s > 1 \end{cases}$$

$$\approx E[s,M] \quad \text{for small s} \qquad (6.33a)$$

Correspondingly, $[E(s,M)]_{M\to\infty}$

$$= -\ell n\,(1-s) - \frac{s\ell n(s)}{1-s} \qquad 0 \leq s < 1$$

$$= -\ell n\,(1-\alpha) - \frac{\alpha \ell n(\alpha)}{1-\alpha} \qquad 0 \geq s \geq 1$$

$$(6.33b)$$

and $H(s, M) \cong H(s, \infty)$ for small s.

6.4 Neural Networks: Simple and Complex

In order to classify a neural network as "simple" or "complex," the relevant inferences pertinent to the above algorithmic derivations can be considered:

- For very small sizes of the neural assembly (microscopic domains) with s << 1, the expected extent of disorderliness is almost independent of the number of assemblies (M) involved.
- For very large sizes of the neural assembly (macroscopic domains) with s >> 1, the expected extent of disorderliness is characterized by the number of assemblies (M) involved.
- The characteristic value of s = 1 bifurcates the system as "simple" or "complex". 'The "simple", microscopically sized assemblies, when grown in large numbers would make the overall neural system as complex. That is, the microscopic assemblies can be regarded as quasiautonomous (simple) subsets, but when grown in large numbers would render the system of neural universe as complex. To model this consideration, the complexity coefficient s can be written as a function of M. Specifically, around s = 1, let $s = (1 - \Delta)$ where $\Delta = A/M \to 0$ as $M \to \infty$ and the constant A remains invariant. Using Taylor's expansion, at s = 1, one has

$$E[s,M] \approx \frac{M}{2}\left(1 - \frac{A}{3}\right) \Big/ \left(1 - \frac{A}{2} + \frac{A^2}{6} + ...\right) \qquad (6.34)$$

and

$$H(s,M) \approx \ell n\,(M+1) \qquad (6.35)$$

The above results correspond to defining a coefficient of neural complexity in a functional form of the type:

$$s = \exp(-A/M)$$
$$s < 1 \quad \text{when } A > 0; \text{ (when the system is simple).}$$

(6.36)

With the aforesaid exponential form of s, the following results can be deduced when $A > 0$:

$$s(e^{-A/M}, M) = \frac{M}{A}[1 - e^{-A}] + \frac{1}{2}[1 - e^{-A}] + \vartheta(A/M)$$

(6.37a)

$$< e^{-A/M}, M> = MF(A) - G(A) + \vartheta(A/M)$$

(6.37b)

$$H(e^{-A/M}, M) = \ln M + U(A) + \vartheta(A/M)$$

(6.37c)

where

$$F(A) = \frac{1}{A} - \frac{1}{(e^A - 1)}, \quad G(A) = \frac{1}{2} + \frac{1}{(e^A - 1) - \dfrac{Ae^A}{(e^A - 1)^2}} \quad \text{and}$$

$U(A) = AF(A) + \ell n\{[1 - e^{-A}]/A\}$. Further, $\vartheta(\cdot)$ represents the "order of $(\cdot)$".

At the critical point of $s = 1$ and in its neighborhood, the mean value of disorderliness depicts the extensive property in respect to the possible number of disordered neural assemblies (M) and this propensity of disorderliness is directly proportional to $\ell n(M)$.

When $s \geq 1$, the exponential law specified by Eqn.(6.22) can be modified as s $= e^{A/M}$ in which case,

$$E[e^{A/M}, M] = \{1 - F(A)\}M + G(A) + \vartheta(A/M)$$

(6.38a)

$$H(e^{A/M}, M) = \ell n(M) + U(A) + \vartheta(A/M)$$

(6.38b)

The aforesaid algorithmic considerations can be adopted appropriately to define a complexity parameter to denote an error-measure of the extent of output deviating from a "guess" or teacher value in respect to an artificial neural network. The complexity measure can be used as an error-measure and backpropagated to train an artificial neural network, such as a multilayered perceptron, via gradient-descent algorithm. The following section describes the relevant methodology.

6.5. Neural Complexity versus Neural Entropy

Considering a multilayer perceptron shown in Fig.6.2, let $\{O_i\}$ represent the set of outputs of the network and the target set sought be specified by a set of teacher parameters $\{T_i\}$. The set of inputs $\{x_j\}$ addressed to the neural network as inputs leads to a weighted sum $X_i = W_{ij} \, x_j$ with W_{ij} being the weighting factor across the interconnection between the ith and j^{th} cell. This summed input is processed by a nonlinear activation function, F_s, so that $F_s(X_i)$ corresponds to the cellular delivery at the output, namely, the output state S_i, as illustrated in Fig. 6.2a.

Now, in order to train the neural network so as to deliver an output O_i that closely matches the target value T_i, conventionally a quadratic error, $\varepsilon = \frac{1}{2}(O_i - T_i)^2$ is considered as discussed in Chapter 1. This error is backpropagated through the network to adjust the weights iteratively by gradient-descent algorithm till such time the error, ε, is minimized (Fig. 6.2b).

Alternatively, one can compare the complexity of the network being trained against the complexity of the system that yields the target, T_i. That is, a large error parameter, ε, indicates that the network under training has a "complexity" extensive enough to veer its performance from delivering an output close to the target value. Or, the complexity of the network is an implicit measure of disorganization in veering the network to achieve the convergence towards the goal.

Explicitly, the measure of disorganization as posed by the complexity of the network can be specified by the entropy associated with the difference error, $|(O_i - T_i)|$. Depicting $|(O_i - T_i)| = d_i$, the corresponding negative entropy is given by Shannon's law:

$$H_{d_i} = -\sum_i p(d_i) \log[p(d_i)] \tag{6.39}$$

where $p(d_i)$ refers to the probability of d_i.

Therefore, referring to Fig. 6.2, the training strategy to be implemented on the network can also be accomplished by maximizing H_{d_i} (by considering it as an error-metric), thereby training the network to learn in getting organized; or, in other words, the network is led toward being of less complexity.

From the above considerations it can be seen that the complexity parameter s indicated in Eqn. (6.22) and the error entropy H_d (of Eqn. (6.39)) correspond to each other on a one-to-one basis. That is, s and H_{d_i} can be matched algorithmically by stipulating that, when H_{d_i} is maximized, the system complexity parameter s is minimized correspondingly.

In order to evaluate H_d explicitly, it is necessary to know the probability density function (pdf) of d_i. Inasmuch as $d_i = (O_i - T_i)$, the pdf of d_i, namely $f_{d_i}(d_i)$, can be written by the following convolution:

$$f_{d_i}(z) = f_{O_i}(z) * f_{(-T_i)}(z)$$

$$= \int_{-\infty}^{+\infty} f_{O_i}(\lambda) f_{(-T_i)}(z - \lambda) d\lambda \qquad (6.40)$$

where f_{O_i} and $f_{(-T_i)}$ represent the pdfs of O_i and $(-T_i)$, respectively.

The above expression (Eqn. (6.40)) can be determined when f_{O_i} and $f_{(-T_i)}$ are explicitly specified. Suppose O_i and T_i are Gaussian processes given by:

$$f_{O_i}(z) = \frac{1}{\sqrt{2\pi}\sigma_{O_i}} \exp[-(z - \mu_{O_i})^2 / 2\sigma_{O_i}^2] \qquad (6.41a)$$

$$f_{T_i}(z) = \frac{1}{\sqrt{2\pi}\sigma_{T_i}} \exp[-(z - \mu_{T_i})^2 / 2\sigma_{T_i}^2] \qquad (6.41b)$$

where μ represents the mean value and σ^2 depicts the variance of the processes involved. Hence, $f_{d_i}(d_i)$ can be written as

(a)

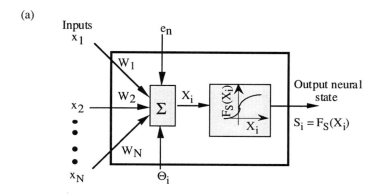

(b)

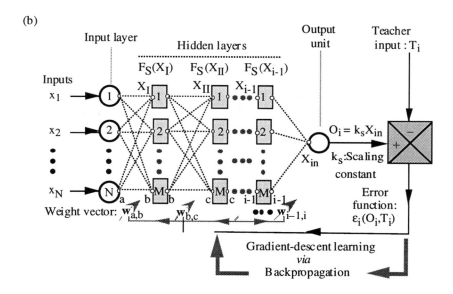

Fig.6.2: (a) Neural cell and (b) artificial neural network (Test perceptron used in the computer simulations)

[*Note:* $\varepsilon_i = (1/2)(O_i - T_i)^2$ represents the square error function; and, $\varepsilon_i = H_{d_i} = -\sum_i p(d_i) \ell n[p(d_i)]$ with $d_i = (O_i - T_i)$ represents the maximum entropy error function]

$$f_{d_i}(d_i) = \frac{1}{\sqrt{2\pi}\,\sigma_{d_i}} \exp[-(d_i - \mu_{d_i})^2 / 2\sigma_{d_i}^2]$$

(6.42)

where $\mu_{d_i} = (\mu_{0_i} - \mu_{T_i})$ and $\sigma_{d_i}^2 = (\sigma_{0_i}^2 + \sigma_{T_i}^2)$. Therefore, H_{max}, which corresponds to maximum value of H_{d_i} is deduced as

$$H_{max} = \frac{1}{2}\ell n\,[2\pi e\sigma_{d_i}^2]$$

(6.43)

Considering the transition region of the system from being simple-to-complex, at $s = 1$, the entropy resulting from the system complexity reaches a maximum value at this transition. It is dictated by the maximum number of subassemblies of the neural complex participating in the learning endeavour. That is, from Eqn. (6.37),

$$\text{Maximum }(H)_{System} = \ell n\,[M + 1]$$

(6.44)

Equating the above relation to the corresponding expression for the entropy deduced in terms of d_i, it follows that

$$\ell n\,(M + 1) = \frac{1}{2}\ell n\,[2\pi e\sigma_{d_i}^2]$$

(6.45a)

or

$$M = [(2\pi e)^{1/2}\sigma_{d_i} - 1]$$

(6.45b)

With the above specifications of the complexity-based entropy of the neural complex being useful as an error-measure, a procedure to train a typical, multilayer perceptron *via* backpropagation mode by gradient-descent method can be elucidated as described in the next section.

6.6. Neural Network Training via Complexity Parameter
In view of the earlier discussions, the strategy involved in the optimization of a neural network towards convergence essentially refers to training it to recognize a set of inputs and yield an output that matches a teacher (or target) function.

Conventionally, this network training is accomplished by the backpropagation mode of an error-metric, deduced by comparing the output and teacher parameters, and adjusting the interconnection weights *via* gradient-descent algorithm (Fig. 6.2). As indicated before, an alternative error-metric compatible for this training scheme is the entropy measure H_{d_i} associated with the error parameter d_i. Maximization of H_{d_i} leads to the output seeking the target/teacher value (Fig. 6.1). This also corresponds to the complexity of the network approaching zero (as measured by $s \to 0$).

For practical implementation of the aforesaid training procedure, it is necessary to evaluate H_{d_i} explicitly in terms of input and target parameters.

For the Gaussian distributed set of inputs considered earlier, the corresponding pdf of the output is given by the transformation:

$$f_{O_i}(O_i) = \left[f_{X_i}(X_i)/F_s'(X_i) \right]_{X_i = F_s^{-1}(O_i)}$$

(6.46)

where X_i is the weighted sum of the inputs and F_s represents the nonlinear activity function of the network.

A stochastically justifiable sigmoid is given by the Bernoulli function [6.1, 6.14],

$$L_Q(v) = g_Q \coth(g_Q v) - h_Q \coth(h_Q v)$$

(6.47)

where $g_Q = (1 + \frac{1}{2Q})$, $h_Q = \frac{1}{2Q}$ and Q is a single parameter which controls the slope of the sigmoid at the origin. The pdf of O_i can be determined explicitly with the following approximation for $L_Q(\cdot)$:

$$L_Q(X_i) = \alpha_o X_i \quad \begin{array}{ll} +1 & X_i > +1/\alpha_o \\ & -1/\alpha_o < X_i > +1/\alpha_o \\ -1 & -1/\alpha_o < X_i \end{array}$$

(6.48)

where $\alpha_o = (2g_Q - 1)/3 = (2h_Q + 1)/3$.

Hence, with the transformation as specified by Eqn.(6.48), $f_{O_i}(O_i)$ is written as

$$f_{O_i}(O_i) = \frac{1}{\sqrt{2\pi}(\sigma_{X_i}\alpha_o/2)} \exp\left[\frac{[O_i - (1/2 + \alpha_o\mu_{X_i}/2)]^2}{2(\sigma_{X_i}\alpha_o/2)^2} \right]$$

(6.49)

from which it follows that $\sigma_{O_i} = (\sigma_X\alpha_o/2)^2$ and the mean value of O_i is $[1/2 + \alpha_o\mu_X/2]$. Therefore,

$$\mu_{d_i} = (\mu_{O_i} - \mu_{T_i}) = (1/2 + \alpha_o\mu_X/2 - \mu_{T_i})$$

(6.50a)

$$\sigma_{d_i}^2 = (\sigma_{0_i}^2 + \sigma_{T_i}^2) = (\sigma_{x_i}^2 \alpha_0^2/4) + \sigma_{T_i}^2 \qquad (6.50b)$$

6.7 Calculation of μ_{T_i} and σ_{T_i}

For the set of Gaussian input values presumed, suppose the values of the teacher functional correspond to the input values. That is, $T_i = V(x_i)$ where $V(\cdot)$ is a known function. Then, the pdf of T_i can be written in terms of the pdf of the input set $\{x_i\}$. Suppose the pdf of x_i is assumed to be Gaussian with a variance $\sigma_{x_i}^2$ and a mean μ_{x_i}.

$$f_{T_i}(T_i) = K\left[f_{x_i}(x_i)/V'(x_i) \right]_{x_i = V^{-1}(T_i)} \qquad (6.51a)$$

The factor K guarantees the axiomatic probability condition that

$$K \int_{\infty}^{T_i} f_{T_i}(z)dz \equiv 1 \qquad (6.51b)$$

Suppose the assumed function $V(x_i)$ is given by

$$V(x_i) = [(x_i)/2 + 1/2] \qquad (6.52a)$$

from which,

$$x_i = (2T_i - 1) \qquad (6.52b)$$

Therefore,

$$f_{T_i}(T_i) = \frac{K}{(1/2)}\left[\frac{1}{\sqrt{2\pi}s_{x_i}} \exp[(x_i\mu_{x_i})^2/2\sigma_{x_i^2}] \right]_{x_i = (2T_i - 1)}$$

$$= \frac{2K}{\sqrt{2\pi}(\sigma_{x_i})} \exp\left[\frac{[(2T_i - 1)\mu_{x_i}]^2}{2(\sigma_{x_i})^2} \right]$$

$$(6.53a)$$

and,

$$2K \left(\frac{1}{\sqrt{2\pi}(\sigma_{x_i})} \right) \int_{-\infty}^{+\infty} \exp \left[\frac{[(2T_i - 1)\mu_{x_i}]^2}{2(\sigma_{x_i})^2} \right] dT_i \;\; = 1$$

(6.53b)

The above condition yields, $K = 1$. Hence, it follows that

$$f_{T_i}(T_i) = \frac{1}{\sqrt{2\pi}\sigma_{T_i}} \exp \left[-(T_i - \mu_{T_i})^2 / 2\sigma_{T_i}^2 \right]$$

(6.53c)

with $\sigma_{T_i} = (\sigma_{x_i}/2)$ and $\mu_{T_i} = (1/2 + \mu_{x_i}/2)$

(6.53d)

Using the explicit values of $(\sigma_{o_i}, \mu_{o_i})$ and $(\sigma_{T_i}, \mu_{T_i})$ in terms of $(\sigma_{x_i}, \mu_{x_i})$ the following relations are obtained:

$$\mu_{d_i} = \left(\frac{1}{2} \right) (\alpha_o \mu_{x_i} - \mu_{x_i})$$

(6.54a)

$$\sigma_{d_i}^2 = \left(\frac{1}{4} \right) (\alpha_{o^2} \sigma_{x_{i2}} + \sigma_{x_i}^2)$$

(6.54b)

Given the set $(\mu_{x_i}, \sigma_{x_i})$ of the input values and assuming the statistical independency of each x_i, then $\mu_{x_i} = \Sigma W_i \mu_{x_i}$ and $\sigma_{x_i}^2 = \Sigma W_i^2 \sigma_{x_i}^2$ when the input set is constituted by gaussian variates. Instead of a Gaussian distributed set of inputs, if the distribion of x_i and its transformed version across the ANN, namely O_i, are assumed to be uniform, the corresponding $f_{d_i}(d_i)$ that can be deduced *via* convolving $f_{o_i}(o_i)$ with $f_{T_i}(-T_i)$ can be shown approximately equal to $C|O_i - T_i|$ where C is a scaling constant. The uniform distribution presumed can be justified under the conditions of *Laplace's principle of insufficient reason*, namely, in the absence of any a priori considerations to the contrary, the outcomes can be taken as equally likely to occur.

Further, as indicated in Eqn.(6.28a), $\mu_{x_i} = [(s_i/(1 - s_i)] - (M + 1)s_i^{M+1}/ [1 - s_i^{M+1}]$, where M can be specified *via* equation (6.47b) as $[(2\pi e)^{1/2} \sigma_{d_i} - 1]$; and the complexity coefficient of the network s_i can be assessed from this set $(\mu_{x_i}, \sigma_{x_i})$. The corresponding error-metric is then given by $H(s_i, M)$, namely,

$$H(s_i, M) = \Sigma \, \ell n \, [(1 - s_i^{M+1})/(1 - s_i)] - \{[s_i/(1 - s_i)]$$
$$- (M + 1) \, s_i^{M+1} / [1 - s_i^{M+1}]$$
$$\equiv - \Sigma \, p(d_i) \ell n \, p(d_i) \qquad\qquad (6.55)$$

In normalized form $H_s(s_i, M)/\ell n \, (M + 1) \rightarrow 1$ as $s_i \rightarrow 1$ which depicts the maximization of the system (negative) entropy with the complexity approaching the transition towards the "simple" regime.

6.8. Perceptron Training: Simulated Results

In order to evaluate the efficacy of the complexity based error-measure (as given by Eqn. (6.57)), in training a multilayered perceptron *via* backpropagation based gradient-descent approach, the network considered in Chapter 4 was adopted in the simulations performed.

The training and prediction phases implemented correspond to those discussed in Chapter 4, except that the backpropagated error (ε_{bp}) used in normalized form corresponds to

$$\varepsilon_{bp} = \left(1 - \frac{H_{d_i}}{H_{max}}\right) \qquad\qquad (6.56)$$

with $H_{d_i} = - \Sigma \, r(d_i) \ell n[r(d_i)]$ where the probability $r(d_i)$ has a pdf given

by $f_{r_{d_i}}(r_{d_i}) = \dfrac{\exp[-(O_i - T_i)^2/2\sigma_{d_i}]}{\sqrt{2\pi} s_{d_i}}$ when x_is are normally distributed. On the

other hand, if x_is are uniformly distributed, $f_{r_{d_i}}(r_{d_i}) = C|O_i - T_i| f_{o_i}(r_{o_i})$.

In training the network using the above backpropagation error-measure, the convergence of the network towards an equilibrium value upon facilitating iterative feedback epochs of the corrputed error-measure depends on the network parameters Q and the learning rate, η [6.14].

Further, the maximum entropy based error function, $H(d_i)$, apart from being represented by the Shannon's formula of Eqn.(6.55), can also be specified by a host of Csiszar's type information-theoretic metrics [6.15]. With respect to the error variate, d_i with a probability of $r(d_i)$ having a distribution (pdf) $f_{r_{d_i}}(r_{d_i})$, typical maximum entropy-based error-measures in the symmetrized form are as follows [6.16]:

Symmetrized Jensen-Shannon measure:

$$H_{JS}^{max}\big[r(d_i)\big]_{sym} = H_S^{max}\big[r(d_i)\big] + H_S^{max}\big[1 - r(d_i)\big] \tag{6.58}$$

where $H_{JS}^{max}\big[z_i\big] = -\sum_i z_i \, log[z_i]$.

Symmetrized Havrada-Charvat measure:

$$H_{HC}^{max}\big[r(d_i)\big]_{sym} = H_{HC}^{max}\big[r(d_i)\big] + H_{HC}^{max}\big[1 - r(d_i)\big] \tag{6.59}$$

where $H_{HC}^{max}\big[z_i\big] = [1/(1 - \alpha)]\{\sum_i[z_i - 1]$ and $\alpha \neq 1$.

Symmetrized Kapur type-1 measure:

$$H_{KP\text{-}1}^{max}\big[r(d_i)\big]_{sym} = H_{KP\text{-}1}^{max}\big[r(d_i)\big] + H_{KP\text{-}1}^{max}\big[1 - r(d_i)\big] \tag{6.60}$$

where $H_{KP\text{-}1}^{max}\big[z_i\big] = -\sum_i z_i \, log[z_i] - [(1 + an)/an]\, log(1 + an)$

$+ [1/an] \sum_i^n (1 + anz_i)\, log(1 + anz_i)$ with $a > 1$.

Symmetrized Kapur type-2 measure of order α and β:

$$H_{KP\text{-}2}^{max}\big[r(d_i)\big]_{sym} = H_{KP\text{-}2}^{max}\big[r(d_i)\big] + H_{KP\text{-}2}^{max}\big[1 - r(d_i)\big] \tag{6.61}$$

where $H_{KP\text{-}2}^{max}\big[z_i\big]_{sym} = [1/(\beta - \alpha)]\, log\left([\sum_i^n z_i^\alpha]/[\sum_i^n z_i^\beta]\right)$ with $(a \geq 1,$ if $b < 1)$; or, $(\alpha < 1,$ if $\beta \geq 1)$.

Symmetrized Kapur type-3 measure:

$$H_{KP\text{-}3}^{max}\big[r(d_i)\big]_{sym} = H_{KP\text{-}3}^{max}\big[r(d_i)\big] + H_{KP\text{-}3}^{max}\big[1 - r(d_i)\big] \tag{6.62}$$

where $H_{KP\text{-}3}^{max}\big[z_i\big] = -\sum_i z_i \, log[z_i] - [(1 + a)/a]\, log(1 + a)$

$+ [1/a] \sum_i^n (1 + az_i)\, log(1 + az_i)$.

Symmetrized Sharma-Mittal type-1 measure:

$$H_{SM\text{-}1}^{max}\big[r(d_i)\big]_{sym} = H_{SM\text{-}1}^{max}\big[r(d_i)\big] + H_{SM\text{-}1}^{max}\big[1 - r(d_i)\big] \qquad (6.63)$$

where $H_{SM-1}^{max}(z_i) = [n^{\alpha-1}/(\beta - \alpha]\{\sum_i^n z_i^\alpha - 1\}[n^{\beta-1}/(\alpha - \beta]\{\sum_i^n z_i^\beta - 1\}$
and $(\alpha \geq 1,\ \text{if } \beta < 1);\ \text{or},\ (\alpha < 1,\ \text{if } \beta \geq 1).$

Symmetrized Sharma-Mittal type-2 measure:

$$H_{SM\text{-}2}^{max}\big[r(d_i)\big]_{sym} = H_{SM\text{-}2}^{max}\big[r(d_i)\big] + H_{SM\text{-}2}^{max}\big[1 - r(d_i)\big] \qquad (6.64)$$

where $H_{SM-2}^{max}(z_i) = [1/(\beta - \alpha]\{\sum_i^n z_i^\alpha - \sum_i^n z_i^\beta\}$ and $(\alpha \geq 1,\ \text{if } \beta < 1);\ \text{or},\ (\alpha < 1,$
if $\beta \geq 1).$

Symmetrized Rényi's measure:

$$H_{RY}^{max}\big[r(d_i)\big]_{sym} = H_{RY}^{max}\big[r(d_i)\big] + H_{RY}^{max}\big[1 - r(d_i)\big] \qquad (6.65)$$

where $H_{RY}^{max}(z_i) = [1/(1 - \alpha]\ell og\{\sum_i^n z_i^\alpha\}$ and $\alpha \neq 1.$

Apart from the maximum-entropy based error-measures indicated above, it is also possible to construct a variety of similar measures using the generalized Csiszàr's functional considerations, as was done in the case of cross-entropy measures discussed in Chapter 4.

The efficacy of maximum entropy based error-measures are indicated above in dictating the performance of an ANN. The results are obtained in respect of the test architecture of Fig. 4.4 and are depicted in Figs 6.3 and 6.6. These results represent a few sample simulations. Specifically, the error measures considered refer to the Jensen-Shannon type (Eqn. (6.58)) and the Rényi's version (Eqn. (6.65)). Further, the parameters considered and adopted to train the test network of Fig. 4.4 are Bernoulli function parameter, $Q = 0.5$; learning coefficient, $\eta = 0.001$; and the momentum factor, $\lambda = 0.9$.

At least 250 iterations were performed in the training and prediction phases and the results were predicted from the mean results of 10 ensemble runs. The test function predicted corresponds to the same \sin(x)\ function considered in Chapter 4.

Figs. 6.3 and 6.5 are the learning curves obtained from which the ability of the network's output to converge is self-evident. Shown in Figs. 6.4 and 6.6 are the predicted results obtained with the two test error measures adopted. Again, the ability of the network to predict the patterns closely can be observed. Increasing the number of iterations can minimize the error associated with the predictions further.

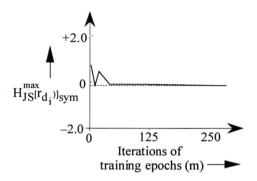

Fig. 6.3: Learning curve obtained from ensemble data on 10 training runs: Dynamics of the symmetrized Jensen-Shannon maximum entropy error measure:

$$H_{JS}^{max}\left[r(d_i)\right]_{sym} = H_S^{max}\left[r(d_i)\right] + H_S^{max}\left[1 - r(d_i)\right]$$

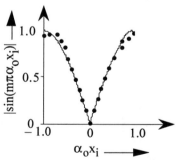

Fig. 6.4: Teacher function (___) and the predicted function (.....) obtained with

$$H_{JS}^{max}\left[r(d_i)\right]_{sym} = H_S^{max}\left[r(d_i)\right] + H_S^{max}\left[1 - r(d_i)\right]$$

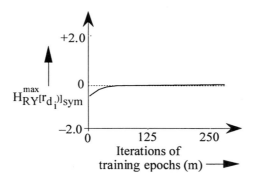

Fig. 6.5: Learning curve obtained from ensemble data on 10 training runs: Dynamics of the symmetrized Rényi's maximum entropy error measure:

$$H_{RY}^{max}\left[r(d_i)\right]_{sym} = H_{RY}^{max}\left[r(d_i)\right] + H_{RY}^{max}\left[1 - r(d_i)\right]; \ a = 0.07$$

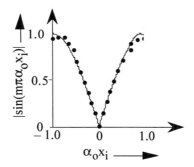

Fig. 6.6: Fig.6.4:Teacher function (____) and the predicted function (.....) obtained with $H_{RY}^{max}\left[r(d_i)\right]_{sym} = H_{RY}^{max}\left[r(d_i)\right] + H_{RY}^{max}\left[1 - r(d_i)\right]; \ a = 0.07$

6. 9 CONCLUDING REMARK

This chapter offers an alternative strategy to use the maximum entropy parameter as an error-metric in lieu of the conventional cost functions such as the quadratic error or cross-entropy error-metrics proposed in Chapter 4. The ability of the proposed cost function in enabling neural network training is illustrated by a simulation pertinent to a multilayered perceptron.

A summary of maximum entropy principles is furnished in Appendix 6.1.

BIBLIOGRAPHY

[1]   Neelakanta, P. S. and De Groff, D.: *Neural Network Modeling Statistical Mechanics and Cybernetic Perspectives* (CRC Press, Boca Raton, FL: 1994).

[2]   Shannon, C. E.: A mathematical theory of communication. *Bell Syst. Tech. J.*, 1948, 47, 143-157.

[3]   MacGregor, R. J.: *Theoretical Mechanics of Biological Neural Networks* (Academic Press, Inc./ Harcourt Brace Jovanovich Publishers, Boston, MA: 1993).

[4]   Siu, K. Y., Roychowdhury, V. P., and Kailath, P.: Rational approximation techniques for analysis of neural networks. *IEEE Trans. Inform. Theory*, 1994, 40, 455-466.

[5]   Haken, H.: *Information and Self-Organization: A Macroscopic Approach to Complex Systems* (Springer-Verlag, Berlin: 1988).

[6]   Curtis, B.: In search of software complexity. In: *Proc. of Workshop on Quantitative Models for Reliability, Complexity, and Cost.* (Washigton D.C., 1979, IEEE Computer Society), pp. 95-106.

[7]   Rissanen, J.: *Stochastic Complexity in Statistical Inquiry* (World Scientific, Singapore: 1989)/

[8]   Landau, L. D. and Lifshitz, E. M.: *Statistical Physics* (Addison-Wesley, Reading, Ma: 1958).

[9]   Ferdinand, A. E.: A statistical mechanical approach to system analysis. *IBM J. Research and Development*, 1970, 14, 539-547.

[10]   Ferdinand, A. E.: A theory of system complexity. *Int. J. Gen. Systems,* 1974, 1, 19-33.

[11]   Leung-Yan-Cheong, S. K., and Cover, T. M.: Some equivalences between Shannon entropy and Kolmogorov complexity. *IEEE Trans. Inform. Theory,* 1978, IT-24, 331-338.

[12]   Li, M. and Vitányi, M. B.: Two decades of applied Kolmogorov complexity. In *Proceedings of the Third Annual Structure in Complexity Theory Conf.*, (Washington DC., June 1988), pp. 80-101.

[13]   Van Emden, M. H.: *An Analysis of Complexity.* (Mathematisch Centrum, Amsterdam: 1971).

[14]   Neelakanta, P. S., Sudhakar, R., and De Groff, D.: Langevin machine: A neural network based on stochastically justifiable sigmoidal function, *Biol. Cybern.*, 65, 1991, 331-338.

[15]   Rumelhart, D. E., and McClelland, J. L. (Eds.): *Parallel Distributed Processing I,* (MIT Press, Cambridge, MA: 1986).

[16]   Watrous, R. L.: A comparison between squared error and relative entropy metrics using several optimization algorithms. *Complex Syst.*, 6, 1992, 495-505.

[17]   Neelakanta, P. S., Abusalah, S., De Groff, D., Sudhakar, R. and Park, J.: Csiszár's generalized error-measures for gradient-descent based optimizations in neural networks using back-propagation algorithm. *Connection Science,* 8,1996, 79.

Appendix 6.1

Maximum Entropy Principles

6A.1 General

There are theoretical perspectives on the maximum entropy concept [6.16] which can be translated to the information-theoretic framework of the neural complex in order to foresee certain computational difficulties which may inhibit the use of generalized measures. However, with the advent of robust computers, compatible for nonlinear optimizations, these measures can be adopted on an ad hoc basis.

The symmetrization procedure indicated in Chapter 4 can be extended to the maximum entropy based measures without any loss of generality so as to guarantee a convergence in the optimization pursued.

Further, the following considerations can be duly taken into account in the maximization of the functionals as appropriate.

While using the simple Lagrange's method of constraint optimization, if the maximizing probabilities come out to be negative, say under a prescribed mean value, then a nonnegativity constraint should be imposed explicitly. This situation would be encountered, for example, when the Havrada-Chavat measure of order 2 is maximized subject to a prescribed mean [6.16].

Nonlinearity and logistic characteristics of complex systems can be perceived explicitly by maximizing (subjected to a set of plausible constraints) the generalized measures of entropy. For example, the models of population growth, innovative diffusion, etc., which exhibit typical characteristics of a complex system, have been comprehended in terms of maximum entropy principles. This chapter indicates the scope of relevant considerations that can be adopted in the pursuit of maximum entropy specific to information organization by the neural complex.

6A.2 Jaynes' Maximum Entropy Considerations

According to this principle pertinent to a stochastical system, from among all those probability distributions which satisfy a given set of constraints, one may choose that probability distribution which maximizes the Shannon entropy associated with the system. This principle is designated as *Jaynes' principle* in order to distinguish it from the other generalized principles. It is widely used in practical applications and provides the impetus to define the other principles *via* suitable exensions of the underlying concepts.

Jaynes' principle [6.17] aims at maximizing uncertainty subject to a set of given constraints in a restricted sense of being confined to Shannon's measure of uncertainty. However, the uncertainty, in reality, is "too deep and complex". Therefore, a single measure may not be adequate under all circumstances towards entropy maximization and/or minimization endeavors. For example, when uncertainty refers to fuzziness of data, or when uncertainty refers only to partially probabilistic and partially nonstochastic situations, the use of Shannon's measure alone becomes questionable. Hence, a set of generalized measures of entropy involving certain unique parameters should be used in lieu of Shannon's measure which applies to a limited, totally probabilistic stochastical system. The generalized principle of maximum entropy, on the other hand, allows a wide choice of distributions and can satisfy a variety of constraints while optimizing the entropy/information.

Contrary to the maximum-entropy-based optimization, the generalized principle of minimum, cross-entropy discussed in Chapter 4 requires one to choose a distribution for which a specified measure of mutual entropy is a minimum for those sets which satisfy a given set of constraints. Both the maximum entropy based as well as minimum entropy-based optimization principles, however, follow the so-called *direct principle* wherein the objective is to determine the maximum or minimum entropy distribution satisfying given constraints. In an *inverse principle,* the goal is to determine the constraints a priori, when the probability is specified.

It is also possible to consider simultaneous maximization and minimization of an entropy functional. When there is a distinct feasibility of classification, one may often decompose the total entropy of the system into " entropy between classes" and "entropy within classes". Each class is, however, required to be as homogeneous as possible. Also, these different classes should be distinguishable among themselves as much as possible. To accomplish this goal, one has to maximize the entropy within the classes and minimize that entropy that prevails between the classes. This leads to the so-called *mini-max entropy principle.*

In addition to maximization of entropy and minimization of cross-enropy, it is also feasible to consider a measure known as *measure of dependence,* D. This is essentially a measure of cross-entropy of the joint distribution of a number of variables related to the product distribution of marginal distsributions.

6A.3 Inverse Entropy Principles

First inverse maximum entropy principle

This principle allows the determination of a set of constraints that will give rise to a given observed or theoretical probability distribution such as Jaynes' maximum entropy probability distsribution. As indicated earlier, the measure of entropy used is the Shannon entropy.

Second inverse maximum entropy principle

This principle is similar to the first principle except that a generalized measure of entropy is used in lieu of Shannon's entropy.

Third inverse maximum entropy principle

This principle permits the determination of a measure of entropy so that the given probability distribution is obtained by maximizing this measure of entropy subject to given constraints. Accordingly, application of this principle may lead to entropy measures other than the Shannon entropy.

6A.4 Maximum Entropy versus Minimum Entropy in Optimization Strategies: A Summary

- For every principle of maximization of entropy, there is a dual principle of minimization of cross-entropy

- Optimization may refer either to the principle of minimization of entropy or the method of maximization of cross-entropy, or both

- For every principle of entropy-based optimization, there are one or more inverse principles where the constraints, measure of information, and *a priori* probability distributions have to be determined uniquely

- There also exist principles which involve simultaneous maximization and minimization of entropy and cross-entropy measures

- Finally, there are also interdependence principles which involve maximization or minimization of cross-entropy of a joint multivariate probability distribution from the multivariate distribution obtained from the product of the marginal density function.

6A.5 Optimization Postulation in a Nut shell

Every probabilty distribution, theoretical or observed, is an entropy optimization distribution. That is, it can be obtained by maximizing an appropriate entropy measure or by minimizing a cross entropy measure with respect to an appropriate a priori distribution, subject to satisfying its appropriate constrants.

Chapter 7

Information-Theoretic Aspects of Neural Stochastic Resonance

J.C. Park and P.S. Neelakanta

Resonance ... an act of intensification or
enrichment ...
- Webster's Collegiate Dictionary

7.1 Introduction

Noise is an ubiquitous companion in the neural complex and imposes an inevitable limiting boundary which often constrains the accuracy of estimating the physical entities associated with the macroscopic realm of spatiotemporal neural activity.

In general, in a *linear system* representation of a noisy medium intended for transmission of information, increasing noise power while holding the other system parameters invariant, would result in the reduction of the signal-to-noise ratio (SNR) in the output. However, in certain *nonlinear* systems considered in the detection, classification and estimation of weak, periodic signals, an optimum signal-to-noise ratio may be encountered in the presence of certain nonzero value(s) of additive noise. Analysis of this counter-intuitive situation reveals that noise is rather, a requisite component in such nonlinear systems— physical and/or biological to achieve an optimum information transfer. This phenomenon has been termed as *stochastic resonance* (SR) [7.1-7.9].

As pointed out by almost every researcher in this area, the terminology *stochastic resonance* is perhaps a misnomer. This is due to the fact that there is no "resonance phenomenon" whatsoever that exists in the conventional sense pertinent to a physical unit tuned to "resonance" by virtue of the associated mass, compliance, and frictional characteristics in the case of mechanical systems; or, equivalently, an electrical system attaining resonance through inductive, capacitive and resistive elements. Resonance depicts in a general sense, the response of a system with a peak in its transfer-function over a well-defined band of spectral range. SR situations, however, depict a resonance phenomenon viewed as a cooperative coupling prevailing between the inherent noise and a low-amplitude, coherent signal which, jointly and optimally excite a nonlinear class detector. Thereby, the cooperatively coupled noise-plus-signal maximizes the information transfer through the nonlinear system. However, it should be noted that without the explicit presence of noise, such SR-dependent, nonlinear systems may not be capable of facilitating a coherent information transfer at all.

The essential system attributes required for stochastic resonance are remarkably simple: A bistable system with an inherently nonlinear transfer function, a small-amplitude coherent signal, noise, and a valid range of thresholds controlling the stable state-transitions. Though a specific type of probability distribution of the noise amplitude is not important, nor is a strict periodicity of the signal mandated, it is assumed that noise power could be several times, or possibly several orders of magnitude greater than the signal power. Consistent with the fact that there are a multitude of dynamical systems, including the biological neuron, which satisfy these broad requirements, it is not surprising that SR manifests itself in numerous real-world systems—physical, biological and anthropogenic. Nature exploits the presence of noise in optimizing information transfer architectures by incorporating SR into many of its information-processing applications wherein small signals are overwhelmed by noise. To gain an understanding of this processing would, therefore, aid many technological pursuits which rely on sensitive detection of signals in the presence of noise and would also enable understanding of the role played by noise in such information transfer endeavors.

In short, noise becomes integral to the operation of certain nonlinear detectors employing SR to detect weak coherent signals. For example, an artificial neural network (ANN) while learning under input conditions of SR, can exhibit a lower rate of information loss during the organizational epochs of the network than the same network learning from input conditions which do not encompass the SR. An interesting realm with an intersection of information theory, signal-processing, and dynamic activities pervades the SR-based neural ensemblages.

7.1.1 The emergence of SR

Existence of stochastic resonance was originally postulated as a mechanism to explain the 100,000 year periodicity of glacial epochs in the Earth's climate [7.1, 7.2]. Prior to the emergence of SR-based theoretical reasoning, it was unclear whether global external force phenomena, such as perturbations in solar radiation intensity due to variations in the Earth's orbital dynamics, or internal processes like feedback between oceanic and atmospheric mechanisms, were the driving entities in triggering the glacial-interglacial climatic transitions. It has been well-known that a small amplitude periodicity in solar radiation caused by a periodic variation in the Earth's orbital eccentricity has a dominant period of 100,000 years. But the climatic energy-balance models used to predict climatic transitions were unable to correlate this insolation forcing mechanism to the observed cycle of interglaciation. However, when the weak, periodic atmospheric-forcing introduced by the solar radiation variations is coupled with a stochastic forcing inherent in the various atmospheric radiative and transport mechanisms in evolving energy-balance equations, the result is a dominant peak seen in the power spectrum of energy-balance solutions. This corresponds to the observed periodic transitions of the Earth's climatic behavior between glacial and interglacial states.

It may be of interest to note that the energy-balance equations, a set of stochastic differential equations, when cast in the form of a one-dimensional climatic potential relation with two stable states, simply refers to the Fokker-Planck version depicting the continuous time evolution of the transition probabilities of a bistable system. As is made evident in Chapter 5, the Fokker-Planck equation plays a central role in understanding the dynamics of the bistable model of biologic neurons as well.

Experimental observation of stochastic resonance can be made from the measurements of a signal-plus-noise driven bistable electronic circuit, namely, the *Schmitt trigger* [7.3]. A Schmitt trigger, as is conventionally known in electronics, is a bistable, dual-threshold device that changes its state when an input stimulus crosses the upper threshold from below, or when the stimulus crosses the lower threshold from above. Stochastic resonance has also been verified experimentally in a bidirectional ring-laser, in which a periodic signal, unless coupled with some noise, would not facilitate correlated laser modulations [7.4]. Further, stochastic resonance has also been demonstrated in many other dynamical systems, such as the single-well potential system, and in the "integrate and fire" dynamics employed commonly to model the response of sensory neurons [7.5]. The theory of SR has also reached a mature state, wherein many of the interdependencies between the external parameters such as the noise and signal intensity, threshold settings, etc. have been experimentally investigated as well as theoretically justified [7.6-7.7].

More recently, stochastic resonance has been verified in a number of important biological processes. These include information-transfer from mechanoreceptor hair-cells situated on the tail of the red swamp crayfish, *Procambarus clarkii* [7.8], cercal sensory neurons of crickets [7.9], signal-transduction across the voltage-dependent ion-channels of cell membranes in neuronal networks prepared from the temporal lobe hippocampal sections of a mammalian brain [7.10, 7.11], and even in the exteroreceptive somatic nervous system [7.12], as well as in muscle-spindle afferents [7.1] of *Homo sapiens*. It is not surprising that many anthropogenic and biological systems which take advantage of the SR phenomena have been identified and there can be little doubt that many more SR-related phenomenologies will be brought to surface in the coming years.

Biological neurons operate in an inherently noisy environment, and the fact that SR has been observed in them indicates that such noise is an integral part of neurocellular activity. Studies pertinent to SR in the neural complex can be viewed in two formal ways. From a signal-processing perspective, SR results in an optimum input/output signal-to-noise ratio at non-zero values of noise intensity. Viewed from the information theory standpoint, this means that there is a significant entropy (counter-information) transfer through the system.

Contemporary studies have applied information measures (direct entropy and/or mutual information) to quantify the noise-induced, maximum information transfer resulting from SR [7.14-7.16]. These results demonstrate that models of a single neuron do exhibit the maximization of information rate at non-zero noise levels. They do not, however, illustrate whether or not cooperative populations of a neural assembly make use of SR in such information transfer endeavors. An answer to this question is attempted here and the influence of stochastic resonance on neural learning in an artificial neural network (ANN) is elucidated.

The detection and transfer of information from neural sensory input to the mammalian brain is known to be encoded in the time-interval sequence between 'firings' of neurons which constitute a train of action-potentials. A statistical examination of these time-interval sequences (or inter-spike intervals) reveals that the temporal coding follows a stochastic process. However, despite the fact that the statistics of action potential trains have been investigated comprehensively [7.17], it is not established precisely how the associated sensory information is encoded. It has been assumed historically that appropriate mathematical models which represent the statistics of time-interval sequences of neurosensory action potentials refer to a *homogeneous Poisson point-process*. These time-interval sequences belong to the family of *modified Poisson point-processes* with the inclusion of temporal refractory effects. These models are inherently *renewal processes*, indicating that they predict independent, identically distributed, inter-event intervals. Such intervals are also uncorrelated implicitly; and therefore, can be described completely in a statistical sense by their inter-event-interval histograms (IIH). However, it has also been shown [7.18] that sequences of inter-event intervals may exhibit positive correlations over long time periods; renewal process based statistical models are rather inadequate for a complete description of such action-potential statistics having correlatory (Markovian) attributes. In fact, if one were to measure the action-potential rate in mammalian auditory nerves, it can be observed that fluctuations in the rate do not often subside to any significant extent even for very long averaging periods; that is, the rate may exhibit sustained fluctuations on all time-scales [7.18]. Therefore, it has been suggested [7.19] that a complete statistical model for the action-potentials should rather be based on a *fractal stochastic point-process*.

Interestingly, regardless of the model employed to describe the temporal attributes of the neural spike occurrences, the presence of inter-event correlations does not alter the outcome of the associated IIH. The IIH can, therefore, serve as a tool to demarcate the presence or absence of stochastic resonance and can be adopted as a simplified statistical tool to model the sensory neuron response.

In this chapter, the information flow through an artificial neural network adapting to the IIH produced by a simulated neuron in response to input conditions of stochastic resonance is studied. Artificial neural networks often make a convenient tool to model the flow of neural information. They are based on the information-processing structure of biological neurons and derive their ability to learn complex-mappings from their massively parallel interconnection structure composed of inherently nonlinear set of neurons. Neural networks, in general, operate in two distinct phases. The first is the training phase, where the network adjusts its internal parameters in response to a set of training data or *rule-bases*. Second is the predictive phase, where the trained network responds to input data and produces a *functional-mapping*.

Neural networks can be classified generally either as *supervised* or *unsupervised* architectures. Supervised paradigms require a "teacher" that produces an error-output in response to the training data, and minimization of the error (or cost-function) directs the learning process which is essentially an adjustment of the network's internal parameters. An unsupervised network does not require an external teacher to generate an error, but relies on a rule-base to adjust its internal parameters in response to the network output during training. Neural networks can be classified further as either feedforward, or feedback types. This classification refers to the direction of the flow of information during the predictive phase of operation.

It has been pointed out by the authors [7.20] that the flow of information during the training phase in an artificial neural network is a competitive endeavor between the positive and negative information available. Positive information relates to the actual learning of the network, namely, its ability to predict the desired outcome with an increasing degree of confidence. Negative information refers to confusing the learning process arising from extraneous parameters or due to the inevitable presence of noise. It is therefore indicated in [7.20] that it is more appropriate to model the information transfer during the learning process in terms of the classical definition of information based on entropy concepts [7.21]. Further, it is shown in [7.20] that the dynamics of information flow during ANN learning correspond to that of a stable, chaotic system with a dominant well-defined attractor basin.

The essence of this chapter is to synthesize the dynamic behavior inherent in stochastic resonance with the information rate dynamics of neural learning. This would facilitate assessing the information transfer characteristics of neural learning conjectured to exploit optimally the stochastic resonance. It is demonstrated that the flow of information during the training phase of an ANN is more informatic when the ANN models a dynamic system exhibiting SR, than in the case of a similar, but non-SR system. That is, a non-SR dynamic results in a larger information-theoretic entropy (network uncertainty) during learning than the SR case.

In order to demonstrate this information-theoretics based result and performance of SR-based neural networks, the remaining sections of this chapter are organized as follows. First, some of the characteristic features of SR are outlined. Next, the statistical characteristics of state-transitions exhibited by biological neurons in response to SR are reviewed; then the Schmitt trigger as a bistable nonlinear detector is described. Lastly, a general threshold parameter is identified to quantify the signal-plus-noise input in relation to the thresholds of the detector. Results of computer simulations which generate state-transition statistics of the Schmitt trigger in response to the signal-plus-noise are presented and contrasted with relevant biological results. Comparison of biological response statistics with those obtained *via* computer simulations are used to establish the presence or absence of SR. Pertinent computer generated response statistics and corresponding detector threshold parameters are subsequently used as training sets for a test ANN. This test neural network and the details of its implementation to provide a mechanism to quantify the flow of neural information during learning is described. Relevant aspects of using an information-theoretic cost-function [7.22] in the backpropagation algorithm adopted to train the network is indicated. The results *vis-a-vis* learning evolutions are then quantified and discussed in terms of the average uncertainty perceived by the neural network during training. Finally, this chapter is concluded with remarks regarding the flow of neural information through systems attempting to maximize the extraction of information from weak periodic signals buried in noise by means of the mechanism of SR.

7.1.2 Characteristics of SR

Perhaps the most significant characteristic and useful attribute of SR is the nonlinear relationship between the output signal-to-noise ratio (SNR) and the input (or channel) noise power. This feature emphasizes essentially the nonlinear nature of a system which exhibits SR. If the system were purely linear, or weakly nonlinear, the addition of noise power to the transmission channel or to the detector input would result in a decrease in the detector output SNR. In the case of systems exploiting SR, the output SNR exhibits a characteristic peak in the range of additive noise power.

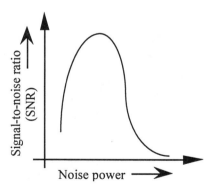

Fig. 7.1: Output SNR as a function of input noise power under stochastic resonance

Fig. 7.1 depicts a typical relation between output SNR and the additive noise under the auspices of SR. At the initial values of additive noise, the output SNR rises sharply and exhibits a peak at a specific non-zero value of the additive noise. As the noise power is increased beyond this regime, the output SNR decays rather rapidly.

Another intriguing aspect of SR is that the signal power alone would be incapable of triggering information-carrying state-transitions of the detector. As mentioned earlier, in SR-based systems, the signal power is assumed to be weak relative to the noise power; and, the detector thresholds are set such that state-transitions would not possibly take place with the signal alone. However, when noise of sufficient magnitude (of characteristic value) is presented, along with the signal, it enables constructive state-transitions towards information-transfer to occur, thereby portraying a significant enrichment in the output SNR.

7.1.3 Physical description of SR

The physical picture of SR can be understood by considering the dynamics of a particle trapped in a bistable potential well. Suppose a particle is confined within the double-well potential depicted as in Fig.7.2.

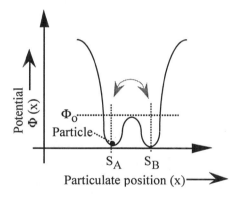

Fig. 7.2: A dual-well potential which governs the dynamics of a particle confined to bistable-states corresponding to the basins of attraction located at S_A and S_B

The potential Φ can be represented as a quadratic function of the form: $\Phi(x) = ax^4 - bx^2$ with stable basins of attraction located at S_A and S_B. The output of this system refers to the time-dependent position of the particle, $x(t)$. The input consists of a superposition of a periodic modulating signal, $\psi(t) = A \cos(\omega_o t)$ and a random forcing function, $\zeta(t)$ which can have arbitrary statistics or probability distribution. For convenience, $\zeta(t)$ can be assumed to be as a zero-mean Gaussian noise. The input forcing displaces the potential-well as a function of time either to the left $(-x)$ or the right $(+x)$ of the mean positions S_A and S_B. If both the modulating signal and the random forcing are absent, a sufficiently damped particle introduced into the potential well(s) would eventually come to rest either at S_A or S_B. Now, if only the random

311

perturbations are introduced with a sufficiently large value of variance, σ^2, the particle will make occasional transitions surmounting the potential-barrier Φ_o and reach the other stable-state. The first mean transition time of such transitions is given by the well-known *Kramers time* [7.23]: $\tau_K = \sqrt{2\pi}[\exp(2\Phi_o/\sigma^2)]/2b$, which depends only on the potential barrier height Φ_o, the noise power (σ^2), and the spatial nature of potential wells at the minima and maximum. Corresponding *Kramers frequency* (rate) is given by $\omega_K = 1/\tau_K$. The underlying assumption for a valid Kramers time is that the probability distribution of a particulate position within a potential well should be Gaussian with a central tendency towards the mean, stable-point in the basin of attraction, so that the particle would remain proximal to the stable-points most of the time. As the noise variance increases, the frequency with which state-transitions occur will also increase quite rapidly at the initial stage, but later at a slower rate when the noise power is sufficient enough to drive the particle to surmount the barrier more easily.

When a small amplitude periodic forcing is added (A ≠ 0), the potential-well is regularly shifted to the left, or alternatively rotated about a fixed center, and then to the right at the frequency ω_o. It is assumed that the periodic amplitude, A, acting without the noise coupling, is incapable of driving the particle over the barrier from one stable state to the next. It is also presumed that the period of the sinusoid is greater than the characteristic intrawell relaxation time of the bistable system.

Now, consider the effect of increasing the noise power from a small initial value. When the noise power is small, so that, the corresponding Kramers time is relatively long, there is enough momentum imparted to the particle to surmount the potential barrier only occasionally. As the noise power increases, the Kramers rate increases. Further, as $\omega_K \rightarrow \omega_o$, the incoherent noise power couples with the periodic signal resulting in larger inter-well transition probabilities than what may occur in the presence of noise alone. As the noise power increases further, the Kramers rate would exceed the periodic forcing and the cooperative feedback between the noise power and sinusoidal will be lost. Thus, from a purely physical perspective, the phenomena of SR can be viewed as a matching of the Kramers rate with a periodic driving force. It is in this sense that there is indeed a "resonance" occurring. That is, a frequency matching between the mean inter-well state transition time is imposed by the stochastic excitation and the weak periodic stimulus, which otherwise is not energetic enough to initiate a state-transition. The portrait above constitutes the basis behind the physics of SR and elaborates the counter-intuitive notion that incoherent noise can couple with a coherent signal so as to excite state-transitions in an optimal manner.

Having outlined the basic development and a physical description of SR, the main question posed in this chapter is the following: Do cooperative assemblages of neuronal population exhibit information transfer characteristics which are supportive of SR? In order to respond to this query, it is necessary to review some basic relations between measured IIH in respect to certain biological events and the characteristics of SR.

7.2 Inter-Event Histograms (IIH) and Stochastic Resonance

7.2.1 Biological characteristics of IIH

As mentioned earlier, the IIH is a valid and plausible statistical tool that can be used to characterize the distribution of neural signaling intervals in biological neurons. Therefore, in order to incorporate biologically meaningful information into a model of information transfer *via* SR, it is appropriate to review the biological characteristics of IIH. Longtin et al. [7.24] have analyzed a measured set of IIH data collected 23 years apart from the auditory nerve fibers of a squirrel monkey and from the primary visual cortex of a cat, both subjected to periodic stimulus. They observed the following substantiative features of biological IIH in response to weak periodic stimuli:

- Response modes are located at integer multiples of the stimulus period
- Response mode amplitudes decay more or less exponentially as indicated by a linear slope of mode amplitudes plotted on a logarithmic scale.

The second feature suggests that the spike rates are governed by rate processes. Longtin et al. [7.24] modeled the sensory neuron information transmission on the basis of bistable Schmitt trigger dynamics. By comparing the resulting state-transition statistics of the bistable model with the experimentally obtained physiological results, it is concluded in [7.24] that noise is a requisite component in order to produce experimentally justifiable IIH when neurons are stimulated by periodic inputs for at least the following two reasons:

- The firing threshold of the system is sustained to remain above that of the periodic stimulus amplitude
- Noise is required to produce a set of response modes at intervals other than the stimulus period.

These characteristic attributes of SR, lend substantial credence to the idea that nature employs SR in the information transfer processing of extero-receptive somatic cells of the mammalian nervous system.

7.2.2 Computer simulations of stochastic resonance

The statistical description of inter-event intervals, which result from the transition dynamics of nonlinear systems exhibiting stochastic resonance, have been verified in both biological as well as electronic systems as discussed above. Further, the double-well potential system considered can be conceived in terms of a bistable, dual-threshold detector which detects information-bearing voltage levels transmitted through a noisy information channel. For double-well potential with a heavily damped particle resident, the system input, in general, corresponds to the superposition of a stochastic and periodic signal which perturbs the double-well potential. The system output refers to the position of the particle as a function of time. Likewise, the bistable electrical system under consideration is a dual-threshold detector exposed to fluctuating voltage input levels, and the output corresponds to a time-record of state-transitions occurring in the detector. This electrical system can also be regarded as an analog of a neural cell. Thus, the bistable, dual-threshold detector, the analog of a somatic nerve cell, receives voltages from a network of synaptic interconnections. The corresponding noisy information channel is represented by the background voltage fluctuations of the summed synaptic inputs to the detector. The detector output is, therefore, analogous to the delays between voltage spikes which constitute an information–bearing message from a neural cell.

The dual-threshold detector that is presently being considered to emulate the inter-event transitions under conditions of stochastic resonance is a Schmitt trigger. Statistics of the inter-event transitions are extracted as depicted in Fig. 7.3. The Schmitt trigger is a dual-state, dual-threshold device. The operational characteristics of the Schmitt trigger are defined as follows: HI and LO represent the two stable-states with the corresponding thresholds α_U and α_L, respectively, where $\alpha_U > \alpha_L$. The state-point logic of the Schmitt trigger is as follows: LO $\rightarrow$ HI, if the input crosses α_U from below and the state is not already HI; and, HI $\rightarrow$ LO, if the input crosses α_L from above and the state is not already LO.

Let the noise considered be zero-mean Gaussian-distributed with a variance σ^2. It is added to a periodic process with a peak-to-peak amplitude of $2A$. Relevant to the application of this signal-plus-noise to the bistable Schmitt trigger, it is assumed that $\alpha_U > A$. The lower threshold a_L is assumed to take a value close to A but below α_U. An appropriate system parameter that encompasses the relevant features of signal-plus-noise applied to a threshold detector is the ratio of threshold excess power, namely $(\alpha_U - A)^2$, to the variance, σ^2, of the noise. Hence, the system threshold parameter γ can be defined as follows:

$$\gamma^2 = (\alpha_U - A)^2/\sigma^2 \qquad (7.1)$$

This single parameter can be used to characterize the power of the excess signal-plus-noise required to initiate a state-transition of the detector relative to that of the noise. In the case when $\sigma \rightarrow \infty$, then $\gamma \rightarrow 0$ and it is pointless to attempt to extract useful information from the statistics of the detector state transitions. If the noise variance is greater than the required threshold excess power, that is, when $(\alpha_U - A) < \sigma$, it follows that $0 < \gamma < 1$. This parametric regime, which encompasses the low signal-to-noise conditions and the detector threshold levels, constitutes a hallmark of stochastic resonance. In the event of the unlikely occurrence that the noise variance and excess power are precisely matched, then $(\alpha_U - A) = \sigma$, with $\gamma = 1$. Lastly, if the transition threshold is set far above the level of signal amplitude in relation to the noise, then $(\alpha_U - A) > \sigma$, resulting in $\gamma > 1$.

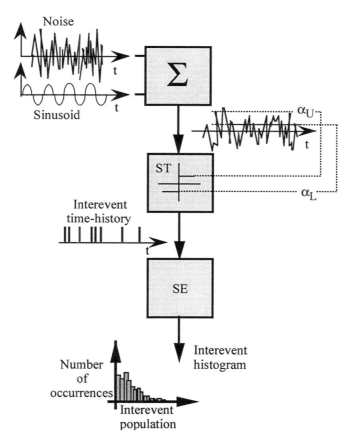

Fig. 7.3: Extraction of interevent statistics from
a Schmitt trigger subjected to a small amplitude sinusoid embedded in noise and
enabling conditions of SR

ST: Schmitt trigger
SE: Statistical enumeration of interevent intervals and construction of IIH

In the computer program implemented, a Schmitt trigger logic produces IIH data for any value of γ using a sine wave plus zero-mean Gaussian, or a uniformly distributed noise. The number of points of signal-plus-noise data

315

which are processed in each simulation is 50,000. Fig. 7.4 is a typical IIH that results from threshold and noise conditions specified by $\gamma = 0.8$.

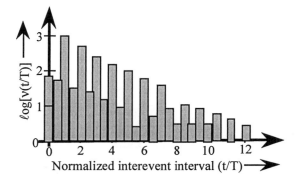

Fig. 7.4: IIH obtained from a Schmitt trigger (with a threshold parameter $\gamma = 0.8$) subjected to a sinusoid plus Gaussian noise

(The interevent intervals (t) are normalized with respect to the sine wave period (T) and ν is the number count of Schmitt trigger state-transitions that occurred.)

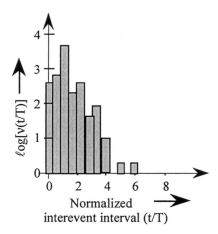

Fig. 7.5: IIH obtained from a Schmitt trigger (with a threshold parameter $\gamma = 0.1$) subjected to a sine wave Gaussian noise

(The interevent intervals (t) are normalized with respect to the sine wave period (T) and ν is the number count of Schmitt trigger state-transitions that occurred.)

It may be noted that the primary response modes are located at integral multiples of inter-event intervals normalized by the period of the sine wave. The amplitude of the modes decay approximately exponentially, in agreement with the statistics of biological IIH as indicated by Longtin [7.19]. The optimum range of γ which results in biologically plausible IIH ranges from 0.7 to 0.9.

When values of γ are less than 0.7, the detector threshold is set slightly above the level of the periodic signal amplitude, and the resulting state-transitions are dominated by the periodic signal as illustrated in Fig.7.5. In the case of $\gamma > 1$, the threshold level is significantly remote from the signal amplitude in relation to the noise power so that the noise fluctuations dominate to control the detector dynamics. The resulting IIH then tends to exhibit amplitudes distributed uniformly across the range of inter-event-intervals shown in Fig.7.6

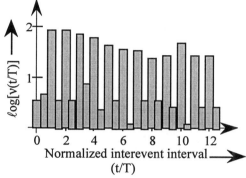

Fig..7.6: IIH obtained from a Schmitt trigger (with a threshold parameter $\gamma = 1.5$) subjected to a sine wave Gaussian noise

(The interevent intervals (t) are normalized with respect to the sine wave period (T) and ν is the number count of Schmitt trigger state-transitions that occurred)

7.3 A Neural Network under SR-Based Learning

In this section, a multi-layer perceptron is described which is implemented to evaluate the learning dynamics involving SR. The perceptron is a supervised feedforward ANN (Fig.7.7). That is, it requires an error-source or cost-function of the network output in response to training data during the learning phase. The network architecture is configured as a multilayered structure. Specifically, the network consists of one input layer, one hidden layer, and one output layer. The input layer contains twenty-six neurons, the hidden layer has 15 neurons, and there is a single neuron in the output layer, as depicted in Fig.7.7.

The input layer serves as a signal-multiplexer and routes the value of each input to each neuron in the hidden layer. Connecting each neuron in successive layers is a trainable weight. The weight between the ith and j^{th} units, denoted as w_{ij}, is a numerical value which is multiplied by the output of the ith unit. This weighted value is then presented as one of the multi-inputs to the jth unit. Each weight is modified during the training process to produce a minimum error output from the network. Neurons in the hidden and output layers perform computations as follows: Let $x_i = \Sigma w_{ij} \chi_j$ 1represent the weighted sum of the multi-inputs X_j. This summed input is processed by the activation function to produce the neurons' output signal O_i.

The activation function provides each neuron with a nonlinear transfer function, so as to allow the processing of large input values devoid of overload, while simultaneously facilitating sensitive response to low-level input activity. The activation functions used in the hidden layers are sigmoidal Bernoulli functions, $L_Q(x)$, with $Q = 1/2$, [7.25] and a linear function in the output layer. The linear output activation functions allow the network output to converge to values outside the ± 1 interval set by the Bernoulli function bounds. Further, the input as well as the hidden layer have an additional bias unit clamped to a fixed value of -1 connected to each unit in the succeeding layer through a trainable weight.

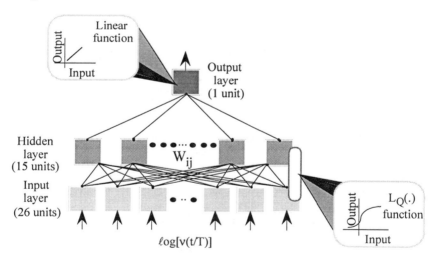

Fig. 7.7: Architecture of a multilayer perceptron implemented to learn the state-transition statistics associated with stochastic resonance

7.3.1 Training phase

The objective of the training process is to allow the network to learn the functional mapping of input data to the desired output vector. This is achieved by presenting to the network repeatedly a known set of input/output pairs (training sets) and adjusting the weights to minimize some measure of error, or cost-function, between the desired output and the computed network output. In

318

the case of the multi-layer perceptron, the conventional error minimization approach refers to the so-called backpropagation algorithm [7.26,7.27]. The fundamental entity used in the weight adjustment process is the error ε_i of the network output O_i at the ith unit, deviating from a target value T_i. It is used to calculate the effective gradient of the weight modification term in the backpropagation algorithm. The effective gradient, δ_j, has two distinct definitions depending on whether or not a target value is available for a particular unit. In the case of network output units, for which a target is known, δ_j is defined as the distance (error) of the jth unit times the derivative of the activation function evaluated at the output value of the ith unit. That is, $\delta_j = (\partial O_i/\partial X_i)\varepsilon_j$ where X_i represents the ith unit input to the activation function. When the unit resides within a hidden or input layer, a target value is not available for computation of the network error ε. In this case, the definition of δ_j is modified such that the product of cumulative effective gradients from the next layer and the interconnection weights are "backpropagated" to these units. In other words, $\delta_i = (\partial O_i/\partial X_i)\Sigma_j\delta_j\, w_{ij}$. With an appropriate expression for the error-gradient in hand, the basic prescription to adjust the weights at the nth training step is given by the well-known *Widrow-Hoff delta rule* [7.26], namely, $w_{ij}(n) = w_{ij}(n-1) + \eta\delta_jO_i = w_{ij}(n-1) + \Delta w_{ij}(n)$ where η is the *learning rate*.

In regions of the error surface where large gradients exist, the δ-terms may become inordinately large. The resulting weight modifications will also be so large as to set extensive oscillations at the network output, thereby preventing convergence towards the true error-minimum. The learning coefficient can then be set to an extremely small value to counteract this tendency; however, this would drastically increase the training time. To avoid this situation, the weight modification can be given a "memory" so that it may no longer be subject to abrupt changes. That is, the weight-change algorithm is given a modified form as $\Delta w_{ij}(n) = \eta\delta_jO_i + \lambda[\Delta w_{ij}(n-1)]$ where λ is known as the *momentum parameter*. If λ is set to a value close to 0, the search in the parameter space will be determined by the gradient accumulated over several epochs instead of a single iteration, improving the stability of the network towards convergence.

In order to train the network as robustly as possible, it is desirable to expose it to an ensemble of training sets during the learning phase. This will avoid incorporation of details into the "memory" of the network that are specific to a particular training pattern. Therefore, the network is trained with L distinct training patterns for each realization of signal-plus-noise and threshold conditions. These L training patterns are presented sequentially to the network at each learning step n.

7.3.2 Cost-function

The cost-function commonly employed in neural network training is the Euclidean distance which is specified by: $\varepsilon_i = (T_i - O_i)$. This error-metric is usually represented in absolute terms as the root of the square error $\varepsilon_{SE} = \sqrt{(T_i - O_i)^2}$ and is referred to as the square-error (SE) cost-function. The SE is solely a measure of the network deviation from the desired goal, and it does not quantify implicitly the information flow dynamics of the learning process. Alternatively, the cost-function can be specified in terms of relative informational entropy (ε_{RE}) between the present state of the network output and the desired network goal [7.18]:

$$\varepsilon_{RE} = \frac{1 + T_i}{2} \ell og \left[\frac{1 + T_i}{b + O_i} \right] + \frac{1 - T_i}{2} \ell og \left[\frac{1 - T_i}{b + O_i} \right] \qquad (7.2a)$$

where,

$$\beta = a^2 + b^2 \left[\left(\frac{1}{\tanh(b\,O_i)} \right)^2 + \left(\frac{1}{\sinh(b\,O_i)} \right)^2 \right] - \frac{2ab}{\tanh(a\,O_i)\tanh(b\,O_i)}$$

$$(7.2b)$$

with $a = (3Q + 1)/2$, $b = (3Q - 1)/2$, and Q is the parameter of the Bernoulli activation function $L_Q(x)$. During learning, weight modifications are made proportional to the gradient of the network error. In the information-theoretic domain, this means that learning takes place concordant with the rate of information flow towards or away from the desired goal, as dictated by the instantaneous stochastic state of the network.

7.3.3 Gross-features of network uncertainty

Since the RE error-measure is an implicit metric of the relative informational entropy, an examination of the neural learning curve under the direction of RE quantifies the flow of neural information during the learning endeavor. Specifically, the value of relative entropy (ε_{RE}) at any discrete learning step n, averaged over the L input training patterns, will quantify the average relative uncertainty of the current network organization:

$$HL(n) = \frac{1}{L} \sum_{l=1}^{L} \varepsilon_{RE}(n) \qquad (7.3)$$

Further, it is necessary to ensure that the network training is truly representing the average information available for processing during learning. Suppose one performs a single training cycle on the network using n (= 1, 2, ..., N) sequential presentations of L input patterns. It would then result in a learning evolution with an under-represented choice of random numbers that were selected to initialize the interconnection weights. In order to avoid this, the network is trained m = 1, 2, ..., M times, with the L training sets presented N times using a different random number seed to initialize the network weights for each of the M runs. The average network uncertainty at a particular learning step, n, for L training sets after M training cycles, is then given by

$$H_M(n) = \frac{1}{L} \sum_{m=1}^{M} H_{Lm}(n) \tag{7.4}$$

This quantity represents the average (informational) relative entropy pertinent to the network goal-directed organization as the network architecture seeks to minimize the divergence of its output from that of the desired goal. It is used to examine the discrete-step temporal evolution of neural learning as the multilayer perceptron learns to recognize the statistics of stochastic resonance.

7.4. Simulation Results

The test perceptron is trained to learn the functional mapping from IIH input to threshold parameter output, for both cases of SR and non-SR discussed earlier in Section 7.2.2. The number of training sets for each threshold parameter realization is $L = 10$. The network was trained $M = 10$ times for each threshold parameter specification with a distinct random number seed in each case. The inputs to the neural network are the amplitude values of the IIH computed from the Schmitt trigger logic corresponding to a fixed value of γ. The inputs are specified at 26 values, covering the range from 0 to 12.5 in increments of 0.5, corresponding to the inter-event intervals normalized by the forcing period. The network output is defined by the respective value of the threshold parameter γ. The neural network was trained via the backpropagation algorithm with the RE error-metric. A learning coefficient of $\eta = 0.0015$ was adopted, along with a momentum term of $\lambda = 0.9$. In order to provide a valid comparison of learning dynamics for different threshold parameter conditions, the random number generator which sets the initial interconnection weights is initialized with the same seed for each of the successive M training cycles at the start of each training run. For example, to compare the learning dynamics for $\gamma = 0.8$ and $\gamma = 0.1$, each of the M training cycles for the distinct values of γ are started with the same random number seed. This ensures that the evolution of network dynamics for each value of γ initiated identically.

Fig. 7.8a compares H_M as a function of training iterations for the case of SR ($\gamma = 0.8$) with those relevant to non-SR statistics where the detector dynamics are dominated by the modulating periodic signal ($\gamma = 0.1$). It can be seen that when SR is present, the (informational) entropy during the early stages of learning is significantly lower than when SR is absent. It is also observed that the SR network configuration completes the learning (cessation of oscillatory/jagged behavior) at $n = 60$, while for the non-SR case, the network convergence is achieved at $n = 70$. There is not a significant deviation from these values up to n

321

= 120 learning steps, corresponding to the maximum extent of the simulations envisaged.

A comparison of information dynamics of the network for the SR case with $\gamma = 0.8$ and for a case characterized by $\gamma = 1.5$ (noise-dominated statistics), is presented in Fig. 7.8b. It is observed that the early learning phase is not significantly different, except for the large uncertainty of the non-SR dynamic at the initiation of training. Both systems have converged to a stable network configuration at n = 60; however, the non-SR system has migrated to a state of larger entropy. The non-SR system is, therefore, less certain about the proximity of its output to the goal than in the SR case.

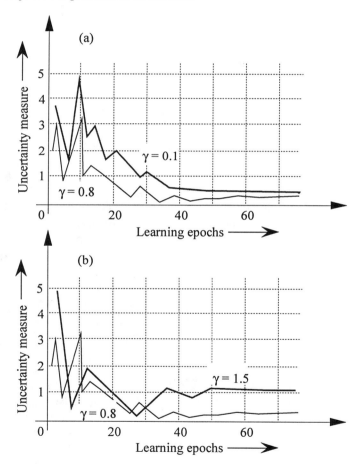

Fig.7.8: Evolution of average information flow (uncertainty measure)
in the neural network
(a)   For threshold parameters of $\gamma = 0.1$ and $\gamma = 0.8$
(b)   For threshold parameters of $\gamma = 0.8$ and $\gamma = 1.5$

Figs. 7.8a and 7.8b demonstrate that the nonlinear detector statistics inherent under conditions of SR result in a more efficient organization of information-flow during neural learning than when SR is absent. Therefore, the learning of a neural network under conditions of stochastic resonance are robustly more informatic (less negentropic) than under similar conditions when SR is not present.

After training the neural network, it is of interest to assay the total extent of relative entropy experienced by the network during the learning process. The neural complex during learning perceives this cumulative average uncertainty. It can be quantified in terms of the RE error-measure corresponding to M distinct ensembles of L input patterns presented over the N discrete-learning steps to the network. This is given by:

$$H_{M,N} = \frac{1}{N} \sum_{n=1}^{N} H_M(n) \qquad (7.5)$$

The average (informational) entropy computed in respect to the learning curves shown in Figs. 7.8a and 7.8b is presented in Table 7.1 for two terminal values of the learning step, N. The cumulative uncertainty of Eqn. (7.5) was computed for both terminal values of N, inasmuch as the question of when the network converges is always open for interpretation. Examination of the early stages of learning, say at N = 50, reveals that the total relative uncertainty for the SR conditions ($\gamma = 0.8$) are approximately one-half of those when SR is absent. This condition remains valid even when N = 70, at which point it can be observed that the network has fully organized in all cases considered (SR as well as non-SR). Therefore, it can be surmised that the informational uncertainty experienced by the neural complex during learning under SR conditions is only about one-half of the uncertainty under non-SR dynamics.

Table 7.1. Total extent of average uncertainty $H_{M,N}$ perceived during the learning phase

Learning steps N	Threshold parameter (γ)		
	0.1	0.8	1.5
50	1.148	0.586	1.009
70	0.844	0.462	1.019

7.5 Concluding Remarks

Stochastic resonance arises in a wide variety of nonlinear detectors exposed to a simultaneous coupling of weak periodic signal and noise. In the case of a bistable system, one of the attributes is the quasiperiodic modulation of the transition probabilities of dichotomous states pertinent to the bistable potential. This indicates that for a neuron modeled as a nonlinear bistable detector, a pulse-train constituted by quasiperiodically modulated state-transitions can serve as a robust representation of proliferating neural events along the time-axis. In a biological sense, this model refers to the firing events triggered by the threshold crossings of membrane potential undergoing a biased random-walk. This biased random-walk or *Wiener process* refers to the excitatory and/or inhibitory synaptic potentials in the presence of intra- and/or extraneural disturbances in a biological neuron. The nonlinear bistable model of a neuron also reflects the robustness of competitive negentropic and posentropic synergism, which leads to a stable attractor point for the neural dynamics, as observed recently by the authors [7.20, 7.22] and others [7.28]. In reference to the ANN, this robustness indicates achieving an isomorphism between the mapping of information flow at the input and output sets of the neural network that preserve certain properties of the input domain set.

The relative information in the domain-space () and the range-space ($\Re$) can be considered in terms of a *differential information density function* defined as the difference between the maximum entropy and the observed entropy. In reference to the network output range, $H_\Re = (H_\Re^{max} - H_\Re^{obs})$. Similarly, the information density over the input domain space can be defined as $H_\vartheta = (H_\vartheta^{max} - H_\vartheta^{obs})$. Within these bounded values of entropy associated with the input-domain and the output-range of the network, the fidelity of narrowing the output-range of information (being close to a target value) can be improved without sacrificing the *variety* associated with it. This is true if the input-domain of information is manipulated within a narrow range of additive noise corresponding to the established signal amplitude and detector threshold conditions. The added noise at the input-domain increases the maximum value of disorder so that the network training robustly self-imposes certain constraints and ordering. As a result, the *S-redundancy* (Shannon redundancy) is introduced implicitly to reduce the range of uncertainty at the output. The domain-to-range concept of stochastic resonance isomorphism, when viewed in Shannon's perspective, indicates that the more negentropic (less informatic) aspect of the input-domains, and less negentropic (more informatic) certainty of the output-range settling in the basin of convergence is a synergistic endeavor. This utilizes both the *randomness* (introduced *via* noise) and the *orderliness* achieved through the network's self-organization efforts. This endeavor is specific to set of detection thresholds which facilitate the maximization of net information-transfer. The added noise allows robustly the search-process of self-organization in the network to find a more compact set of less cardinality which can be accommodated in the global minimal basin of attraction.

Stochastic resonance in both real and artificial neural networks refer implicitly to an adaptive scheme of information-processing. As indicated in Finley's simulation studies [7.29], neural networks require a "parametric tuning" in order to force them to display a robustness in converging towards a steady-state. Otherwise, the network's initial behavior would typically either degenerate rapidly to zero activity, or would tend to an unstable dynamic with wide oscillatory fluctuations in neural firing. Considering stochastic resonance, this "parameter tuning" is introduced *via* a narrow range of additive noise with respect to the set detection thresholds. The noise so introduced influences the spatiotemporal neural activity and maximizes the information transfer through the network. The result is an efficient convergence towards network stability.

In summary, considerations of information-theoretics applied to neural network learning so as to recognize the state-transition statistics of a nonlinear, bistable detector system under the conditions of stochastic resonance are presented in this chapter. The information-flow dynamics are examined in terms of an information-theoretic cost-function defined by the relative (informational) entropy associated with an ensemble of training sets averaged over the temporal evolution of training cycles. The network architecture adopted consists of a multilayer perceptron evolving under the guidance of the backpropagation algorithm. For the purpose of emulating stochastic resonance (SR), a Schmitt trigger logic is used as the nonlinear detector. It generates state-transitions exhibiting SR in response to a sine-wave signal superimposed with Gaussian noise. The output statistics of the Schmitt trigger are used to train the multi-layer perceptron towards recognizing the extent of SR present in the detector state-transition dynamics. It is demonstrated that information flow dynamics, under the conditions of stochastic resonance, are inherently more informatic (or less negentropic) than in cases wherein the state-transition statistics are dominated by non-SR conditions. The SR or non-SR conditions are dictated by higher or lower signal-to-threshold ratios, respectively. Stochastic resonance in relation to biological neurons, as indicated in this chapter, offers an insight into feasible aspects of deliberately incorporating the SR phenomeno1 3y in ANNs in order to achieve more robustness in information processing tasks.

Bibliography

[7.1]   Benzi, R., Parisi, G., Sutera, A., and Vulpiani, A.: Stochastic in
          resonance climate change, *Tellus*, **34** , 1984, 10-16.

[7.2]   Nicolis, C.: Stochastic aspects of climatic transitions - response to a
          periodic forcing, *Tellus*, **34**, 1982, 1- 9.

[7.3]   Fauve, S., and Heslot, F.: Stochastic resonance in a bistable system,
          Phys. Lett., **97**A, 1983, 5-7.

[7.4]   McNamara, B., Wiesenfeld, K., and Roy, R.: Observation of stochastic
          resonance in a ring laser, *Phys. Rev. Lett.*, **60**, 1988, 2626-2629.

[7.5]   Weisenfeld, K. and Moss, F.: Stochastic resonance and the benefits of
          noise: from ice ages to crayfish and SQUIDs, *Nature*, **373**, 1995,
          33-36.

[7.6]   McNamara, B., and Weisenfeld, K.: Theory of stochastic resonance,
          Phys. Rev. A, **39**, 1989, 4854-4869.

[7.7]   Gammaitoni, L., Marchesani, F., Menichella-Saetta, E. and Santucci,
          S.: Stochastic resonance in bistable systems, *Phys. Rev. Lett.*, **62**, 1989,
          349-352.

[7.8]   Douglass, J. K., Wilkens, L., Pantazelou, E. and Moss, F.: Noise
          enhancement of information transfer in crayfish mechanoreceptors by
          stochastic resonance, *Nature*, **365**, 1993, 337-340.

[7.9]   Miller, J. P. and Levin, J. E.: Broadband neural encoding in the cricket
          cercal sensory system enhanced by stochastic resonance, *Nature*, **380,**
          1996, 165-168.

[7.10]   Bezrukov, Sergey M., and Vodyanoy, I: Noise-induced enhancement
          of signal transduction across voltage-dependent ion channels, *Nature*,
          378, 1995, 362-364.

[7.11]   Gluckman, B. J., Netoff, T. I., Neel, E.J., Ditto, W. L., Spano, M. L.,
          and Shiff, S.J: Stochastic resonance in neuronal network from
          mammalian brain, *Phys. Rev. Lett.,***77**, 1996, 4098-4101.

327

[7.12] Collins, J. J , Imhoff, T. T., and Grigg, P.: Noise-enhanced tactile sensation, *Nature*, **383**, 1996, 770.

[7.13] Cordo, P, Inglis, J. T., Verschueren, S., Collins, J.J., Merfeld, D. M., Rosenbaum, S, Buckley, S., and Moss, F.: Noise in human muscle spindles, *Nature*, **383**, 1996, 769-770.

[7.14] Heneghan, C., Chow, C. C., Collins, J. J., Imhoff, T. T., Lowen, S. B., and Teich, M.C.: Information measures quantifying aperiodic stochastic resonance, *Phys. Rev. E*, **54**, 1996, R2228 - R2231.

[7.15] Bulsara, A. R., and Zador, A., Threshold detection of wideband signals: A noise induced maximum in the mutual information, *Phys. Rev. E*, **54**, 1996, R2185 - R2188.

[7.16] Neiman, A. Shulgin, B., and Anischenko, V.: Dynamical entropies applied to stochastic resonance, *Phys. Rev. Lett.*, **76**, 1996, 4299-4302.

[7.17] Srinivasan, S. K. and Sampath, G., Stochastic Models for Spike Trains of Single Neurons. In: *Lecture Notes on Biomathematics*, Vol. 16 (Springer Verlag, Berlin: 1972).

[7.18] Teich, M. C., Johnson, D. H., Kumar A. R. and Turcott, R. G.: Rate fluctuations and fractional power-law noise recorded from cells in the lower auditory pathway of the cat, *Hear.Res.*, **46**, 1990, 41-52.

[7.19] Teich, M. C.: Fractal character of the auditory neural spike train, *IEEE Trans. Biomed. Eng.*, **36**, 1986, 150-160.

[7.20] Neelakanta, P. S., Abusalah, S., Sudhakar, R., De Groff, D., Aalo, V. and Park, J.C.: Dynamic Properties of neural learning in the information-theoretic plane, *Comp. Syst*, 9, 1995, 349-374.

[7.21] Park, J. C., Neelakanta, P. S., S.Abusalah, De Groff, D. and Sudhakar, R.: Information-theoretic based error-metrics for gradient descent learning in neural networks, *Comp. Syst.*, 9, 1995, 287-304.

[7.22] P. S. Neelakanta, Abusalah, S., De Groff, D., Sudhakar, R., and Park, J.C.: Csiszar's generalized error measures for gradient-descent-based optimizations in neural networks using the backpropagation algorithm, *Connection Sci.*, **8**, 1996, 79-114.

[7.23] Kramers, H. A.: Brownian motion in a field of force and the diffusion model of chemical reactions, *Physica,***7**, 1940, 284.

[7.24]   Longtin, A., Bulsara, A., and Moss, M.: Time-interval sequences in bistable systems and the noise-induced transmission of information by sensory neurons, *Phy. Rev.Lett,* **67**, 1991, 656-659.

[7.25]   Neelakanta, P. S., Sudhakar, R., and De Groff, D.: Langevin Machine: A neural network based on a stochastically justifiable sigmoidal function, *Biol. Cybern.*, **65**, 1991, 331-338.

[7.26]   Wasserman, P. D., *Neural Computing* (Van Nostrand Reinhold, New York, NY: 1989).

[7.27]   Müller, B. and J. Reinhardt: *Neural Networks* (Springer Verlag, Berlin: 1990).

[7.28]   van Vreeswijk, C. and Sompolinsky, H.: Chaos in neuronal networks with balanced excitatory and inhibitory activity, *Science*, **274**, 1996, 1724-1726.

[7.29]   Finley, M.: "An experimental study of the formation and development of Hebbian cell-assemblies by means of neural network simulation", *Tech. Rept 08333-1-T* (Dept. of Computer Sciences, Univ. of Michigan, Ann Arbor, MI: 1967).

Chapter 8

Neural Informatics and Genetic Algorithms

P. S. Neelakanta and T.V. Arredondo

*"For cells, as for societies,— time's arrow points
ultimately to the dust of entropy"*
... Albert Einstein

8.1 Entropy, Thermobiodynamics and Bioinformatics

Poised robustly with bioenergetics, the other facet of bioentropy, namely, informatic-biology embarks at the most primitive level—the breaking of universal genetic-codes in neural cells. Though tossed by the tempest of intra-extraneural disturbances, it sails triumphantly across the ocean of neural complex in search of far-reaching shores of information-content pertinent to the giant-molecules of the living systems. This formidable effort is paddled and steered skillfully through the morsels of nucleodites. With the amino-acids at the helm of these cybernetic efforts, this bioinformatic voyage converges at its terminal phase, to the haven of macroscopic information-management which docks the gross features of complexity in the life systems.

Thermobiodynamic perspectives of entropy (S), the cybernetic attributes of bioenergetics and the notions of information theory (due to Shannon and Weaver [8.1] and Khinchin [8.2]) are the three equilateral planes which enclave the hypersphere of informational biology (Fig.8.1).

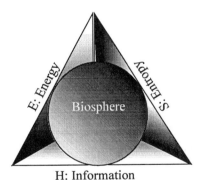

H: Information

Fig. 8.1: Hypersphere of informational biology

The multidimensional perception of bioinformation is the same as the negentropy concept for information (H = − S) in the sense of the improbability of specified ordered states of the biological system—either at the microscopic, neucleoditic level or at the macroscopic, cerebral level.

As discussed in the earlier chapters, certainty, predictability, or an organized state of order is the negative of S (entropy) associated with the physical and/or biological systems and can be specified quantitatively by the celebrated Shannon's definition of probabilistic information, namely, $H = - S = - \Sigma p_i \ell og(p_i)$. Further, the fundamental concepts of thermodynamic states in biological systems, as specified by the equations of state and the extents of energy (ultimately convertible into mass) constitute the entropic explanations in alternative ways which keep the bioinformation "bits together with their energetics".

Pertinent to the perpetual complexity in living systems, the posentropy (positive entropy) faces a struggle with negentropy or information acquired through assimilation of energy. For example, *photosynthesis* is a complex performance in the living system and depends on "informing" and energizing the actions   perceived from the irradiant source of electromagnetic (light) energy. The process of photosynthesis is *semantic* and *pragmatic*. It refers to a meaningful, message-bearing action executed at the molecular level. The associated information governs the matrix of molecular receptors and excitation energy in reducing the uncertainty among the degrees of freedom in the expected modes of iterations of a photosynthetic cycle. The gamut of information in biological processes includes deliberating the informatics of the energy involved, coded guidance of the interaction(s), and the regulatory aspects as well as the cybernetic endeavors of the process itself. The associated semantic or meaningful information-content can be specified in terms of the molecular bonds, the weights of interactions (stored as memory bits), and by the capacity of the channels of neural transmissions.

Conceptually, the flow of matter-energy in a system is inseparable from the proliferation of the coexisting entropy or information. Such flows in biological systems stemming from the metabolic activities and regulatory considerations are dictated by C^3I protocols[1] as observed by one of the authors elsewhere [8.3]. Further, like any communication system, the bioinformatic flow at all conceivable levels would encounter "transduction, interpretation, discrimination, threshold behavior, phases with meaningful lags and leads, space and time-constants, frequency-bandwidth and concepts of conflict, noise and antinoise".

Specifically, the cellular system is well-known as an ensemble of information-processing units, wherein the visible neural units function as sensory transducers, and the interconnections between the neurons store the learned data through their correlatory weights. There also exists a mechanism of decision-making and related functions in the neural informatic setup in its role as a pattern recognizing system.

It is not surprising that when the core biological entities, namely, *genes*, function as per a set of definitive, algorithmic trends with an associated

---

[1] C^3I : Command, communication, control and information.

informatic profile, the cellular units which hold these genes could be presumed to follow similar informational trends in their self-organizing endeavors, as well.

The brain, representing the information center of biological systems, has the capability of information storage and management, data-banking and the related processing functions. Its neuronal cells receive information and store it by molecular biochemical means. In this process, "information sees the *axon* as a gate, where under the management of the nervous system, the incoming information is self-organized by association with the necessary *synapses*, organized for a network searching function and aimed at a *consensus*". The self-organization amounts to the neural transmission maintaining the spatiotemporal quality of the signal by counteracting the intra- and/or extraneural disturbances (noise).

The gross features and functions of biological entities denote an *extensive* representation. The corresponding *localized* characteristics, at least intuitively, can be analyzed by the considerations of physical sciences as governed by mass, momentum, energy, entropy, and action. Hence, the principles of thermodynamics and the dogmas of stochastical mechanics play a vital role in managing the mass-energy considerations and the order-disorder statistics of the biological systems as in the physical systems.

As detailed in Chapter 1, entropy is a thermodynamic concept and thermodynamic information decides the number of choices or alternatives posed by the uncertainty (or entropy) involved as a result of order-disorder conflict existing naturally in a system. *Order* in a system presents a negentropy as it is generated, stored, transcribed, and copied (retrieved). It "informs" the system to organize towards an objective function. Hence, it constitutes an information. In contrast, the associated *disorder* tries to offset the system's objective and, therefore, constitutes a posentropy. The order-disorder conflict is a thermodynamical tug-of-war.

The negentropy (or information) is measured by the $\ell og_2(\cdot)$ function where the argument $(\cdot)$ is decided by the choices or alternatives which result from the system's state of order-disorder. The unit of information is the bit and 1 bit = $\ell og_2(2)$ refers to the information content resulting from 2 (binary) choices. Thus, depending upon the complexity of the system, the push-pull mechanism of order-disorder dictates the choices, which are enumerable. For example, a mammalian cell uses about 2×10^{10} bits of information and the information associated with a protein molecule is about 100 to 200 bits, which is typically equivalent to a written paragraph fully packed with messages.

The "Siamese twins" relation between entropy and energy enables the computation of energy equivalence of the bits of information. That is, if E is the energy consistent with the thermodynamical indicator, namely, the temperature, T, then the *Boltzmann's energy relation* indicates that:

$$E = (k_B T)\log_e(N) \quad \text{ergs} \tag{8.1}$$

where N is the number of choices and k_B is the *Boltzmann constant* equal to 14×10^{-16} ergs per degree. At $T = 298° K$ and $N = 2$, $E = 3 \times 10^{-14}$ ergs.

Elementary principles of probability theory and concepts of uncertainty indicate that as entropy decreases, the uncertainty decreases, since the system becomes more ordered in predictable states.

Uncertainty in a system can be estimated by considering the statistical probabilities involved as elaborated in Chapter 1. That is, a measure of information is functionally dependent on probabilities leading to *Shannon's law*, $H = - S$ (negentropy) $= - \Sigma \; p_i \log(p_i)$ bits, where p_is represent the relative numbers or probability of occurrence of the choices.

Entropy (S) links the transforming (but conserved) energy and information, consistent with the change of state(s) involved. It is specified by the *Boltzmann's entropy relation*, namely,

$$S = k_B \log(\Omega) \tag{8.2}$$

where the Boltzmann constant k_B links the thermoentropy (or temperature) associated with the thermal energy of the various states. Further, $\log (\Omega)$ describes changes encountered in a system having a constant (conserved) energy and mass.

In the perspectives of thermobiodynamics, the *First Law of Thermodynamics* can be reiterated to confirm the cent-percent transformable aspect of mass $\leftrightarrow$ energy while retaining their conservative feature intact. The *Third Law of Thermodynamics* is also validly perceived in the biological system maintaining the tranquillity of the state of rest when the thermoentropy is zero, that is, when $T = 0°K$. The *Second Law of Thermodynamics*, however, warrants a philosophical explanation when applied to living systems. When applied to a "self-acting mechanism", this law stipulates that the mechanism can transform only a fraction of available energy (due to the inherently prevalent bound-energy) of the system. That is, as transformation progresses with time with the conserved (constancy of) energy and mass, it will result in a positive trend in entropy. However, as organisms grow up to the point of infirmity or death with time, self-organizing themselves with more ordered states, the prevailing improbable arrangements (or disorder) are replaced by more probable arrangements (or order). That is, contrary to the Second Law of Thermodynamics which stipulates an increase in entropy in actual systems with forward progression of time, the living system shows a negative trend in entropy. Therefore, it appears that the *Second Law of Thermodynamics* becomes applicable only with a time-reversal or backward progression in time in reference to living systems.

This "arrow of time describes the irreversibility of physical events as entropy gives time a direction—toward more products and arrangements". Organisms "want" to organize—grow and live or make negative entropy. That is, a designated objective towards the self-arrangement of the system is rendered. There is a less inherent probability for the resultant state. If all life stopped, and decay occurred, disorder would increase (positive entropy) and this would stop the negative entropy trend. This amounts to a thermobiodynamic description of death [8.4].

However, this divergent trend in negentropy contained by the constancy of the energy in the universe is considered and amounts to the available energy shrinking to zero, suggesting a catastrophe—"like a battery running down to zero". But, it does not happen.The bioproductivity continuously replenishes energy *via* synthesis (which is similar to charging the battery), so that growing negative entropy in living systems continues until death occurs. The negative entropy is against the arrow of time and is restricted only by the tiny "island of biosphere", which is a subset of larger surroundings wherein posentropy is perpetually divergent.

It was indicated in Chapter 1 that between a system composed of nonlinear, mutually interacting dynamical entities and its environment, there exists an informational interaction with an associated differential entropy $\Delta S(t)$. This is the sum of the internal differential entropy $\Delta S_i(t)$ of the system and the external differential entropy $\Delta S_e(t)$ of the environment.

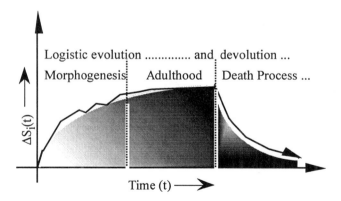

Fig. 8.2: Evolution and devolution of $\Delta S_i(t)$ in differentiating organisms

In complex biological systems, the evolution of internal entropy is fast during the embryogenetic phase (*morphogenesis*) of differentiating organisms. This logistic growth, however, levels off at a steady-state value corrresponding to the attainment of adulthood, wherein a balanced profile of negentropy *versus* posentropy prevails.The internal differential entropy drops monoto-nically after a certain stage of adult-life indicating a progress towards the death process, that is a devolutionary action sets in (Fig. 8.2).

"Turning back the hands of the time" appears as "what is life from the point of view of physics". The Second Law of Thermodynamics seems to pass a "death sentence" and enforces it mercilessly in the "inanimate world which is dead in advance". Life, as Alekseev [8.4] points out, "suspends this sentence and takes advantage of the fact that the verdict is passed without any fixed term of execution". It is a universal law of biology, according to Bauer as quoted in [8.4], that "the balanced state of the inanimate systems is stable". Surprisingly, the unbalanced state of animate systems is also stable. It could be that the animate systems carry *free-energy* which may be liberated under certain conditions so as to support the unbalanced state. Living systems produce posentropy and ultimately approaches a state of maximum entropy (death). The unbalanced state is supported by extracting negentropy from the environment. That is, metabolic functions release perpetually posentropy and extract negentropy. Thus negentropy is the "extract of order" so as to "inform" the organism to enhance its regularity—or, as Wiener said, enzymes could be the *Maxwell's metastable demons* which devour the entropy!

The concept of entropy changes (ΔS) can be specified at a temperature T in terms of *Gibbs* free-energy (ΔG) relation given by

$$\Delta G = (\Delta \mathscr{H} - T\Delta S) \tag{8.3a}$$

where $\Delta \mathscr{H}$ is defined as the change in the *enthalpy*.[2] The above relation (Eqn. (8.3a)) can be interpreted as follows:

Change in enthalpy($\Delta \mathscr{H}$) = Change in free-energy (ΔG) +
$\qquad\qquad$ T × Change in entropy (ΔS) $\qquad\qquad$ (8.3b)

That is,

Change in total energy = Change in useful energy +
$\qquad\qquad$ T × Change in fixed energy of organization

$\qquad\qquad\qquad\qquad\qquad\qquad\qquad\qquad\qquad$ (8.3c)

---

[2] *Enthalpy:* Whenever energy changes are encountered (for example, as in a chemical reaction), it is convenient to assume each participating material as having a certain heat-content or *enthalpy*, $\mathscr{H}$ (Greek *enthalpein,* to warm in). A heat change in a reaction is termed as a change in enthalpy, $\Delta \mathscr{H}$.

In addition to a change in enthalpy, every bio-, physico- or chemical change would involve a change in the relative disorder of the constituent atoms, molecules, or ions involved. A measure of this disorder refers to *entropy*, S of the system.

In essence, a part of the enthalpy change is involved in changing the relative disorder or entropy of the system, and a part is associated with the useful work. The maximum amount of useful work that can be done by any process at a constant temperature and pressure refers to ΔG.

The entropy change informs implicitly the changes in the states of the system. That is, more information is required to specify a disordered system because of the random distribution of its entities. Hence, the concept of entropy is applicable as informational entropy concurrent to its general meaning as the state of the disorder or thermodynamic entropy. Therefore, by formalizing a correspondence between Boltzmann's relation on entropy and Shannon's expression for information measure, H, namely,

$$S \Leftrightarrow H: \ [\ k_B log(\Omega)] \Leftrightarrow [- S = - \Sigma \ p_i log(p_i) \ ], \qquad (8.4)$$

it is clear that $log(\Omega)$ represents how a jaggered system turmoiled between order and disorder can be quantified by a dimensionless number counting the modes of (such ordered and disordered) arrangements in the system; whereas, information is a matter of statistical (relative) enumeration of the chances of achieving a specific arrangement in the same system. The entropy of the jaggered system decreases as the information specifies the reduction in the uncertainty of (disordered) arrangements.

Biological systems are intrinsically information-rich. This can be inferred from the following expression on the information-content, I ($\equiv$ H), in knowing whether a particular biological entity (say, a cell) is alive [8.5]:

$$I = [\ (\mathcal{E}_a/k_BT) - log_2(X) + log_2 (Z)] \ \ \text{bits} \qquad (8.5)$$

where, for the ensemble of a living system under consideration, there are Ω possible states and X of these correspond to confirm that the system is a living one. Z is the *partition function* which is the normalizing factor in the evaluation of the probability (p_a) given by

$$p_a = [X exp(- \mathcal{E}_a/k_BT)]/Z \qquad (8.6)$$

where p_a is the probability that the ensemble entity is alive under isothermal conditions and $\mathcal{E}_a$ is the energy level that all living entities of the ensemble strive to attain.

Using the thermodynamic relations between *Helmoltz free-energy* and the partition function, equation (8.5) can be rewritten as

$$I = [\ (\mathcal{E}_a - U)/k_BT] - [k_B log_2(X) - S]/k_B \qquad (8.7)$$

where U is the internal energy, and S is the entropy.

The first term in equation (8.7) depicts the difference in energy between the living system and an equilibrium system of the same temperature. Further, the living systems are also energy-rich by virtue of constant "charging" from say, photosynthesis. The difference $(\mathcal{E}_a - U)$ cannot sustain in thermal modes lest it would transfer to the reservoir as kinetic energy. Therefore, $(\mathcal{E}_a - U)$ sustains in the form of potential energy (of molecules) leading to a stable energy-rich state.

The second term of Eqn. (8.7) namely, $(S/k_B) - \log_2(X)$, is always positive since $(S/k_B) > \log_2(X)$. The reason is as follows: In the biosphere, the number of living entities (posing the value X) is always smaller than the number of possible states.

The ordered state of a biological system would decay "to the most disordered possible state unless work is performed constantly to restore order in the system". This work is done in accordance with the Second Law of Thermodynamics following the principle of the heat-pump. The source and sink of heat is provided by the exterior ambiance of the biosphere, namely, earth, sun, etc.

The growth process involves consumption of nutrients, emergence of new cells and excretion of waste products. This process is accompanied by heat exchange to and from the reservoir (or stored energy); and, the corresponding entropy change can be written as

$$\Delta S_{Total} = [\Delta S_{System} + \Delta S_{Reservoir}] > 0 \qquad (8.8)$$

The total entropy change is always positive since the process is irreversible. The positive attribute of the total entropy vis-à-vis growth of cells, is, therefore, consistent with the Second Law of Thermodynamics.

It may be of interest to study the extent to which Shannon's entropy concepts and Boltzmann's perspectives on thermodynamical entropy be addressed cohesively. As indicated in Chapter 4, Kapur, in his studies on the measures of uncertainty in mathematical programming and physics, has indicated that Shannon's entropy leads to relevant aspects of Boltzmann distribution of statistical mechanics, in concurrence with the corresponding studies by Jaynes. However, Shannon entropy fails to link with Bose-Einstein and Fermi-Dirac entropies of quantum mechanics!

8.1.2   Informatics of DNA molecules

Having grazed the concepts of energetics, entropy and information of global biological systems, the specific attributes of information-theoretics at the biological molecular level can be considered [8.6].

A living system, viewed at the most primitive entity where a meaningful deciphering of information content can be done, refers to the information content of *DNA molecule*. DNA (or *deoxyribonucleic acid*) is a giant, linear, polymeric molecule existent in all living systems. It consists of two polynucleodite chains wound helically about a long central axis. The bases on opposite chains are joined through *Hydrogen* bonds with specified constraints.

The linear polymeric bonds of DNA are based on 4 sub-units, namely, the bases *Adenone* (*A*), *Gauanine* (*G*), *Cytosine* (*C*) and *Thymine* (*T*), linked in a chain through *Phosphate* (*P*) and *Deoxcyribose* (*R*) bridges as shown in Fig.8.3.

Fig. 8.3: A typical DNA chain

The base sequences exhibit statistical ordering which can be specified as vectors posing magnitude and directional attributes. Characteristic to every individual living system, the sequences of bases along its DNA chain are unique and represent meaningful, message bearing statistics. That is, a negentropy value can be ascertained from the sequential features of the DNA chain. Such a negentropy value denotes the characteristic genetic information possessed by the living system of the DNA considered.

The genetic information contained in the cell is almost stored entirely in a coded form as actual sequences of bases. This information constraint the cell in what order the amino-acids should be strung together in the protein constituents. This is done indirectly, first by transcribing the information in sections to an intermediate polymer called *messenger RNA* (or mRNA) which is similar to DNA except that the sugar residue **R** is replaced with a slightly different one, namely Ribose **R′**; and, the corresponding four bases are **A, G, C** and *Uracil* (**U**), replacing *Thymine* (**T**) of the DNA.

DNA sequences represent very long strings—of the order of 10^4-10^9 base pairs. Therefore, they are stochastically viable for representation in the information-theoretic plane by virtue of their extremely large number.

Mathematically, DNA represents a linear sequence of symbols and exactly duplicates this sequence each time the molecule reproduces itself. This confirms the role of DNA as the site of information storage on the genetic and hereditary characteristics of the biological species. The information learned and stored in the brain has been largely identified and characterized in terms of the spatiotemporal attributes of the neuronal firings at a macroscopic level. However, it has not been confirmed and correlated with the genetic information storage strategies governed by the DNA and/or RNA, *Ribonucleic acid* at the sub-cellular level.

A life system stores and processes the information necessary for its own reproduction. The associated intricate information-processing is facilitated by the DNA and RNA strands. First, the information stored in the base sequence of a DNA strand is copied by the formation of a strand (a messenger) of RNA using the single DNA strand as a template. That is, a process of *transcription* takes place which is analogous to the DNA *replication process*. This occurs at a specific site on one strand of DNA known as the *transcription initiation site* which has a characteristic base sequence.

The transcription proceeds through a specific chemical pairing (termed as *base pairing)* between **G** and **C** (or **C** and **G**) and between **A** and **U** and **T** and **A** as illustrated below (Fig. 8.4):

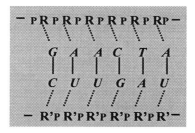

Fig. 8.4: Base-pairing

The RNA enzyme flows along the DNA strand producing a stretch of single strand RNA until it reaches a second characteristic sequence of bases which carries information that instructs the transcribing operation to halt. The RNA molecule is then set free. Thus, the mechanics of transmission are primarily chemical and can be summarized as follows.

Upon the transcription performed *via* base-pairing, the mRNA behaves as a machine-controlling instruction tape to a special protein synthesizing system in the cell, where it is translated into a set of amino acid sequences of the proteins which are produced *in situ.* Each amino-acid gets coded for a triplet of bases, and this triplet coding scheme is called the *genetic code.* For mammals, the genetic information content realized *via* genetic coding is in the order of 10^{10} bits. The transcription process is an information retrieval technique from the memory units of the DNA. In order to quantify the negentropic characteristics of the DNA, the following questions should be answered:

- Are the bases in the DNA chain independent events?
- Does the occurrence of any one base along the chain alter the probability of occurrence of the base next to it?
- What is the conditional probability that when a base A occurs, it will be followed by A or any other base designated at T, C or G ?

Starting with the first question, studies reveal that the DNA chains are not independent. The divergence from independence results from a linear ordering of the bases along the chain which reduces the entropy, deviating from the maximum entropy.

Equiprobable occurrence of bases would yield a maximum negentropy (information) denoted by H_{max}. Suppose the divergence from this equiprobable state is specified as D_1. Further, due to nonindependency, the actual information content would deviate from the maximum value H_{max}. Suppose D_2 represents the divergence due to nonindependent status. Then (D_1+D_2) is the total divergence from the maximum state. This can be designated as the *stored information*, I_s because (D_1+D_2) denotes the total extent to which the entropy has been lowered from the maximum entropy state. This is equal to the extent to which information can be replenished or stored. It is also known as *information density*, I_d. Shannon referred to this is as the *redundancy*.

The salient aspects of DNA information can be summarized as follows [8.5]:

- DNA information is inherently redundant
- It represents, at least a *first-order Markov source* output
- DNA information can be regarded as the output of an ergodic source
- Total divergence from the maximum entropy state is composed of D1 and D2 as indicated earlier. The indices (D–indices) which specify the relative contributions of D_1 and D_2 to the total divergence are defined as:

$$RD_2 = D_2 / (D_1 + D_2) \qquad (8.9)$$

- Shannon's *redundancy factor*:

$$R = 1 - [H_m/\ell og(a)] \qquad (8.10)$$

where a is the number of states (or events) of the statistics involved and $\ell og(a)$ corresponds to the maximum entropy of the sequence of equiprobable, independent elementary events I =1,2,...,a; H_m is the *entropy of the first-order Markov* chain; and $H_m/\ell og(a)$ is defined as the *relative entropy*

- D_2 is an *evolutionary index* which separates the higher organisms such as vertebrates from lower organisms
- *R-indices*: Vertebrates have achieved higher R values by holding D_1 relatively constant and increasing D_2. Lower organisms, on the other hand, may achieve higher values of R primarily by increasing D_2. That is, D_2 is increased by a divergence in the base compositions from the uniformly distributed characteristics and marching towards a central tendency area at the primitive state of evolution of the vertebrates.

- $(R, D_1) \Leftrightarrow (R, RD_1)$ and $(R, D_2) \Leftrightarrow (R, RD_2)$ can be regarded each as a set of coordinates defining a vector

- *Shannon's Second Theorem* and genetic information transmission: Consistent with a typical communication system consisting of a source of information, a noisy channel and receptacle which receives the information, one can consider modeling the genetic information transmission as follows:

  1. *Source output*: Base sequence of the DNA in encoded form
  2. *Received message*: Amino-acid sequence in protein
  3. *Noisy channel*: Mechanics of protein synthesis involving transcription of the base sequence of DNA into mRNA and the translation of mRNA on the Ribosome.

According to *Shannon's Second Theorem*, despite the presence of noise, under certain conditions, it is possible to transmit the message without error and without loss of message rate, if the message is encoded appropriately at the source.

In biologistic language, Gatlin [8.5] translates Shannon's Second Theorem as follows: "It is possible, within limits, to increase the fidelity of the genetic message without loss of potential message namely, provided that the entropy variables change in the proper way, namely, by increasing D_2 at relatively constant D_1".

Vertebrates can accomplish such a source encoding. That is the reason they are referred to as "higher" organisms.

- *Information spaces*: Pertinent to genetic information, the information space is decided by the vector spaces of $(R,D_1) \Leftrightarrow (R,RD_1)$ and/or $(R,D_2) \Leftrightarrow (R,RD_2)$. In "DNA space", the RD_2 vector describes "the path by which the system is required to climb if it is to optimize the conflicting elements of variety *versus* reliability". It is an optimum path and Gatlin [8.5] suggests that evolution refers to the organisms ancestral to vertebrates climbing this path.

8.2 Genetic Code

The specific regions of information involved in making proteins in the DNA are constituted by a stretch of nucleodites or genes. These distinct regions of the DNA (or the genes) correspond to distinct bits of information. Thus, the stretch of DNA, that is, the trace of genes, is analogous to a string of alphabets constituting a message as indicated earlier. In the DNA, the nucleotide pairs correspond to genetic letters or alphabets and a set of such alphabets constitute a *gene*. The organized combination of genetic letters (namely, *A, T, G* and *C*) constitute a specific *genetic code*. Studies have revealed that genetic code is read as triplets of nucleodites. Triplets in DNA that correspond to amino acids are called *codons*. For example, four that correspond to amino acids are called *codons*. For example, four nucleodites in RNA, taken three at a time, can form 64 combinations of which 61 correspond to amino acids in protein; and the other 3 are stop (termination) codons.

Gene expression is the process by which the information in DNA is converted to protein. This process involves the "central dogma" of molecular biology specified by:

- *Transcription*: The cell uses the information in a particular gene to direct the formation of a single-stranded RNA.
- *Translation*: The mRNA uses the information it contains to direct the formation of a specific protein.

Gene expression is a controlled process. Gene regulators, called *repressors*, can block transcription so as to control (attenuate) the gene expression. In the repression process, the gene regulator protein blocks RNA synthesis. An opposing event, *gene activation*, is also possible via binding of special proteins called *entrances*.

There is also a feasibility of *DNA replication*. Appropriate base-pairing complementarily allows an exact copying of DNA leading to DNA replication. In this way, two DNA molecules may appear from one.

Occasional errors made in DNA replication would lead to *genetic mutation*. The mutation, when occurring may lead to an organism which is different from its parent. Such an altered organism is called a *mutant*. In information-theoretic point of view, mutation is a sudden change in the information content of the hereditary substance.

8.2.1 Biological cells, genetic mapping and genetic algorithms

The biological organization is built on the cellular architecture. The cell is essentially a membranewhich offers a physical (geometrical) boundary and provides an interface with its environment. Another characteristic of cells is the presence of genetic material organized into one or more chromosomes, each of which consists mainly of complex DNA.

Chromosonal DNA is an unbranded chain of nucleodite pairs and, as such, the genetic information in a chromosome represents a linear sequence of genes. The genetic information can be localized or "mapped" on a chromosome. Since the gene is a binary encoding parameter, a chromosome — an array of genes— can be regarded as a computer algorithm.

Each chromosome can be associated with a cost-function which represents the relative characteristic of that chromosome. Suppose there is a set of chromosomes with a high cardinality and each element of this set bears a cost-function. Based on the values of these cost-functions, chromosomes can be ranked in a descending order from the most-fit to the least fit. A truncation of this ranked list at the bottom eliminates the unwanted chromosomes, leaving the subset of superior species to survive.

The survived set of genes perform cross-over mating to reproduce enough chromosomes to offset the discarded chromosomes so as to keep the gene population constant. They map their genetic features in their offsprings. Some of the offspring could, however, be mutants which will be discarded in the ranking-by-cost-function procedure iterated.This iteration continues until a "solution" yielding a set of desirable chromosomes is realized.

The aforesaid process, involving selection, rejection, mutation, etc. which is present in biological systems can be mimicked to formulate a global numerical-optimization technique. That is, patterned after the natural processes of genetic recombination and evolution, global optimization procedures that can handle a large number of discrete parameters have been algorithmized as computer codes. Such algorithms are known as *genetic algorithms* (GAs). These constitute a subset of search-algorithms adopted conventionally in optimization problems in reference to cybernetic efforts in complex systems, as indicated in the following sections.

8.3 Search Algorithms

8.3.1 Cybernetics, complex systems and optimization

Cybenetics, in general, refers to the art of maneuvering complex systems through optimization strategies in order to achieve a purposeful objective [8.7]. Essentially, cybernetics deals with the human efforts pertinent to the endeavors of complex systems in order to realize an objective goal in the same manner as machines extend their capabilities to aid and steer the human hands in tooling a job. Cybernetic efforts, therefore, call for a comprehensive interplay of theoretical methods, analytical techniques and computational strategies so as to find an optimum combination of an assemblage that constitutes the complex system conceived to achieve a specified goal.

Accordingly, a computational method of ascertaining optimal (or suboptimal) input parametric options so as to realize a solution towards minimizing certain desired requirements (for example, cost), and/or maximizing certain other considerations (such as performance) is addressed in the following sections. In a complex system, it may be required that more than a single objective criterion be satisfied. Hence, the optimization pursued here is designed to meet a set of multiple objectives. A typical, triple-objective case study is considered and analyzed thereof. Relevant computation is based on a genetic search algorithm deployed in an artificial neural network (ANN) and cast in the information-theoretic (IT) plane.

In reference to complex systems, the associated cybernetics could be governed by anywhere from a few to several factors. A cohesive analysis and strategic evolution of such systems, therefore, are pertinent to the so-called "large systems" constituted by several interactive subsystems. Further, such a complex system represents an integrated unit enclaving a network of sub-units, each linked to one or more sub-units. It may cover a sizable spatial region and include both temporal and spatial stochasticity.

The cybernetics of complex systems has a *structure* constituted by the hardware of the constituting sub-units, a *connective configuration* of sub-units, and a *state* governed by the input-output response of the system. Such a state would also include the conditions of the system specified by: (i) Normal operating (static) conditions, (ii) abnormal or fault state(s), and (iii) transient or steady-state attributes in a given time-frame.

In the context of designing complex systems, it is therefore necessary to enunciate a strategy which includes cohesively a set of mixed, sub-unit characteristics which meet the near- and long-term performance of the system behavior in an efficient and reliable manner at the optimal cost-functional outlays. Hence, the objective of cybernetic efforts towards such a strategy is to realize an overall technique that maximizes (or optimizes), (i) desired/satisfactory system outputs, (ii) a robust performance, (iii) risk-tolerance in handling scenario with stochastical attributes and, (iv) offer minimal cost-functional characteristics.

Indicated in the following paragraphs is an optimization strategy based on a genetic algorithm applied to an artificial neural network in the information-theoretic plane. The optimization technique is posed as a paradigm to model and evaluate a complex system. The projected methodology is described with an illustrative numerical example.

A typical optimization pertinent to a complex system can be envisaged to maximize the system performance consistent with the stochastical variations in the environment of the system. Such performance optimization depends largely on the additions/subtractions of variables involved and changes in the system parameters intended towards dynamic updating.

Further, as mentioned earlier, the optimization may be aimed at minimizing certain cost commitments and/or maximizing certain other performance indices. In many situations, such cost-minimization and performance-maximization could have polemical influence on the system's endeavor towards achieving the said optimization(s). More so, a complex system may warrant a temporal (time-dependent) strategy to predict the futuristic system performance using the stochastical data collected over a stretch of time.

A method of implementing such a design, in general, would call for a dynamic programming method. Such dynamic evaluation can, however, be done to a meaningful extent only with sufficient historical data collected on the stochastical fluctuations and upon system performance.

In summary, optimization warranted in the design/evaluation of complex systems is conceived to realize optimal or sub-optimal cost outlays plus an acceptable level of overall performance with a multiple set of constraints imposed on the system. For example, in an industrial complex, more capital, operational and maintenance (O&M) expenses are warranted if a high degree of performance is anticipated. Typically, the three functional relations indicated in Fig. 8.5 can depict cost-factor versus performance index of a complex system.

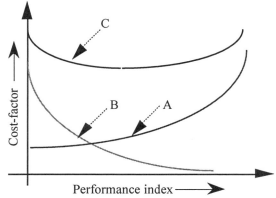

Fig. 8.5: Variation of cost against performance in a typical complex system

Cost-profiles:
    A: High-performance implementation
    B: Low- performance implementation
    C: Net cost versus performance

Curve A in Fig. 8.5 denotes an explicit, monotonic increase in cost when the system performance requirements are enhanced. If, on the other hand, a low performance is facilitated by the system, it would reflect implicitly an excess cost to meet and compensate for the consequences of low performance. This excess cost would diminish (Curve B) when the system performance is improved. Hence, the resultant (total) cost *versus* the level of performance (Curve C) would show a minimum.

Achieving the minimal cost with an acceptable extent of performance under a set of time-invariant, multiple constraints is the objective goal of the problem under discussion. This optimization can be realized by an optimally (or, at least suboptimally) selected set of input options from several available alternatives with multiple constraints imposed on the operation of the system. These input options and constraint sets are to be searched and selected by an optimization algorithm from among several, feasible and constrained alternatives.

The optimization technique described here is based on an information-theoretic-based genetic search algorithm implemented on an artificial neural network. The cost-functions used thereof, refer to meeting three objective criteria imposed at appropriate stages of the algorithm applied, both during training and prediction sessions of the neural network architecture. One of the cost-functions adopted is an Euclidean distance-measure and the other two are information-theoretic distance-measures. The following sub-section gives a description of a typical complex system and the related optimization problem with the associated cost-functions. The implementation considerations are presented subsequently.

8.3.2 Problem description

The optimization vis-á-vis the design aspects of a complex system, being considered here as an illustration, refers to embarking upon a scheme with the following parameters:

C_I: Value of the initial cost

C_p: Value of the ongoing (recurring) cost

K: Maximum number of time-units $(1 \leq k \leq K)$

L: Maximum number of subunits $(1 \leq \ell \leq L)$

S_V: Salvage value of each subunit

M: Maximum number of possible alternatives depicting finite number of design optimizations feasible $(1 \leq m \leq M)$

N: Maximum number of constraining factors in each feasible design optimized $(1 \leq n \leq N)$

There are three objectives posed to the designer. They are as follows:

(i) The first objective refers to minimizing the total cost, namely, F_{kmn} which includes the initial cost plus ongoing cost minus salvage value of subunits. That is,

$$F_{mn}(L, K) = \left[\sum_{\ell=1}^{L} \left\{ \sum_{k=1}^{K} \left[C_I \right]_{kmn} + \left[C_p \right]_{kmn} - \ell \times \left[S_v \right]_{kmn} + \left[C_c \right]_{kmn} \right\} \right]$$

(8.11)

where m = 1, 2, ..., M; n = 1, 2,..., N; and, C_cs depict the cost-factors resulting from the constraints imposed on the system. Thus, the first objective is to achieve F_{kmn} close to a desired target value, F_T.

(ii) The second objective is concerned with the subcosts buried in F_{kmn}. Suppose F_{kmn} itself is constituted by a set of r $(1 \leq r \leq R)$ subcosts, namely, $\{\Phi_r\}$ such that

$$F_{kmn} = g\left\{ \sum_{r=1}^{R} b_r \Phi_r(k, m, n) \right\}$$

(8.12)

where g represents a functional relation, which for simplicity, can be taken as a linear function so that g(x) = Gx and G is a constant of proportionality which can be set normalized to unity. Further, R depicts the maximum number of subcosts involved, $\Phi_r(\cdot)$ is the value of rth subcost and b_r is a binary coefficient, 0 or 1 weighting the rth subcost.

The subcosts indicated are usually the costs of achieving a specified performance level in each of the r independent subunits of the system which decide cumulatively the overall cost of the system. The value of the binary coefficient b_r being 1 or 0 denotes the importance of the cost of the rth subunit, namely, Φ_r in either contributing to or not influencing at all, the overall cost respectively.

The subcosts can be sequenced according to their diminishing importance. Ideally, an optimization strategy should take into account all the R subcosts involved. However, a specific complex system may not suffer unduly if at least one or two priority types of subcosts are omitted. Accordingly, a target binary-string of $\{b_{Tr}\}_{r=1,2,...R}$ can be constructed with one or more b_{Tr} values being zero as necessary. In other words, a Φ_r with a coefficient $b_{Tr} = 1$ specifies that the rth subcost is significant and should not be precluded. If, on the other hand, $b_{Tr} = 0$, this r^{th} subcost is small and the system envisaged can survive even without it. Such zero-valued binary coefficients should be placed towards the least significant bits of the bit-stream represented by $\{b_{Tr}\}$ so that the priority ranking envisaged in sequencing the subcosts is preserved. The second objective can, therefore, be specified in reference to the binary coefficients, $\{b_r\}$. The outcome of optimization should be such that the binary-string $\{b_r\}$ of the outcome should have a minimum distance when compared with a target-string $\{b_{Tr}\}$.

(iii) The third objective is pertinent to constructing $F_{mn}(K, L)$ meeting n constraints posed for the mth input alternative. These constraints may, for example, refer to the system organization and each can also be associated with a cost.

In Eqn. (8.11), the net influence of n constraints on F_{kmn} is specified by an additive term $[C_c]_{kmn}$. That is, C_c can be obtained by combining the cost due to n constraints, assumed to be independent. Written explicitly,

$$[C_c]_{kmn} = f\left\{\sum_{n=1}^{N} a_n [C_n(k,m)]\right\} \qquad (8.13)$$

where f represents a functional relation and in a simple case, f can be a linear function so that

$$[C_c]_{k,m,n} = D \sum_{n=1}^{N} a_n [C_n(k,m)] \qquad (8.14)$$

where D is a constant of proportionality which can be set normalized to unity. Further, a_ns denote weighting coefficients on the individual constraint-based cost involvements namely, C_ns. The set $\{a_n\}$ can again be a binary-string and C_ns can be sequenced in the order of their importance (from C_1 through C_N). In the corresponding bit-map* of a_ns, if the least significant bits are zeros, the associated constraint-induced costs are presumed to be negligible. The objective in relation to $[C_c]_{kmn}$ is that the bit-strings (a_ns) realized should be of some (stochastical) minimum distance with respect to a target string, a_{Tn}.

In short, the three objectives discussed above towards optimization can be written concisely as follows: Suppose $F_T(k, \ell, m, n)$ is the target value of revenue requirement in implementing the conceived planning, and,

$$F_T(k, \ell, m, n) = \sum_{r=1}^{R} b_{Tr} F_r(k, \ell, m, n) \qquad (8.15a)$$

The first objective function can be specified by the following Euclidean distance-measure:

$$\left| F_T(k, \ell, m, n) - F_{k\ell mn} \right| \le \epsilon_1 \qquad (8.15b)$$

where ϵ_1 is a permissible error. Alternatively, the Euclidean distance-measure can be replaced by a root mean-square error arising from the differences between $F_T(\cdot)$ and $F_{k\ell mn}$.

The bit string $\{b_{Tr}\}$ can be compared with $\{b_r\}$ via Hamming-distance [8.8] to predict the closeness of one against the other. Hamming-distance is obtained by performing a modulo-2 operation between $\{b_{Tr}\}$ and $\{b_r\}$ and counting the number of 1s in the result. Denoting the Hamming-distance pertinent to b_r and b_{Tr} as $Hd_b[.]$,

$$HD_b\left[\{b_T\} \text{ and } \{b_r\}\right] \le \epsilon_2 \qquad (8.16a)$$

specifies the second cost-function to be minimized in order to realize the optimization with a permissible error value of ϵ_2.

---

* *Bit-map*: A string of binary numbers. In discrete optimization problems, a set of bit-maps is used to represent conveniently a collection of information. In such a string of binary numbers, the bit value 1 may signify an affirmation of certain information and the bit value 0 may signify negation of certain information.

Now, comparing the binary strings $\{a_n\}$ and $\{a_{Tn}\}$ would yield another Hamming-distance, HD_a. Hence,

$$HD_a\left[\{a_{Tn}\} \text{ and } \{a_n\}\right] \leq \in_3 \qquad (8.16b)$$

refers to a third cost-function minimization criterion set by a choice of the error, $\in_3$.

Alternatively, an optimal bit-map of a_n can be realized by generating a bit-string of length n, where, as indicated earlier, its most-significant bit to the least-significant bit sequence represents the descending order of priority of the constraints involved. The optimality of bit-map is verified by checking a certain fitness value associated with the bit string. That is, for a given string generated randomly and used, if the fitness value surpasses a certain acceptable limit, the bit-map of the string is taken as the optimal representation of the constraints being imposed. A method of information-theoretics based fitness-check [8.21] is adopted in the present study and will be described later.

The triple-objective optimization described above is illustrated *via* a block schematic presented in Fig. 8.6.

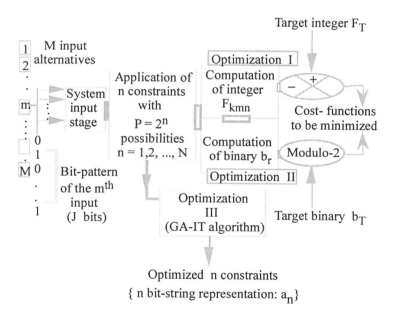

Fig. 8.6 : A triple-objective optimization of the complex system

Suppose the M input alternatives are such that each input set is constituted by J elements. The optimization task projected in Fig. 8.6 refers to deciding μ input options (out of the M possible alternative design sets) such that each of the μ input-options yields an output which satisfies all three objectives stipulated above. If any one of the objectives is not met, the search is pursued further via iterations until all the successful input options which yield results commensurate with the triple objectives envisaged are sieved out.

Optimization methods, as discussed above, are useful in general to identify a good set of optimal or suboptimal design solutions of a problem with a specified set of constraints within a reasonable amount of time. These methods should also accomodate a specified objective or a set of multiple objectives. The outcome of a particular search performed under specified constraints is compared with a desired solution (or set of targets) iteratively until the outcome represents an acceptable estimate of the desired objective(s). For the purpose of such a comparison, a functional relation, or a set of relations, is used to determine the closeness of the outcome(s) to the objective criterion or multiple criteria.

There are number of attractive optimization algorithms available in practice [8.9]. Genetic algorithm is a subset of these candidates. A few methods of traditional optimization, including the GA, are outlined below for comparison purposes. The functional relation indicated above is known as the *cost-function* or the *objective function*[3] of the search process.

8.3.3 Integer programming

Integer programming [8.9] refers to numerical methods which explore discrete search spaces in an attempt to find optimal solutions. A typical approach is the "hill-climbing" method. It begins with a random solution for a problem specified with a set of constraints and the cost-function is evaluated thereof. Then, a control parameter is changed to obtain a new value of the cost-function. If this new value of cost-function is better than the initial value, the corresponding control parameter is retained and the initial one is discarded. This process is iterated until the evaluated value of the cost-function shows no further improvement. This indicates that a local minimum is reached.

---

[3] *Cost-function/objective function*: In optimization problems, the outcome of a particular search performed under a set of specified constraints is compared with a desired solution iteratively until the outcome represents an acceptable estimate of the desired value. For the purpose of such comparison, a functional relation is used to determine the closeness of the outcome to the objective criterion. This functional relation is known as the *cost-function* or *objective function* of the search process.

8.3.4 Simulated annealing

Simulated annealing [8.10] optimizes a function by performing a directed random search based on statistical mechanics considerations. Similar to a metal attaining a low energy state when heated up and cooled slowly in a controlled fashion, a solution can be improved by perturbing it and then slowly reducing the perturbations. Initially, the probability of accepting a perturbation which worsens the solution would rather be high. However, as the perturbation size reduces, the said probability would diminish.

8.3.5 Hybrid method

The hybrid method refers to a mixed strategy in which two or more optimization methods can be combined. For example, integer programming and simulated annealing can be used together. Likewise, simulated annealing or interger programming can be used jointly with the GA.

8.3.6 Genetic algorithms

Genetic algorithms [8.11-8.14] were developed almost as a parallel strategy to simulated annealing. Their efficacy in optimization problems is comparable to that of simulated annealing. As discussed earlier, they are based on the mechanisms of *adaptation* in biology. Adaptation is the act or process of making modifications to an organism to increase its *fitness* to a particular environment. *Natural selection* is a mechanism by which adaptation occurs in nature. In natural selection, individual species within a population which are well-suited to their environment have a greater chance of producing offspring than individuals within the population that are less fit. Conceptually, parents with a high degree of environmental fitness will pass information to each of their offspring in terms of *genes*. Each offspring is, therefore, made up of *genetic information* derived from both parents.

Diversity, or the stochastical distance in the genetic information, is determined by the differences between the *genetic codes* (also known as *strings*) between the parents. Another mechanism that influences adaptation is the possible random *mutations* which cause the genes in an individual to change. This may alter the fitness of an individual to increase; thus improving its chances to participate in reproduction. If the random mutation causes the fitness to survive of an individual to decrease, then the probability of participation in the reproduction process by that individual would decrease according to natural selection. Consequently, the genetic alterations may not proliferate into subsequent generations.

Based on the underlying concepts of adaptation, GAs perform directed random searches through a given set of alternatives with the aim to find the 'best' option subjected to a given set of criteria of goodness. These criteria are required to be expressed in terms of either a single objective function or by a gamut of multiple objective criteria, commonly referred to as fitness functions. Further, GAs require that the set of alternatives to be searched through is finite. If it is required to apply them to an optimization problem where the requirement on the finite-size of alternatives is not saisfied, some alternatives should be discarded via predesign considerations on an ad hoc basis and only a finite size of alternatives should be used in the search. It is also necessary that the set of alternatives should be coded in strings of some specific (finite) length which consists of symbols from some finite alphabet. (for example, as a binary-string). These strings represent the chromosomes; and the symbols constituting the strings are the *genes*. The set of genes is known as the *gene pool*. In a biological sense, the chromosomes are the carriers of genetic material in the nucleus of a cell.

Genetic recombination and the evolution process provide the basement in structuring the genetic algorithms. The genes are the building blocks of genetic algorithms which facilitate a form of a directed random search based on natural selection as mentioned before. In a simple form, a gene can be a binary encoding of a parameter in a GA.

The genetic algorithm begins with a large list of random chromosomes. Every chromosome has an associated cost-function assigning a relative merit to that chromosome. The objective criteria of the optimization involved stipulate the algorithm for the cost-functions which can be evaluated for each chromosome. The chromosomes are then ranked from the most fit to the least fit according to their respective cost-functional values.

In respect to using GA as an optimum search method, the initial population of chromosomes refers to creating an arbitrary set of solutions with an objective to search for an optimum version of the set. In order to accomplish this, future generations are determined probabilistically by selecting fit-parents and combining them to bring forth elite children (or offspring). Selecting fit-parents corresponds to discarding unacceptable chromosomes and leaving the superior species—a subset of the original list. Hence, genes which survive become parents and by swapping some of their genetic materials, they produce a set of new offspring. The parents reproduce enough to offset the discarded chromosomes. Thus, the total number of chromosomes remains constant after each iteration. The offspring (children) may also undergo mutation which may cause small, random changes in a chromosome. Mutation thus enhances the emergence of more permutable offspring with distinct characteristics. Cost-functions are again evaluated for the offspring of the mutated chromosome, and the process is repeated. The algorithm stops after a specified number of iterations terminated by a stop-criterion, or when an acceptable (optimal or suboptimal) solution is obtained.

8.3.7 Paradigm of genetic algorithms

The procedural steps in conceiving a genetic algorithm are as follows. The algorithm begins by defining a chromosome as an array of parameter values to be optimized. If the chromosome has N_{par} parameters (pertinent to an N-dimensional optimization problem), given by $P_1, P_2, \ldots P_i \ldots, P_{Npar}$, then the *chromosome* is defined by the following array:

$$\text{Chromosome} = \{P_i\}; \quad i = 1, 2, \ldots, N_{par} \qquad (8.17)$$

Each chromosome has a cost-function found by evaluating a function, ϕ at P_1, $P_2, \ldots P_i \ldots, P_{Npar}$. The cost-function as mandated by the objective aspects of optimization, can be represented by a functional relation set forth by the elements of the chromosomal set, $\{P_i\}$. Or,

$$\text{Cost} = \phi[P_1, P_2, \ldots P_i \ldots, P_{Npar}] \qquad (8.18)$$

The parameters, P_n, can be either discrete or continuous. If the parameters are continuous, some limits need to be placed on them. Alternatively, they should be restricted to assume only a handful of possible values. One way to limit the parameters is to encode them in a binary sequence, such as:

$$q_n = \sum_{m=1}^{M_n} B_w[m]2^{1-m}Q_{max} \qquad (8.19)$$

where,

q_n	= Quantized version of P_n
M_n	= Number of quantization levels for q_n
B_w	= Array containing the binary sequence representing q_n
Q_{max}	= Largest quantization level
	$\Rightarrow$ Half the largest possible value of q_n

In general, the binary-encoded parameter, q_n, need not be mathematically related to P_n, as in Eqn. (8.19). Instead, q_n may just represent some value of P_n. For instance, if P_n represents eight values of a specific parameter (with a dimension/unit or dimensionless) then q_n has the following representation:

$$
\begin{array}{lllll}
q_n & = & 000 & \Rightarrow & 1 & \text{unit} \\
q_n & = & 001 & \Rightarrow & 2 & \text{units} \\
\cdot & & \cdot & & \cdot & \\
\cdot & & \cdot & & \cdot & \\
& \cdot & & \cdot & & \\
q_n & = & 111 & \Rightarrow & 8 & \text{units}
\end{array}
$$

$$(8.20)$$

The implementation of genetic algorithms described here only works with the binary encoding of the parameters, and not with the parameters themselves. Whenever the cost-function is evaluated, the chromosome must first be decoded. An example of binary-encoded chromosome that has N_{par} set of parameters each encoded with $N_{pbit} = 10$ bits, is given by:

Chromosomes = [111001001 0011011111 ... 0000101001]

$$q_1 \qquad\qquad q_1 \qquad\qquad\qquad q_1$$

$$1 \qquad\qquad 2 \qquad\qquad ... \ N_{par} \qquad\qquad (8.21)$$

Substituting this binary representation into Eqn. (8.19) yields an array of quantized versions of the parameters. This chromosome has a total of $N_{gbit} = N_{pbit} \times N_{par}$ bits.

Upon devising a scheme to encode and decode the parameters, a list of random chromosomes is generated. Each chromosome has an associated cost, calculated from the cost-function of Eqn. (8.18).

Table 8.1: A list of 8 random chromosomes and associated costs

#	Chromosome	Cost as per Eqn. (8.18)
1	011110010001	22
2	111001100101	17
3	111110001010	21
4	000101010111	128
5	001010100001	102
6	101010010101	16
7	111111110000	27
8	110011001100	33

Table 8.2: Ranking and selection of chromosomes

#	Chromosomes	Cost	Keep top 50 % ?	Keep costs less than 50 %?
6	101010010101	16	Yes	Yes
2	111001100101	17	Yes	Yes
3	111110001010	21	Yes	Yes
1	011110010001	22	Yes	Yes
			Discard	
7	111111110000	27		Yes
8	110011001100	33		Yes
				Discard
5	001010100001	102		
4	000101010111	128		

Suppose for a list of $N_{chro} = 8$ random chromosomes, the associated costs determined by Eqn. (8.18) could be as, for example, listed in Table 8.1. The next step in the algorithm is to rank the chromosomes from the best to worst. Assuming that a low cost is the preferred choice or the problem objective, the ranking of the chromosomes (with the cost functions as in Table 8.1) can be done as shown in Table 8.2.

Pursuant to ranking, a set of unacceptable chromosomes can be discarded. The criterion of acceptability is user-defined. Typically, the top N_x are kept (where N_x is even), and the bottom ($N_{chro} - N_x$) can be discarded. As an example, if 50% of the chromosomes are discarded, then chromosomes 1, 2, 3, and 6 are kept, while chromosomes 4, 5, 7, and 8 are discarded (Column 4 in Table 8.2).

An alternative strategy is to let the cost meet a specified level. As an example, should the cost be less than 50, then chromosomes 1, 2, 3, 6, 7, and 8 are kept, while chromosomes 4 and 5 are discarded (Column 5 in Table 8.2).

Suppose the top 50% costs are kept. After ranking and discarding the chromosomes, the next step is to pair the remaining 50%, that is, $N_{chro}/2$, chromosomes for mating. In general, any two chromosomes can mate as a stochastical choice. For computational convenience, some possible approaches are to pair the chromosomes from top to bottom of the list, pair them randomly, or pair them 1 with $N_{chro}/2$, 2 with ($N_{chro}/2) - 1$, etc. Once paired, new offspring are formed from the pair-swapping genetic material. The genetic algorithm may also use the *elitist* population concept in the reproduction strategy. That is, a certain percentage of each generation is reserved for the *elite class*. When creating children of the next generation, the best members of the current population will be copied directly into the elite class. As an example, suppose chromosome 6 is paired with 2, and 3 is paired with 1, from Table 8.2. Now, a random cross-over point is selected. The binary digits to the right of the cross-over point are between bits 5 and 6. The new chromosomes are formed as illustrated below (Fig. 8.7). After the surviving $N_{chro}/2$ chromosomes pair and mate, the list of $N_{chro}/2$ parents and $N_{chro}/2$ offspring results in a total of N_{chro} chromosomes the same number of chromosomes as at the start.

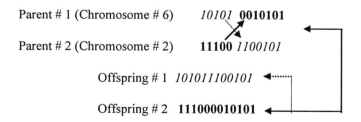

Fig. 8.7: Pair-swapping to yield a new offspring

At this juncture, random mutations may alter a small percentage of the bits in the list of chromosomes. For mutation, a bit is selected randomly from the total number of (N_{chro} x N_{gbit}) bits in all the listed chromosomes and implemented by changing a "1" to a "0" or vice versa. Increasing the number of mutations increases the freedom of the algorithm to search outside the current region of parameter space. This flexibility on the search-space becomes more important as the algorithm begins to focus on a particular solution. Typically, on the order of 1% of the bits may mutate per iteration. Mutations do not, however, occur on the final iteration.

At the termination of the mutation process, the costs associated with the offspring and mutated chromosomes are calculated and the process is repeated. The number of generations which evolve thereof, depends on whether an acceptable solution is reached; or, when a specified number of iterations is exceeded. After a while, all of the chromosomes and the associated costs may become the same except for those that are mutated. At this point, the algorithm should be stopped.

8.3.8 Genetic algorithm based optimization

In summary, there are essentially five components of a GA:

♦ Representation of solutions by chromosomes
♦ Initializing the population of chromosomes
♦ Establishing an *evaluation (fitness) function* to rate the solutions
♦ Prescribing a set of *genetic operators* to alter the composition of chromosomes during reproduction
♦ Setting appropriate parameters for the algorithm in terms of population size and applying the genetic operators.

Shown in Fig. 8.8 is a flow chart depicting the procedural aspects of optimization involving the GA pertinent to a specific problem. The procedure commences with assigning a set of parameters specific to a given problem. The next step involves generating a random chromosome population. This randomly chosen population is then subjected to an evaluation criterion so as to decide on the fitness of each chromosome. The chromosomes are ranked according to their fitness values evaluated. After inferior chromosomes are rejected, the rest are allowed to mate via pair-swapping. If the mating results in an elite chromosomal string, the procedure is stopped. Otherwise, a slight mutation is introduced to realize a new population, which is then subjected to fitness evaluation and so on. The procedure may also be terminated if a certain stop criterion has been met.

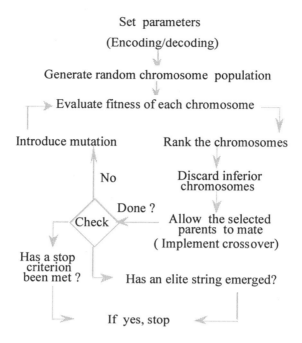

Fig. 8.8: A GA-based optimization procedure

8.4 Simple Genetic Algorithm (SGA)

Genetic algorithms were formulated by John Holland [8.15, 8.16] and others at the University of Michigan. Relevant studies were focused on abstracting and explaining the adaptive processes of natural systems so as to design artificial system software. Holland demonstrated that genetic algorithms can search for complex fitness spaces in order to find solutions to a variety of constrained optimization problems. The GA has the ability to find a good solution in complex and noisy environments, also known as *robustness*. This corresponds to an organism surviving in many different types of environments.

As mentioned earlier, the robustness of genetic algorithms is greater than that of other search techniques, such as gradient and continuity-based search methods. In general, calculus-based search techniques may not find solutions in noisy environments with discontinuous functional attributes, such as the one depicted in Fig. 8.9. Whereas in such environments, GAs have shown better performance.

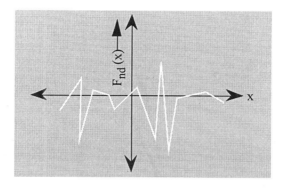

Fig. 8.9: Noisy and discontinuous function: F_{nd} (.)

As indicated before, a genetic algorithm has a population made up of several different members (chromosomes). These members of the gene pool compete with each other for the ability to reproduce. Such an ability to reproduce is determined by the fitness function. In this competitive reproduction exercise, there are several steps that must be completed. These steps include the following: Reproduction and fitness attribution, crossover, and mutation, as described briefly below.

8.4.1 Fitness attribution and reproduction

A type of genetic algorithm commonly known as the *Simple Genetic Algorithm* (SGA) uses an encoding scheme in order to create a string (usually in ASCII binary format) of a specified length to represent each member of the generic population distinctly. In the context of the present optimization problem, a population string refers to the bit-pattern a_n considered earlier.

Fitness of a population string can be decided by comparing the population string a_n with the corresponding target-string, a_{Tn}, and the number of bit-errors (BE) in a_n vis-à-vis a_{Tn} is determined. The *fitness-factor* $(FF)_p$ of pth-string is then evaluated as follows: $(FF)_p = (BE/TB)_p /[\_{p = 1,2, ..., CC} (BE/TB)_p]$ where TB refers to the total number of bits in the string and CC is the number of strings—a global population of chromosomes considered.

The *reproduction* strategy is based on a simple, random (parent) selection mechanism which determines those strings that get to reproduce. As stated earlier, the parent-selection strategy adopted shoud offer more reproductive chances, on the whole, to those population members that are the most fit. That is, parent-selection corresponds to embarking upon a pair capable of yielding a child-string (or child-*schema*) which correlates well with strings having high fitness, and will occur often, 'like the genes or *alleles* of a positive biological survival trait". A common method of implementing this is based on the so-called *roulette-wheel selection* principle [8.14] described below.

The randomness associated with a roulette-wheel, in general, corresponds to a uniform probability distribution in which the occurrences of every pointer-position on the wheel in a spin have equal probabilities. Thus, there exists a statistical attribution of equally likely occurrence of pointer-positions on a roulette wheel. In using the roulette-wheel principle towards parent-selection, however, a weighting strategy is followed. The occurrences of strings can be weighted by a "prorated pie-chart" attribute introduced deliberately to the pointer-positions on a roulette wheel (Fig. 8.10). In other words, these pies are allocated to the population members with each slice proportional to the member's fitness. Selection of a population member to be a parent can be viewed as a spin of the wheel, with the winning population member being the one in whose slice the roulette spinner ends up. The weighting or prorating of the pies is set proportional to the fitness-factor of a member in the selection process of the GA adopted, as described below.

Consider a hypothetical set of CC = 10 chromosomes (indexed as #1, #2, ..., #10) representing a population group (strings) and each has a fitness-factor obtained over a set of evaluations, say, 90. Table 8.3a depicts the population/string set and the average of the fitness-factor for each string obtained over 100 evaluations.

Table 8.3a: A set of string population and the associated fitness values

String #	Fitness value	Cumulative fitness
1	9.6	9.6
2	10.0	19.6
3	8.0	27.6
4	9.8	37.4
5	7.5	44.9
6	10.3	55.2
7	8.6	63.8
8	8.8	72.6
9	9.0	85.6
10	8.4	90.0

Further, shown in Table 8.3b is a row of 10 random numbers picked from a random number generation in the interval 0 to 90. Now, ten weighted pies on the roulette wheel can be constructed each corresponding to the indexed string in Table 8.3a which has a running total of fitness-factor as listed in Table. 8.3a greater than or equal to the random number generated as presented in Table 8.3b. That is, the ten pie-sizes on the roulette wheel in Fig. 8.10 are prorated representations of the random numbers in Table 8.3b.

360

The pairs of strings which will be allowed to crossover towards reproduction (as described below) are selected from Table 8.3b consistent with the criterion set on the minimum fitness value, say, as 25. The other strings will be rejected and will be replenished with those having higher fitness values emerging from the crossover between the selected pairs.

Thus, the parent-selection *via* roulette wheel technique described above the promotes the reproduction of the fittest population members by biasing each member's chance of selection in accordance with its evaluation.

Table 8.3b: A set of chosen string population using the weighted roulette wheel

Random #	40	10	56	20	88	65	73	90	28	45
Chosen String #	4	1*	6	2*	9	7	8	10	3	5

[*Note: These two strings will not be selected for reproduction since their fitness values are below the minimum fitness value chosen (25)]

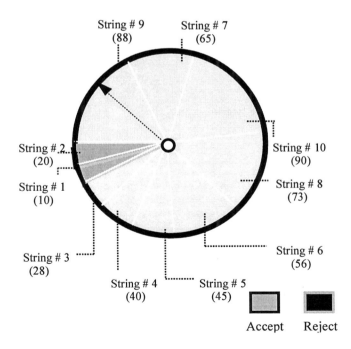

Fig. 8.10: Roulette wheel with weighted "pies" and assignment of strings to each pie

8.4.2 Crossover

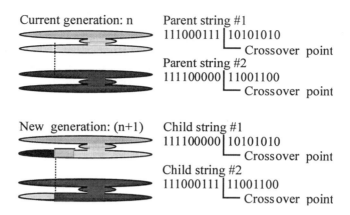

Current generation: n   Parent string #1
111000111 | 10101010
└─ Crossover point
Parent string #2
111100000 | 11001100
└─ Crossover point

New generation: (n+1)   Child string #1
111100000 | 10101010
└─ Crossover point
Child string #2
111000111 | 11001100
└─ Crossover point

Fig. 8.11: An example of simple, single-point crossover

As stated earlier, crossover is the way in which strings exchange information during reproduction. It can be based on a random selection of crossover points for each of two parents. That is, an exchange of genetic information between a pair of parent strings (via crossover) can be set to occur at certain selective crossover points in order to generate two new "children" strings.

Indicated in Fig. 8.11, is a simple crossover facilitated by selecting a single crossover point. The function of the crossover is to cause chromsomes (strings) created during reproduction to differ from their parents.

8.4.3 Mutation

In effecting the mutation, bits are selected randomly from each string in the population. Changes are made as depicted in Figure 8.12 at these randomly selected bits in each string depending on a given probability of mutation, usually of a very small value. The bit selected for mutation is modified from 1 to 0 and *vice versa*.

Generation: n
Selected string #1
Selected bit: 5
Bit-map of string #1: 01101011011000100101
Generation: (n+1)
Bit-map of string #1: 01101011011000000101

Fig. 8.12: An example of mutating a bit-map of a selected string

362

In short, a genetic algorithm has a population made up of several different members. These members compete with each other for the ability to reproduce. Such an ability to reproduce is determined by a *fitness function*. In order to reproduce, there are several steps that must be completed and these steps include: fitness evaluation, reproduction, crossover, and mutation.

8.5 Genetic Algorithms and Neural Networks

Genetic algorithms thus essentially represent search algorithms useful in optimization strategies. They are based on the mechanics of natural genetics and mimic the processes in natural evolution.

GAs have also been adopted successfully in training the artificial neural networks towards synthesizing appropriate network structures, as well as to optimize the values of learning parameters.

Artificial neural networks were developed as paradigms to represent biologically motivated approaches to machine learning. The relevance of using genetic algorithms and various ways of applying them appropriately in the study of neural networks came into being as alternative strategies to search for new avenues in neural network designs, explore the space of possible neural network architectures, and uncover the feasibilities of building appropriate theories and modelings to cast the artificial neural networks as images of real neuron interconnections.

Specifically, genetic algorithms have been applied to the problem of neural network design in respect to the following:

- To train a network of known architecture
- To apply constraints to the connection-matrix of networks trained *via* backpropagation
- To discover the size, structure and learning parameters of a network trained with genetic algorithms

Pertinent to the above, the role played by GA is essentially a strategy set forth in search of an optimum performance of the neural network. The practical aspects of using genetic algorithms vis-à-vis ANNs are, however, restricted to networks with a limited number of neural units. Otherwise, the search-space may increase exponentially with the size of the network. Nevertheless, a variety of neural network optimization problems using the genetic algorithms have been developed.

8.5.1 Neural networks: A review

a) Biological neural networks: In human beings, the neural complex is embodied by the brain, the spinal cord, and a massively interconnected web of neurons. The neural complex has the function of processing a vast amount of information, brought to it in the form of electrical impulses gathered from the senses (perceptive responses) such as smell, sight, hearing and touch. This processing enables the neural complex to make decisions based on available information.

The neural complex is made up of about 10^{10} individual cells known as neurons and about 10^{14} interconnections between them. The interconnections are known as *synapses* and can range in number between 1000 to 10000. Each biological neuron has one or more outputs called axons and many inputs called *dendrites* (Fig. 8.13). The interconnections between the neurons can mediate *excitatory* and *inhibitory* effects. An excitatory connection induces the neuron to fire, while an inhibitory connection tells the neuron not to fire. The conglomeration of inputs to a neuron determine whether the neuron should fire or not.

b) Artificial neural networks: Artificial neural networks are modeled after the neural complex present in most living organisms. Artificial neural networks were first modeled mathematically on the basis of a logical (0 and 1) representation of the neurocellular activity by McCulloch and Pitts [8.17]. That is, the McCulloch-Pitts model is a two-state device which corresponds to two possible output states. Similar to implementing digital computers, it is convenient to represent these two possible states of a neuron in the binary format as 1 or 0 (Fig. 8.14). The output of the ANN is determined by a set threshold.

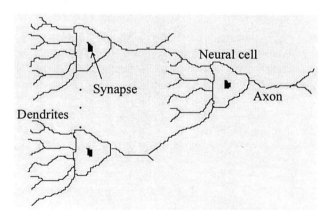

Fig. 8.13: Biological neural network

364

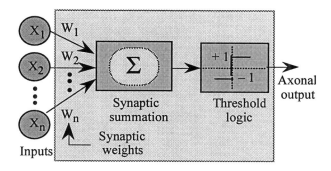

Fig. 8.14: McCulloch-Pitts' model of a neuron

If the weighted sum of inputs is greater than the threshold, then the output is declared as 1, otherwise it is 0. A complex issue in neural networks is that of determining the optimum set of interconnection weights and the architecture of interconnections between neurons to produce an optimum output for a given set of inputs. The extent of the correctness of the output decides the quality of the performance of the network.

A neural network is composed of many neurons, which due to their interconnections and the associated weights, can produce an optimum response to a set of inputs (Eqn. (8.15)). The inputs to the network are received by a set of neurons, usually known as the sensory array. The set of neurons that process the input information is called an associative array. Once the information is processed, an output is produced by a single output neuron (Fig. 8.15). Consider a model where the jth neuron has several inputs X_i each of which is multiplied in turn by an interconnection weight, W_{ij}. The sum of the products of $(X_i W_{ij})$ is then used to determine if the neuron should produce an output according to a specific threshold level. This output is given by

$$NET_j = \sum_{i}^{n} W_{ij} X_i \qquad (8.22)$$

The associative array is normally called the *hidden layer* because it is not visible at the inputs or at the outputs. A neural network composed of a single layer of artificial neurons connected by weights to a set of inputs is known as a *perceptron* [8.18].

8.5.2 Neural network training algorithms

Neural network learning has two basic forms: *supervised* and *unsupervised*. In supervised learning, the output pattern and a reference pattern, known as the teacher pattern, are compared. The most popular method of supervised learning is the *backpropagation* method [8.18]. In the backpropagation technique, by comparing the patterns of the output and that of a supervising reference, the neural weights are adjusted to train the network. That is, by comparing an output O_j with an expected output T_j (known as the *teacher value* or the *supervisory reference*) for a particular input, a quadratic cost-function **E** (**W**) is obtained as follows:

$$E_{wN}(\mathbf{W}) = (1/2) \sum_{n=0}^{N-1} [T_{N_j} - O_{N_j}(\mathbf{W})]^2 \tag{8.23}$$

$$O_j = f_{Sj}(\text{NET}_j) \tag{8.24a}$$

$$\delta_j = (T_j - O_j) f'_{Si}(\text{NET}_j) \tag{8.24b}$$

where $f_{Sj}(.)$ refers to a nonlinear *activation function* and the prime denotes differentiation with respect to the argument.

The network applies δ_i and a *learning rate variable* η to modify each weight in the weight array (Eqn. (8.25)) via backpropagation. The neural network then uses the modified weights, as given in Eqn. (8.25), and delivers an output. A corresponding updated error value is then evaluated. This procedure continues until the set of weights in the neural network produces a desired output with the error tending to a minimum value. In each iteration of this procedure, the average error diminishes indicating a convergence. If the weight change in each iteration is proportional to the error itself, then it leads the system to settle at a stable-state, with the error being sufficiently small. In neural dynamics, this refers to reaching a *basin of attraction:*

$$\Delta W_{ij} = \eta O_i d_j \tag{8.25}$$

In unsupervised learning (*Hebbian learning*), the network is trained with a set of input patterns but without a reference (the teacher) pattern available for comparison. The network learns by comparing the values of the input and output at each neuron. The strength of the neural weight will increase according to the product of the input and output vectors. There are also other considerations involved, such as the self-weight (W_{ii}) elements of the weight array need to be zero. Otherwise, the implementation of the learning algorithm can suffer chaotic oscillations. In addition, the W_{ij} storage matrix must be symmetric. Precisely, the Hebbian learning algorithm can be stated as follows: Given a set of inputs {**v**} and a set of outputs {**u**} each weight can be computed by calculating the sum of their product over all states {**s**}. This procedure can be described by the following relation:

$$W_{ij} = 1 \qquad \text{if} \quad \sum_s v_i^s u_i^s > 0$$
$$= 0 \quad \text{otherwise} \qquad\qquad\qquad (8.26)$$

8.5.3 Application of genetic algorithms to artificial neural networks

An automated mechanism of obtaining optimum performance in ANNs refers to a method of using genetic algorithms to optimize the interconnection weights and interconnections for a set of neurons. That is, each member in the population is regarded as a neural architecture (a collection of interconnected neurons) that will compete with other neural network architectures so that the genetic algorithm will find the fittest neural network for a given problem.

For example, a neural network with the supervised learning algorithm of backpropagation can be embedded into a genetic algorithm as follows. A neural network computes its output value for a given set of inputs and then uses the backpropagation learning algorithm to modify the weights in the ANN. A fast convergent technique (such as, the *conjugate-gradient method*) can be used in the backpropagation towards weight-adjustment. The inputs to the ANN characterize the different possible realizations and the output characterizes the objectives and constraints of the optimization pertinent to the problem and specified *via* the value(s) of cost-function(s). By presenting the network at its input, a set of example realizations/patterns, the supervised learning algorithm with backpropagation facilitates "training" of the network towards recognizing and predicting the extent to which an unknown pattern (given at its input) resembles the characteristic features of a desired pattern. The trained network stores these desired features through the weights of its interconnections in its learning/training phase. Knowledge gained by the network during the learning phase is encoded in the interconnecting weights. Having trained the network, the task is to find/search optimal patterns among possible realizations which, when presented at the input, yield an output meeting the constraints specified *via* cost-function criteria.

This search can be effected by a SGA. That is, a SGA can be used to generate/simulate a set of possible realizations which are then presented to the trained network. In other words, the network architecture is used to encode each member of the GA. For each encoding, the network predicts the fitness, as dictated by the cost-function, and hence, the GA proceeds to determine the fittest string among the various alternatives that corresponds to the pattern memorized (stored) in the learning phase. If an optimal solution is not found with the input set presented, the GA is geared to produce a new population *via* crossovers and mutations as indicated earlier. The SGA approach indicated here can be generalized for various ANN strategies with different learning methods used conventionally.

In summary, knowledge acquisition by ANNs is accomplished generally in one of the three ways:

♦ *Unsupervised learning*: A self-organizational strategy in which the learning does not rely on the ANN being provided with a desired response to the input or a performance score of the network's response. Instead, the ANN (self)-organizes its knowledge internally into categories which appear to have associated or collective properties.

♦ *Reinforcement learning*: This relies on a "teacher" grading the network's output responses to the training patterns. A high grade results in synaptic weight reinforcement, a low grade results in an adjustment of the weights to determine whether the grade can be improved.

♦ *Supervised learning*: Here, the network is provided with the desired response to the training input patterns.

There are four major ANN learning algorithms properly used:

1. *Hebbian* and *non-Hebbian learning algorithms:* This corresponds to the unsupervised learning method. Hebbian law suggests that if neuron A consistently stimulates neuron B while B is generating an output signal, the weight between A and B will increase in magnitude. That is, during the training of the ANN, neuron B becomes more sensitive to the stimulus applied to neuron A. This learning is handicapped by: (i) unbounded increase of the weights; and (ii) it does not permit negative (inhibitory) weights between neurons. There are also improved versions of a simple Hebbian learning algorithm.

2. *Widrow-Hoff learning algorithm*: This is the first generation of the supervised learning method implemented on ANNs. Here, the errors in the neural network are minimized using *steepest-descent optimization* and an added bias weight.

3. *Perceptron learning algorithm*: This is based on applying a linear threshold logic concept. Under this learning algorithm, the output of the perceptron is 1 if the value of the weighted sum of the inputs exceeds a certain threshold; otherwise it is 0. If the desired output response is 1 (0) and the actual output is also 1 (0), no changes are made to the synaptic weights of the inputs. When desired output is 1 and the actual output is 0, some adjustment (increase) on the weights of those inputs of value 1 is done. If the desired output response is 0 and the actual output is 1, adjustment to decrease the weights of those inputs of value 1 is carried out. While perceptron learning treats Boolean signals, Widrow-Hopf learning handles non-Boolean signals.

4. *Backpropagation learning*: This refers to recurrent networks with feedback connections or loops described briefly in the previous

subsection. In this learning method, the error ε perceived at the output of the network in respect to the actual output versus a desired (teacher) value is propagated backwards through the network to update the interconnection weights. This approach differs from the perceptron learning and/or Widrow-Hoff method in the type of threshold or activation function used. Backpropagation learning requires a differentiable function such as a sigmoidal nonlinear activation function.The other two methods employ hard-limiting thresholds. The backpropagation algorithm makes use of robust optimization to solve an unconstrained nonlinear optimization.

A cost-function $\mathbf{E}(\mathbf{W})$ is constructed by normalizing the sum of the output errors over the entire training set. That is,

$$\mathbf{E}(\mathbf{W}) = (1/N) \sum_{N=0}^{N-1} \varepsilon_N \qquad (8.27)$$

where N is the number of patterns in the training set and ε_N is the output error for the Nth training pattern.The function $\mathbf{E}(\mathbf{W})$ represents the objective function which is minimized with respect to the independent variable $\mathbf{W}$ depicting the vector of the network's weights. It is specified as a mean-square error function [8.18] explicitly as indicated earlier (Eqn. (8.23)). Alternatively, information-theoretics-based distance-measure [8.19,8.20], namely, maximum entropy and/or minimum entropy types, are specified in lieu of the quadratic cost-function as discussed in Chapters 4 and 6. The information-theoretic criteria have more impact on the weight changes than the squared-difference criterion when the activation is close to 0 and the target activation is 1 or vice versa.

The conventional method of implementing an ANN with the inclusion of GA is as follows:

- Conceptualization and formulation of the optimization problem: Specification of the contraints and the objective function
- Incorporation of the optimization problem in components that can be represented by a particular type of ANN
- Training the ANN with a set of realizations by presenting them as the inputs to the network; that is, altering the architecture to yield maximum performance of the ANN
- Generating and searching through possible realizations *via* GA to find optimal realizations. The trained network is used to predict the value of objective function, dictated by the constraints of the problem, of the possible combinatorial realizations.

The use of GAs in ANN applications with different learning schemes for a variety of optimization problems is well known. However, application of information-theoretic concepts in such problems has been addressed rarely. Therefore, indicated in the following sections are the considerations and implementation methodologies to blend the principles of information theory into the ANN-based optimization problems where the genetic search algorithms are deployed. The salient aspects of the strategy pursued are as follows:

♦ The ANN architecture uses a supervisory GA to do a genetic search of optimal neural inter-connection weights

♦ In the GA indicated, reproduction is facilitated using the concepts of information theoretics

♦ Further, fitness determination is done via the *Hamming distance* principle.

8.6 Information-Theoretics of Genetic Selection Algorithm

8.6.1 Genetic information

The performance of genetic algorithms, like any other learning system, depends crucially on the informatic representation of the task involved. Mostly, the efforts in using genetic algorithms refer to how a task is represented in the form of genetic algorithms. Less is done to generate and exploit the informatic structure vis-à-vis the task involved and match it with that of a genetic algorithm.

Genetic algorithms are based on evolutionary processes on a low stratum and have been considered as a tool in the optimization of tasks which pose a tendency towards evolution. As mentioned before, they have seldom, however, been used in representing the information-theoretics of the tasks and the optimization protocols involved.

There are theories on genetic algorithm (such as *demand theory*) which deal with information conservation pertinent to give regulatory (positive and negative) processes as decided by the mutations involved. Again, they have not, per se, been considered in the information-theoretic domain in order to describe the associated evolutionary considerations *vis-a-vis* the entropy of the given regulatory process.

All biological activities involve informational perspectives. Life itself, for example, "may be defined operationally as an information processing system—a structural hierarchy of functioning units—that has acquired through evolution the ability to store and process the information necessary for its own reproduction". The aforesaid definition is relative to the environment of the life systems and the biotic components involved. Hence, every genetic functional change can be associated with certain information-

370

content. The uncertainty associated with such changes should, therefore, depict the notional entropy of the life complex system.

Genetic information is the key word that links the steps involved in the gene functions that are fused into the genetic algorithms. The hierarchy of instructions or programs built in the natural construction, growth and mutational aspects of living systems indicate a master code of genetics with an associated-entropy in Shannon's sense. Even at a primitive level, how much information is stored in the base sequence of a given DNA structure becomes an inquisitive query. A body of mathematical knowledge and probabilistic attributes that goes with the stochasticity of DNA changes by brute force the views of Shannon, Weaver [8.1] and Khinchin [8.2] to be embraced by the algorithms built on genes.

In view of the above considerations, it is attempted here to translate the existing concepts of genetic algorithms to the information-theoretic space and use the resulting genetic operators as optimization tools in training neural networks. Representing the associated neural input-to-output characteristics in transinformational perspectives is also attempted. Blending genetic algorithms and neural functional attributes in the information-theoretic plane is a conceptual experiment which lays the carpet for this chapter. The relation between entropy and information considered cohesively in the earlier chapters, and reviewed in an earlier section, renders an exploration of the implications on bioinformatic *versus* entropy paradigms. In the following sections, the logical considerations and methods of approach in viewing genetic algorithms in the information-theoretic framework are detailed, Also, the use of such information-theoretic based genetic algorithms in neural network applications is indicated *via* simulations [8.21].

8.6.2 Use of IT plus GA in neural networks

If genetic algorithms are cast in the information-theoretic plane, will the resulting format of the formulations facilitate a set of "neural algorithms" representing the informational organization of the nervous system?

An answer to the above query can be sought by first observing the success of genetic algorithms which have already been deployed and specified in a noninformation-theoretic perspective and applied to train the neural networks. Though genetic algorithms have been demonstrated successfully as optimization schedules for neural architectures, there are no available works in which GAs have been presented in the information-theoretic framework. Hence, no neural training has been attempted in the information-theoretic plane with genetic algorithms as the means of optimization search towards the global minimum.

In the following paragraphs a maiden attempt is, therefore, made to study the genetic algorithms in the information-theoretic plane and use the transformed algorithms to achieve neural network training and optimizations.

Relevant experiments involve developing a test ANN wherein the optimization can be tried with different sets of control parameters. Such control parameters include an entity based on the concept of entropy as conceived in the field of information theory. That is, for example, the mutual entropy or information-distance (Kullback-Leibler-Jensen distance measure [8.19, 8.20]) between a pair of genetic candidates can be considered in the reproduction process of the genetic algorithm and adopted as a *selection-constraint* parameter. In using this method, proper parameter selection is mandated to realize an optimal network performance. This concept, as described, could be useful for neural network based pattern-recognition strategies. The essence of this method relies on fusing the concepts of information-theoretics and the principles of genetic algorithms as a combined paradigm for artificial neural network optimization problems [8.21].

Commesurate with the use of information-theoretics plus genetic algorithms for neural network optimizations, the two possible avenues are as follows:

⇒ Developing an ANN with a *feedforward topology* (such as *Hopfield's associative memory model*), and introducing weight adjustments on the basis of the GA. Further, the IT principles are used to pick the elitist population for reproduction, as well as to decide the statistical distance between the ANN predictions and the target values. In this supervisory mode of operation, the ANN has an heteroassociative property. That is, an arbitrary set of input patterns are paired with another set of output patterns and the output pattern, in turn is compared with a target to elucidate the objective criterion of the problem.

⇒ Developing and training an artificial neural network with a backpropagation architecture towards optimizing the connection-weights and using the genetic algorithm so as to generate a new set of binary inputs or *bit-map* for presentation as inputs to the trained network so as to establish their relative fitness. Under a supervisory mode of operation, the cost-function to be optimized can be specified by any one of the host of minimum or maximum entropy functions described in Chapters 4 and 6. These functions implicitly correpond to information-theoretic measures.

In reference to the above implementations, the use of IT considerations, in essence, come into play explicitly while envisaging the following:

◊ Introducing information-theoretic concepts into the GA by adopting a genetic selection method towards the genetic reproduction process, that is, in the crossover strategies

◊   Appropriately implementing an information-theoretics based technique in the training algorithm by using the so-called "distance measure" as the cost-function to decide the relative *fitness* of the outcomes of the ANN.

8.6.3 Information-theoretics aspect of the GAs for ANN training

As discussed earlier, a conventional method of automating neural network architectures is to combine two adaptive processes: Genetic search through a selected network architecture space, and implementing backpropagation learning so as to evaluate the selected architecture. For example, cycles of learning in individual strings can be nested within cycles of evolution in the population as observed in biological systems. Each learning cycle presents the neural network, an instantiation of a particular network architecture with the set of input-output pairs defining the task envisaged. The backpropagation learning algorithm then compares the network's actual outputs with the desired teacher values, and modifies the network's connection weights so that it performs the desired input/output mapping task more accurately. Each evolution cycle processes one population of network designs according to their associated fitness values, computed during the learning cycles, and yields an offspring population of more highly adapted network designs.

Unlike this traditional methodology, the method of neural learning indicated here refers to a supervised form of learning combining GA and information-theoretic principles. It is a supervised form of ANN learning in the sense that the ANN is presented with both inputs and a desired (target) output. The neural network computes the output for a given set of input vectors and compares this output with a target value. This comparison can be made on a quadratic or difference error basis. Alternatively, a statistical distance metric can be adopted, if the output and the target being compared are stochastical in nature and are represented by their probability distributions. Any difference observed thereof, is normalized and stored.

The GA can be brought into play to use these stored (normalized) values of the difference (error) to compute the fitness for each string representation of the ANN for the set of inputs presented. In the event that the objectiveness is not met with as indicated by the computed value of fitness, a new generation of strings is realized through the GA by performing crossover and mutation. Again, in the crossover phase, the information theory can be used to verify the extent of stochastical difference (distance) between each pair of strings intended for reproducing. If the strings are "informationally" far off, that is, if their entropy distance is above a certain tolerable distance value, they are discarded. Only those strings which meet the minimum distance criterion are retained. The new (retained) set of strings are then stored back into the ANN as weights creating a new architecture thereof, and the procedure is iterated [8.21].

The method described above, is consistent with the techniques of a so-called *feedforward neural network* as will be introduced later. In its implementation, the system also uses a sigmoidal (logistic) nonlinearity (Fig. 8.15) as the neuronal activation function. In using this activation function, a threshold value is specified which must be surpassed in order for a neuronal output to be produced.

In its implementation, the neural network uses a collection of neurons and each neuron yields an output when its weighted sum of inputs exceeds a certain threshold. A logistic nonlinearity is used as an activation function to decide the bounded output for a given weighted sum of inputs to a neuron.

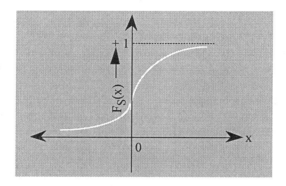

Fig. 8.15: Sigmoidal logistic function: $F_s(x)$
(Typical $F_s(x)$: $1/[1 + \exp(-x/T)]$ whereT is the pseudotemperature)

Typically, a sigmoidal function such as $F_s(x) = 1/[1 + \exp(-x/T)]$ or its variations [8.3] can be used as the nonlinear activation function. Here, T represents the "pseudotemperature" which governs the slope of the activation function chosen. By using an activation function, a threshold value is imposed which must be surpassed in order that a neuronal output is realized.

8.7 A Test ANN Architecture Deploying GA and IT Concepts

GAs have been adopted successfully to train the artificial neural networks (ANNs) towards synthesizing (configuring) appropriate network structures, as well as optimizing the values of learning parameters. Originally, artificial neural networks were developed as paradigms to represent biologically motivated approaches to machine-learning. The relevance of using genetic algorithms and various ways of applying them appropriately in neural networks came into being as alternatives as described earlier in this chapter.

Specifically, genetic algorithms have been applied to the problem of neural network design in respect to the following:

- For generating a number of alternatives or example patterns to be used as inputs to train a network of known architecture
- To apply constraints to the connection-matrix of networks trained *via* backpropagation using GA based weight generations
- To identify the size, structure and learning parameters of a network trained with genetic algorithms

8.7.1 Test ANN for simulation studies

The feedforward topology considered here for simulation studies does not require backpropagation; there are no feedback or recurrent connections. Connections through weights extend from the outputs of neurons, source nodes, to inputs of other neurons, destination nodes. This feedforward architecture enables mapping of an input pattern into an output pattern.

Introducing a nonlinear matrix operator Γ, the mapping of input space $\mathbf{x}$ to output space $\mathbf{o}$ implemented by the network can be expressed as: $\mathbf{o} = \Gamma[\mathbf{Wx}]$, where $\mathbf{W}$ is the weight matrix (connection matrix). Even though the feedforward network has no explicit feedback connection, when x is mapped into o, the ouput values are compared with the target's information, which provides the desired output value, and the corresponding error-signal is employed to let the the network's weights "adapt" themselves iteratively *via* genetic algorithmic steps until the output is classified to conform to a desired data.

The knowledge required to map input patterns into an appropriate classification at the output is thus embodied by the interconnection weights. Initially the weights appropriate to a given problem domain are, unknown. Until a set of applicable weights is found, the network has no ability to deal with the problem to be solved. The process of finding a useful set of weights refers to *training* as indicated before. Training begins with a training set consisting of specimen (teacher) inputs with associated outputs that represent a correct classification or pattern. Training the network involves moving from the training set of neurons placed "visibly" at the input, to a set of interconnection weights which correctly "classifies" the training set vectors, at least to within some defined error limits, in yielding desired output patterns. In effect, the network learns what the training set has to teach it. If the training set refers to "good examples" and the training algorithm is effective, the network, after learning from those examples (or getting trained), should then be able to correctly classify those inputs not belonging to the training set. Thus, an ANN application involves two distinct phases. The first phase is the *training phase* wherein the network weights are adapted to reflect the problem domain. In the second phase, known as the *prediction* or *operational phase*, the weights frozen in the network as a result of the training phase would enable the prediction of the correct classification at the output when a set of test data (real-world data) is presented at its input.

ANNs can be considered as *associative memory networks* when the weights are determined in such a way that the network can store a set of *pattern associations*. Each association refers to an input–output vector pair, x:o. If each vector o is the same as the vector x with which it is associated, then the network is called an *autoassociative memory*. If the os are different from the xs, the network is called a *heteroassociative memory*. In both cases, the network not only learns the specific pattern pairs that were used for training, but also is able to recall the desired response pattern when given an input stimulus that is similar, but not identical to the training input.

The architecture of ANN adopted in the present study corresponds to a feedforward heteroassociative memory model shown in Fig. 8.16. This system can resolve the neural architecture in respect to the associative data-structure pairs, namely, **node_inputs** and **node_outputs** specified at the input and output of the neurons respectively [8.21]. Resolving the algorithmic implies evolving a set of configurable architectures, or the way in which the neurons are interconnected with a set of distinct interconnecting weights, each of which is capable of mapping an input pattern that can be associated with a distinct output classification.

Further, the ANN of Fig. 8.16 uses combined GA and IT principles as described in the following. The test neural network (Fig. 8.16) uses a set of 7 interconnected neurons with 15 links between them. Each link is associated with an eight (8) bit weight. The total of 120 bits (15 weights _ 8 bits/weight) represents the bit-map, a_n, depicting the binary (yes/no) coefficients of the constraints imposed on the problem as indicated earlier. Each bit string, (bit-map) a_n represents a configured ANN architecture. Further, a bit-string so configured, encoded randomly with a set of $n = 120$ bits, is next subjected to optimization by the genetic algorithm so that its optimized version depicts the optimally weighted (configured) architecture of the network.

The weights in their binary format (each with 8 bits) correspond to the output range of each neuron, denoting integer values between +127 and –128. As mentioned earlier, this integer value is obtained by subjecting the sum of the products of all the inputs times the weights to a logistic (nonlinear) function which sets a threshold. An output is realized only when the threshold is surpassed. This ouput value is then normalized to fall in the range between +127 and –128 (in the present context).

Specific to the output neuron, it is biased presently to take only the positive side of the normalized range (of +127 to – 128). That is, a scaling is done so as to specify the output of the network in terms of only positive integers (0 to +128). This scaling thereby simplifies expressing the positive values of the network's output, F_{kmn}, in a binary format, $\{b_r\}$. Hence, a comparison of these formats of the neural output, namely, F_{kmn}; $\{b_r\}$ can be done with the corresponding dual formats of the target value, namely, F_T, specified in the range between 0 and +128 and its associated binary string, $\{b_T\}$.

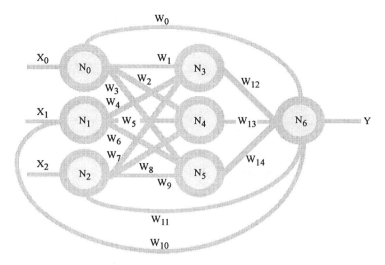

Fig. 8.16: Test ANN: Feedforward heteroassociative memory model

In the present study, the test ANN of Fig. 8.16 also uses information-theoretic concepts to prevent essentially similar sub-strings from being exchanged during the crossover in the early stages of neural network optimization. This is enabled as stated before, with an entropy (information) diversity/distance metric as a check-parameter. That is, the entropy diversity-check used compares the entropy between the sub-strings before they are allowed to cross-over.

8.7.2 Two-point crossover

Having an architecture randomly configured as an initial step, the cross-over principle described earlier can be used next to generate the a_n strings iteratively till an optimum string that conforms with the desired fitness value and yields a desired output (close to a target value) is realized. This optimized bit-string, a_n, by satisfying the criterion set-forth by a fitness value, accommodates the third objective concerning the implementation of n constraints on the problem as specified by C_c for any given m^{th} input alternative.

The crossover mechanism adopted here refers to *two-point crossover*. This is slightly different from the single-point crossover depicted in Fig. 8.11. That is, two points are selected between the lengths of strings to be crossed over (instead of a single-point), and the bit-contents of the entities between the substrings are exchanged to generate a new offspring.

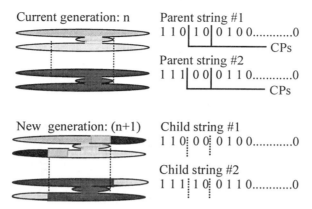

Current generation: n

Parent string #1

1 1 0| 1 0| 0 1 0 0............0
└──────── CPs

Parent string #2

1 1 1| 0 0| 0 1 1 0............0
└──────── CPs

New generation: (n+1)

Child string #1

1 1 0¦ 0 0¦ 0 1 0 0............0

Child string #2

1 1 1¦ 1 0¦ 0 1 1 0............0

Fig. 8.17: An example of two-point crossover operation on two strings of the set {a_n} (String length: 120 bits; CPs: Crossover points taken at 3rd and 5th bits from the most-significant bit)

Shown in Fig. 8.17 is the relevant methodology pursued. Two parent strings say, #1 and #2 in Fig. 8.17 represent two acceptable constraint targets, each of which has two demarcations (or crossover points) as indicated. Each string, a_n, in this example, is taken as 120 bits long depicting 120 possible inclusions of constraints. [*Note:* In the bit-map of a_n, the most-significant bit being 1 refers to a retaining requisite on the associated constraint which is considered as important. The order of importance of 120 possible constraints decreases proceeding from the most-significant to the least-significant bit, as explained in the text. Any bit taking the value 0 signifies that the associated constraint is not needed in optimizing the solution.]

Changing the bits across the demarcations offer several combinations or alternative implementations of the constraints. As a first step, the parent strings (#1 and #2) are constructed such that the bit-maps have the most-significant bit to the least-significant sequencing that represents the prescribed constraints set in the order of their merit or priority. That is, the most significant bits of parent strings #1 and #2 (corresponding to coefficient a_n) are taken as 1, emphasizing that the associated constraint C_1 cannot be ignored. Likewise a_2, a_3,...,a_n are specified for the strings #1 and #2.

Crossover enables the creation of newer generations (of child-strings) depicting several alternatives of constraints and, hence, those constraint-strings which facilitate meeting the overall triple-objectives can be listed.

8.8.3 Information-theoretic (diversity) check for crossover pair-selection

An information-theoretic diversity check is proposed here to calculate the entropy for each of the two substrings that are candidates for cross-over. For this purpose, Jensen-Shannon (JS) relative entropy or Shannon's direct entropy measure can be used. They are determined for each string as follows. Suppose a string of length n has 1s totalling v and 0s totalling $(n - v)$. Let $p_i = v/n$ and $q_i = (1 - p_i) = (n - v)/n$. The JS relative entropy is then given by Eqn. (8.28) and the Shannon's entropy (negentropy) is specified by Eqn. (8.29):

$$H_{JS} = \Pi_1 p_i \ell og(p_i/q_i) + \Pi_2 q_i \ell og(q_i/p_i) \qquad (8.28)$$

$$H_S = - K p_i \ell og(p_i) - K q_i \ell og(q_i) \qquad (8.29)$$

where K is a constant and the coefficients, $(\Pi_1$ or $\Pi_2) < 1$; also, $\Pi_1 + \Pi_2 = 1$.

If the difference in negentropy (information-content) computed *via* Eqn. (8.29) for any two substrings is less than a specified value (threshold), then these two candidates are rejected and replaced by two new candidates. Alternatively, if the JS-entropy (a distance-measure) evaluated for any two substrings is more than a specified value (threshold), again the two candidates are rejected and replaced by a pair of fresh candidates. If the JS or Shannon entropy condition stipulated above between any two substrings is sufficient, then the strings are allowed to crossover and reproduce.

8.7.3 Cost-function calculation

As discussed earlier, let the computed neural output in its integer form be represented by F_{kmn} and its binary format by $\{b_r\}$. Suppose this output of the neural network is compared with the target specification in both formats, namely, F_{kmn} value against the integer target value F_T and in terms of the distance between the target bit-maps $\{b_{Tr}\}$ and $\{b_r\}$. These comparisons refer to determining (i) an error-measure between F_T and $F_{k,m,n}$ and (ii) the Hamming distance [the Modulo-2 (XOR) difference] between the target $\{b_{Tr}\}$ and the neural output $\{b_r\}$. Computation of the Hamming distance enables the determination of bit differences between the compared entities, as in the conventional applications of Hamming distance in error-correction codes. The higher the distance, the greater the difference between the two maps. The greater the difference in bits, more is the deviation of the bit-map of the neural output from that of the target bit-map. Hence, the corresponding cost-functional attribute will be less desirable.

8.8 Description of the Algorithm

A flow-chart of the genetic neural network optimization pursued is shown in Fig. 8.18. The following steps indicate the procedural stages of computation involved thereof. The computer code commences with a specified set of parameters as inputs required for the computation:

Step 0: Display/Print t h e specified parameters: Mutation probability, crossover probability, number of generations (which corresponds to the number of iterations set equal to 2000), retention of the elite string: ON/OFF, implementation of entropy-based check: ON/ OFF, a limit-check on entropy-diversity value, entropy-diversity threshold (set at 0.5), normalizing coefficient for the entropy-diversity threshold (set at 100), minimum acceptable fitness (set at 0.8), neuronal activation threshold (set at the output as 20), activation gain (namely, a pseudo-Boltzmann temperature T, taken as 3.25)

The next part of the algorithm refers to calling **initialize_population**. That is, during this initialization, a population of strings is generated. This is done with an **encode** function that encodes the 120 bit strings considered.

Step 1: Initialize the neural interconnecting weights (15) of the nodes (10): This corresponds to setting up the first generation of the GA.

Step 2: Construct the global population data: Construction of a weight-matrix (15 x 8) for the GA population or chromosomes (taken here as 50).

Step 3: Call the random number generator (0 - 32277): An integer-valued random number is picked and normalized to fall in the range 0 - 128 and used to initialize the population weight.

Step 4: Specify the population in binary strings (8 bits per weight x 15 weights).

Step 5: Increment the generation (iteration) counter,if the iteration is warranted (total iterations = 2000).

* Once the population is initialized, the code enters into an iterative loop with the maximum number of iterations (generations) specified as in **Step 5.**

* The fitness of the generated strings is evaluated in the **eval_fitness** function. This function first decodes each of the strings into a binary number which is stored as the population data structure.

* Then, the function **eval_max_min** is called. It evaluates the average, maximum and minimum of the population fitness. (**Fitness**, here refers to a factor representing the fraction of number of bit-errors observed in a given

380

bit string relative to a target bit string). The results are stored in **pop_data** structure.

* The **statistics** function is then called to display/print the statistics on this population which includes the strings and their fitnesses.

> **Step 6: Evaluate** fitness factors of the entire **(randomly) generated population.**

> **Step 7: Evaluate** the average, maximum, and minimum fitness factors of the entire population.

> **Step 8: Save** the maximum fitness and the corresponding weight string.

* The algorithm next uses the random roulette-wheel scheme to formulate the reproduction scheme. The roulette wheel is a randomizer function that produces pseudo-random numbers up to 32277 and powers of two up to 536870911.

* The computation relevant to roulette-wheel selection is done as follows: To decide the roulette-wheel selection, the **selection** function is run. In this function, a random number from 0 to 99 is picked. The string that falls in a specified weighted pie range of the wheel is selected.

* Then, **crossover** and **mutation** are next called to handle the crossover and mutation functions respectively. The **crossover** function swaps information from each pair of strings that has been selected by the **selection** function. It picks two random spots on the length of the strings to swap the genetic information.

* The **mutation** function imposes the mutation randomly on each binary character of the string by calling the **flip** function. Changing (mutating) 1 to 0 or 0 to 1 of the string characters is done *via* a biased coin-flip strategy. Mutation is done with a probability P_M "to change" and with a probability $(1 - P_M)$, "not to change". Here, P_M refers to the specified (small) mutation probability.

> **Step 9: Select** pairs of strings from the population (50/2= 25) that will reproduce as per the weighted roulette-wheel selection. **Steps 8 and 9** refer to introducing the "elitist reproduction" strategy.

> **Step 10: Perform crossover** between selected string pairs. This is based on selecting pairs of string which have a minimum statistical distance measure.

Step 11: Introduce mutation on the entire string population with the specified mutation probability.

Step 12: Evaluate fitness of the new population realized from the mutated and crossovered population.

Step 13: Evaluate the average, maximum, and minimum fitness factors for the entire population.

* Fitness of the new strings is then evaluated *via* the **eval_ population_fitness** function. Also, a **check** on the retention of the elite string is done. If an elite string does not emerge from the last generation, one of its kind is inserted in a random location in the population. When the loop is exited, corresponding to the triple cost-functions/objectives envisaged, the display or the print-outs made refer to integer value (F_{kmn}), binary representation (b_r) of the output and the optimal a_n binary string (of 120 bits), and the associated fitness value:

(a)   Optimal integer value of the output (F_{kmn}) is decided by the error-metric assaying the difference between the output an the target integer values

(b)   Bit-map of the output string, b_r corresponds to the minimum Hamming distance between the binary format of the output and that of the target values

(c)   Bit-map of the optimal string a_n is decided by the best fitness value corresponding to the elitist child-string yielding the desired outputs of (a) and (b)

Step 14: If the weight string saved previously in **Steps 8** and **9**, corresponding to the best fitness, is not observed in **Step 13**, find a random location among the 50 strings of the new population. Copy to this location the weight string corresponding to the best-fitted one realized in steps 8 and 9. This refers to introducing the "elitist reproduction" strategy.

Step 15: Go back to **Step 5** for next iteration, if, (i) either a string a_n of desired fitness is not realized; or, (ii) minimum cost-function pertinent to F_{kmn} against F_T is not obtained; or, if (iii) a desired Hamming distance between b_r and b_T is not achieved; otherwise, **exit.**

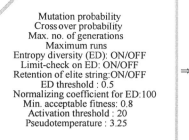

⇒ **Parameter specifications**

Mutation probability
Cross over probability
Max. no. of generations
Maximum runs
Entropy diversity (ED): ON/OFF
Limit-check on ED: ON/OFF
Retention of elite string:ON/OFF
ED threshold : 0.5
Normalizing coefficient for ED:100
Min. acceptable fitness: 0.8
Activation threshold : 20
Pseudotemperature : 3.25

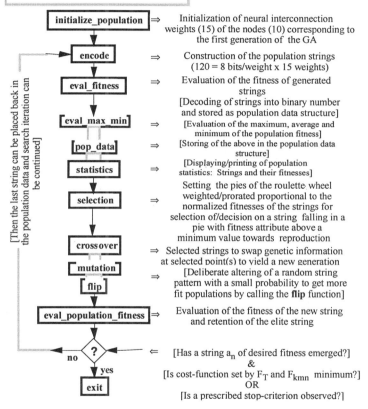

initialize_population ⇒ Initialization of neural interconnection weights (15) of the nodes (10) corresponding to the first generation of the GA

encode ⇒ Construction of the population strings (120 = 8 bits/weight x 15 weights)

eval_fitness ⇒ Evaluation of the fitness of generated strings
[Decoding of strings into binary number and stored as population data structure]

eval_max_min ⇒ [Evaluation of the maximum, average and minimum of the population fitness]

pop_data ⇒ [Storing of the above in the population data structure]

statistics ⇒ [Displaying/printing of population statistics: Strings and their fitnesses]

selection ⇒ Setting the pies of the roulette wheel weighted/prorated proportional to the normalized fitnesses of the strings for selection of/decision on a string falling in a pie with fitness attribute above a minimum value towards reproduction

crossover ⇒ Selected strings to swap genetic information at selected point(s) to yield a new generation

mutation ⇒ [Deliberate altering of a random string pattern with a small probability to get more fit populations by calling the **flip** function]

flip

eval_population_fitness ⇒ Evaluation of the fitness of the new string and retention of the elite string

? ⇐ [Has a string a_n of desired fitness emerged?]
&
[Is cost-function set by F_T and F_{kmn} minimum?]
OR
[Is a prescribed stop-criterion observed?]

no

yes

exit

[Then the last string can be placed back in the population data and search iteration can be continued]

Fig. 8.18: Flow-chart of optimization using GA- IT based ANN strategy

8.9 Experimental Simulations

The experiments (simulations) indicated here demonstrate the implementation of the test ANN which uses the GA learning strategies and IT concepts to evaluate its performance. The test architecture emulates an ANN configuration that corresponds to a specific input (teacher) pattern meeting triple-objective criteria, as demonstrated by a pair of desired cost-functional requirements of the output (F_{kmn}, b_r) and a string-pattern (a_n) depicting the configured weights of the neural interconnections with a specified fitness value. The input vector is a subset of several possible alternatives of binary string patterns available to the network at its input.

The strategy followed in the experiments is to configure a network with a given set of parametric values. Upon initializing the network with the given parameters, training is implemented with a set of teacher inputs. After observing a converging trend of the results (such as the fitness value) in a *single training run**, the optimization is exercised with an ensemble of multiple runs. Such an ensemble of multiple runs is preferred both in the training and prediction phases in lieu of a single run of iterations, since multiple runs would yield robust results in comparison with those of a single run. Thus, the relevant dynamics of the network's performance in seeking the convergence would be less jagged over the epochs of iterations.

In the prediction phase, the performance of the trained network is assessed by posing it with a distinct set of target values (other than those used in the training phase) and with input alternatives (μ out of m) based on the selection made from the observations during the training phase. The μ alternatives used are those input patterns (out of m) which showed a promising trend in meeting the multiple objectives contemplated during the training exercise.

The SGA adopted in the training and learning phases uses 120 bit strings with a population of 50, probability of cross-over = 0.5, probability of mutation = 0.001 or 0.005, and includes an elitist reproduction method—meaning that the best string always gets to reproduce. The iterative generations of the ANN are set at 2000. In summary, two cases of parameters are considered and adopted in the test runs (of training and learning) as indicated below:

---

* *Single run*: This refers to a single run of the network with a given set of parameters, input data structure and ouput target value(s) *versus* iteration epochs.

An *ensemble set of runs* corresponds to the dynamics of repeated (multiple) runs yielding an ensemble average of a network's performance towards convergence.

Case 1: The crossover probability = 0.5, the mutation probability = 0.001, and information-theoretic strategy is used in the crossover decisions, via entropy-check implementation.

Case 2: The crossover probability = 0.5, the mutation probability = 0.005 and information-theoretic strategy is again used in the crossover decisions, via entropy-check implementation.

It may be noted that only one parameter, namely, the mutation probability, is distinct in these two cases. The target values considered refer to a set of desired numbers. Learning is implemented with eight different alternative input (teacher) strings each of 3 bits designated as X_1, X_2, and X_3 in Fig. 8.16 and detailed in the following section.

8.9.1 SET T: Training bit-maps

The input bit-maps considered in the training and/or prediction phases are any subset of eight (2^3) alternative combinations of the following 3-bit strings:{ 000, 001, 010, 011,100, 101,110,111}.

Triple objective functions:
1. $E_{HD} \Rightarrow HD(b_T, b_r)$: Hamming distance criterion; that is, E_{HD} = Minimum number of 1's in $b_T \approx b_r$ where b_r is the output binary string
2. $E_{TO} = (|F_T - F_{kmn}|/ F_T)$ X 100% fi Fractional percentage error between the target (F_T) and output (F_{kmn}) integer values
3. Third objective function refers to the realization of an optimal bit-map of a_n (not specified in Table 8.2 or 8.3)

Table 8.4: Training phase results on network's output (Case # 1)

m	Input teacher Bit-map	Target values Integer (F_T)	Binary (b_T)	Network output Integer (F_{kmn})	Binary (b_r)	Bit errors (E_{HD})
1	**000**	0	00000000	0	00000000	0
2	001	25	00011001	29	00011101	1
3	010	49	00110001	120	01111000	3
4	**011**	71	01000111	67	01000011	1
5	100	90	01011010	126	01111110	2
6	101	106	01101010	122	01111010	1
7	**110**	117	01110101	121	01111001	1
8	**111**	125	01111101	125	01111101	0

The bit strings used as input (teacher) bit-maps for training the test network while adopting the ANN refers to a set of bit patterns shown in Table 8.2. These are designated as SET T. The target values corresponding to each input string are the integer values indicated in Table 2. It should be noted that in the event the input binary pattern is $(X_1X_2X_3) \Rightarrow 000$, the output is always equal to zero. The triple-objectives of the network being trained, as indicated earlier, are:

(1) To match the integer value in each row of the teacher structure for the example patterns depicted in Table 8.2 with the (integer) output of the network using a cost-function criterion.

(2) To realize a minimal (Hamming) distance between the binary coefficients of the target value and those of the network output.

(3) To get a child-string a_n (representing 120 possible constraints) with a desired fitness value that yields the optimized outputs (1) and (2).

Table 8.5. Training phase results on network's output (Case #2)

m	Input teacher Bit-map	Target values Integer (F_T)	Binary (b_T)	Network output Integer (F_{kmn})	Binary (b_r)	Bit errors (E_{HD})	E_{TTO} %
1	000	0	00000000	0	00000000	0	0
2	001	25	00011001	29	00011101	1	16.0
3	010	49	00110001	49	00110001	0	0
4	011	71	01000111	71	01000111	0	0
5	100	90	01011010	127	01111111	3	41.1
6	101	106	01101010	124	01111100	3	17.0
7	110	117	01110101	117	01110101	1	0
8	111	125	01111101	125	01111101	0	0

A rule of thumb is that one would like at least as many examples to train a network as there are weights in the network. Ideally, the number of examples should be 5 to 10 times the number of weights. However, only eight input patterns are considered here for illustrative purposes and are presented to the test network. Further, the training and prediction phases are performed twice, corresponding to two different parametric attributes indicated above, in order to confirm the replicative efficacy of the algorithm developed in offering consistent results under varying parametric conditions.

Listed in Tables 8.4 and 8.5 are the results of the training results. In Table 8.5, the fitness results perceived in the trainings are presented.

Table 8.6: Training phase: Parameters and results on maximum fitness obtained with ensemble runs using SET T teacher and target data

Genetic algorithm implementation	Case #1	Case #2
Parameters		
Probability of mutation	0.001	0.005
Probability of crossover	0.500	0.500
Entropy-check	ON	ON
Number of iterations	2000	2000
Number of ensemble runs	100	100
Fitness results		
Maximum fitness obtained	0.8857	0.8857
Iteration at which maximum fitness string appeared	1071	96
Run at which maximum fitness string appeared	5	31

Table 8.7: Selection of optimal input bit-map alternatives from the training phase results

Input teacher bit-map	Number of times the bit-map yielding optimal outputs
000*	4
001	1
010	1
011*	4
100	0
101	0
110*	4
111*	4

* Selected input patterns (totaling, $\mu = 4$) out of total M = 8 alternatives

The performance of network training vis-à-vis the input patterns used, as summarized in Tables 8.4 and 8.5, indicate that 4 out of 8 input (teacher) patterns train the network effectively for the set of chosen targets under two different cases of parametric attributes envisaged. That is, the training phase demonstrates that among the 8 input bit-map alternatives presented to the network along with the triple criteria, namely, (i) the difference error, $\varepsilon_1 \leq |F_T - F_{kmn}|$ expressed as a fractional percentage (= $E_{TO}\%$) with respect to F_T being less than or equal to 10%; and simultaneously, (ii) the Hamming distance ($\varepsilon_2 = HD_b$) between the binary string (b_r) of the output and that of the target (b_T) being less than or equal to 1, and, (iii) offering a bit-string (a_n) representing an optimal coefficient pattern of constraints imposed with a specified fitness value), at least 4 input bit-maps enable the network to converge yielding optimal/suboptimal solutions under different parametric conditions.

Learning dynamics

The convergence trends of the network towards optimization in the learning phase, pertinent to a single as well as multiple (ensemble) runs performed, are presented in Figs. 8.19a through 8.19d via learning curves which depict the dynamics of learning—that is, the variation of fitness-factor of $\{a_n\}$ with respect to the number of iterations performed.

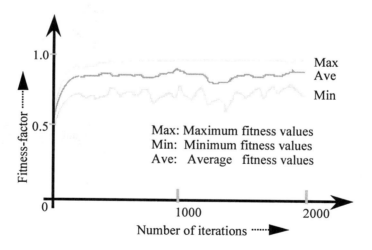

Fig. 8.19a: Case # 1: The convergence dynamics of the test network: Fitness *versus* iterations (single-run)

388

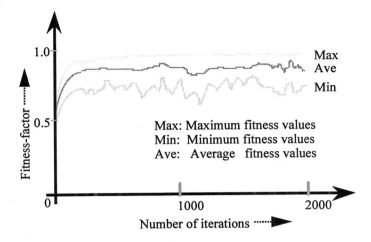

Fig. 8.19b: Case # 2: The convergence dynamics of the test network: Fitness *versus* iterations (single-run)

The learning curves plotted in Figs. 8.19a-8.19b refer to single-run computations and Figs. 8.19c and 8.19d correspond to simulations with ensemble runs respectively for the two different cases of parameters envisaged.

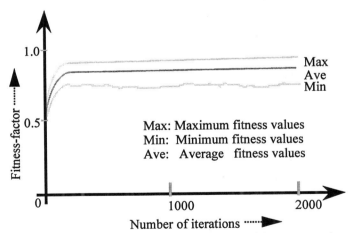

Fig. 8.19c: Case # 1: The convergence dynamics of the test network: Fitness *versus* iterations (ensemble-runs)

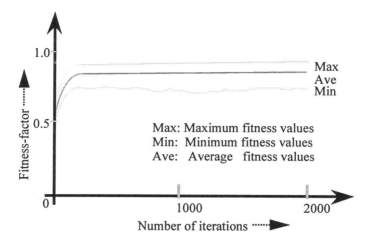

Fig. 8.19d: Case # 2: The convergence dynamics of the test network:
Fitness *versus* iterations (ensemble-runs)

From the dynamics of convergence depicted in Figs. 8.19a and 8.19b in
respect to single-run experiments, the approach of the network's performance
towards convergence as a function of the iterations implemented shows a
jagged characteristic. However, the iterations repeated over several ensemble
of runs allow a smooth convergence of the network, as can be observed from
the results shown in Figs. 8.19c and 8.19d. In both cases, however, the
implementation of the GA based algorithms had not failed in letting the
network to learn and seek the basin of attraction leading to the optimization
sought. Further, in all the cases of ensembled runs, the trend toward
convergence has resulted with the knee of the learning dynamics toward
convergence being around 200 iterative epochs.

8.9.2 SET:P - Prediction phase

Omitting the trivial case of input-pattern, namely, 000, the other
optimal/suboptimal input bit-maps are tested in the prediction phase on an
ensemble run basis to establish their viability in yielding optimal outputs.
This satisfies the dual objective criteria with a corresponding bit-string a_n
satisfying a prescribed extent of fitness value (third objective) and, thereby
depicting an optimal bit-map of the n constraints imposed. Presented in
Tables 8.8 and 8.9 are the simulation results obtained with the selected input
patterns: SET P - { 010, 011, 111} for target values distinct from those used
in the training phase. The fitness results pertinent to the prediction phase
simulations are indicated in Table 8.10.

Table 8.8a: Prediction phase results on network's output (Case # 1)

m	Input teacher Bit-map	Target values Integer (F_T)	Binary (b_T)	Network output Integer (F_{kmn})	Binary (b_r)	Bit errors (E_{HD})	E_{TTo} %
3	011	67	01000011	66	01000010	1	1.5
4	110	111	01101111	111	01101111	1	16
8	111	77	01001101	77	01001101	0	0

Table 8.8b: Optimal bit-maps of a_n corresponding to the prediction phase results (Case # 1)

Input bit-map	Bit-map of a_n (120 bits long)
011	111011110.....0
110	111110110.....0
111	101111110.....0

The bit-maps of a_n shown in Table 8.8b and 8.9b are each 120 bits long. In the optimization envisaged, however, only the first 8 significant bits were considered, representing a variation of eight possible constraints (N = 8) on the problem, and the rest of the 112 bits are set as zero. The binary coefficients (a_n) refer to nonexisting constraints or constraints presumed to be not influencing the cost optimization. The two-point cross-over is done at the third and fifth bits from the most-significant bit end.

Table 8.9a: Prediction phase results on network's output (Case # 2)

m	Input teacher Bit-map	Target values Integer (F_T)	Binary (b_T)	Network output Integer (F_{kmn})	Binary (b_r)	Bit errors (E_{HD})	E_{TTO} %
3	011	67	01000011	66	01000010	1	1.5
4	110	111	01101111	111	01101111	0	0
8	111	77	01001101	76	01001100	1	1.3

391

Table 8.9b: Optimal bit-maps of a_n corresponding to the prediction
phase results (Case # 2)

Input bit-map	Bit-map of a_n (120 bits long)
011	111111100.....0
110	111111010.....0
111	101111110.....0

Table 8.10: Prediction phase: Parameters and results on maximum fitness obtained
with ensemble runs using SET P
teacher and target data

Genetic algorithm implementation	Case # 1	Case # 2
Parameters		
Probability of mutation	0.001	0.005
Probability of crossover	0.500	0.500
Entropy-check	ON	ON
Number of iterations	2000	2000
Fitness results		
Maximum fitness obtained	0.8857	0.8571
Iteration at which maximum fitness string appeared	229	160
Run at which maximum fitness string appeared	57	47

8.9.1 Remarks on test optimization

The following is an outline summarizing the essence of using a combined GA and information-theoretic (IT) based strategy towards neural network optimizations:

- Using GA in the IT-plane for ANN optimization is a feasible alternative to the conventional optimization techniques.
- The results on the learning characteristics obtained here imply that the present algorithm is a viable complement to the traditional ANN learning optimizations.

- Blending the heuristics behind information and entropy formulations appropriately with the GAs adopted in ANN optimization strategies is commensurate with the biological perspectives of the neural complex.
- The scope of the relevant studies on ANN optimization with Gas, together with the IT concepts as indicated here, essentially enclaves the possibility of using the concept of mutual entropy as a statistical measure to distinguish a pair of genetic candidates set in the reproduction process. That is, the mutual entropy information distance has been adopted as a genetic constraint selection parameter. It refers to a distance measure that can be adopted in GA to determine a string's fitness and identify those strings which are remotely similar.
- The second IT consideration used here is pertinent to deploying the Hamming distance concept so as to decide the statistical distance between the desired target and an output binary bit-map, as a part of minimizing the cost-functional objectives.
- In essence, the theory of GA and the processes of genetic selection which are invariably considered in the parametric domain in traditional efforts pertinent to ANN optimizations are viewed differently in IT perspectives in the present study.
  The proposed algorithm is implementable in conventional ANN architecture (such as a feedforward associative memory model); and, for the supervised learning strategy adopted, the test-network has yielded results meeting the triple objectives envisaged. Higher resolution would be feasible with larger sized networks.
- The optimum fitness based network configuration, expressed in terms of optimally configured binary-string, a_n, is decided largely by the number of nodes used in the network architecture and is constrained by the network size.

In reference to the studies performed, the following are still open questions which offer promising topics for future research:

- A rigorous analysis of genetic processes in the IT plane
- Feasibility aspects of IT-based GA for various ANN architectures involving supervised or unsupervised learning with feedforward or backpropagation methods
- Analysis of the effects of mutation and/or crossover probabilities on the genetic process and on the corresponding performance of a network's convergence.
- Possibilities of introducing fuzzy attributes to the GA.

8.10 Concluding Remarks

Apart from devising a strategy to implement and include the concepts of GA and IT cohesively in ANNs, the present study indicates a method of applying the relevant algorithm to a practical problem with the attributes of a complex system. Pertinent studies indicate that a GA-IT based ANN technique enables a triple-objective cost-optimization involving minimization of the overall cost under the constraints imposed by: (1) the subcost outlays and (2) system considerations. This method facilitates the selection of compatible alternatives of the inputs from several available input-conditions with the cost and system constraints imposed. The complexity of the cybernetics involved is tactfully imbedded into the algorithm *via* informatic considerations and principles of genetic algorithms. This effort bears the novelty wherein the informatics based cost-functions are used in the application of GA in the ANN training and prediction phases.

Bibliography

[8.1]   Shannon, C. E. and Weaver, W,W.: *The Mathematical Theory of Communication* (University of Illinois Press, Urbana, IL: 1949)

[8.2]   Khinchin, A. Y. : *Mathematical Foundations of Information Theory* (Dover Publications, Inc. New York, NY: 1957)

[8.3]   Neelakanta, P. S., and De Groff, D.: *Neural Network Modeling: Statistical Mechanics and Cybernetic Perspectives*, (CRC Press, Boca Raton, FL: 1994)

[8.4]   Alekseev, G. N.: *Energy and Entropy* (Mir Publishers, Moscow: 1986).

[8.5]   Gatlin, L. L.: *Information Theory and the Living System* (Columbia University Press, New York, NY: 1972).

[8.6]   Drelica, K.: *Understanding DNA and Gene Cloning: A Guide for the Curious* (John Wiley and Sons, Inc., New York, NY: 1984)

[8.7]   Wiener, N.: *Cybernetics* (MIT Press, Cambridge, MA: 1948)

[8.8]   Oh, S. H, Yoon,T. H. and Kim, J.C.: Associative-memory model based on neural networks: Modification of Hopfield Model, *Opt. Letts.*, 1988, 13, 74-76

[8.9]   Nemhauser, G.L. and Wolsey, L. A.: *Integer Programming and Combinatorial Optimization* (John Wiley and Sons, Inc., New York, NY: 1988)

[8.10] Davis, L. (Ed.): *Genetic Algorithms and Simulated Annealing* (Morgan Kaufmann Publishers, Inc., Los Altos, CA: 1987)

[8.11] Goldberg, D. E., *Genetic Algorithms in Search, Optimization, and Machine Learning* (Addison-Wesley, Reading, MA: 1989)

[8.12] Goldberg, D. E.: *Genetic Algorithms* (Addison Wesley, New York, NY: 1989).

[8.13] Koza, J. R.: *Genetic Programming* (MIT Press, Cambridge, MA: 1992).

[8.14] Davis, L. (Ed.): *Handbook of Genetic Algorithms* (Van Nostrand Reinhold, New York, NY: 1991)

[8.15] Holland, J. H: *Adaptation in Natural and Artificial Systems* (University of Michigan Press, Ann Arbor, MI: 1975)

[8.16] Holland, J. H: Genetic algorithms, *Scientific American*, July 1992, 66-72.

[8.17] McCulloch, W.W. and Pitts, W.: A logical calculus of the ideas inminent in nervous activity. *Bull. Math. Biophys.* 5 (1943), 115-133.

[8.18] Rumelhardt, D.E. and McClelland, J.L.: *Parallel distributed Processing,* Vol.1 (The MIT Press, Cambridege, MA: 1987

[8.19] Park, J. C., Neelakanta, P. S., Abdusalah, S., De Groff, D., and Sudhakar, S.: Information-theoretics based error-metrics for gradient-descent learning in neural networks. *Comp. Syst.* 9, 1995, 287-304.

[8.20] Neelakanta, P. S., Abusalah, S., Sudhakar, R., De Groff, D. and Aalo,V.: Dynamic properties of neural learning in the information-theoretic plane, *Comp.Syst.*, 1995, 9, 349-374.

[8.21] Arredondo, T. V.: *Information-Theoretics Based Genetic Algorithm: Application to Hopfield's Associative Memory Model of Neural Networks.* M.S.E. Thesis, Department of Electrical Engineering, Florida Atlantic University, Boca Raton, FL, May 1997.

SUBJECT INDEX

V
Vagueness, 113, 135
Variety, 10, 15, 42

W
Weber-Fechner law, 2